STELLAR EVOLUTION

STELLAR EVOLUTION

Proceedings of an international conference,
November 13-15, 1963,
sponsored by the Institute for Space Studies
of the Goddard Space Flight Center, NASA

Edited by
R. F. Stein and A. G. W. Cameron
Institute for Space Studies, Goddard Space Flight Center
National Aeronautics and Space Administration
New York, New York

 **Plenum Press · New York · 1966**

Library of Congress Catalog Card Number 65-25285

PREFACE

On November 13–15, 1963, the Institute for Space Studies of the Goddard Space Flight Center, National Aeronautics and Space Administration, was host to an international group of astronomers and physicists gathered to discuss the evolution of the stars. This was the fifth in a continuing series of interdisciplinary meetings on topics in space physics held at the Institute. The conference was organized by B. Strömgren of the Institute for Advanced Study and A. G. W. Cameron of the Goddard Institute for Space Studies.

During the last century there has been a gradual development of the empirical science of spectroscopy and of its application to astronomy. Stellar spectra were classified according to the character of the lines which appeared within them. Gradually it was realized that the stellar spectral classes were closely related to the surface temperatures of the stars. When it also became possible to determine the absolute luminosities of the stars, the way was open for Hertzsprung and Russell to take the important step of plotting stellar luminosities against spectral classes, which established the fundamental diagram bearing their names.

With the growth of astrophysics in the present century, it was natural to wonder how the structures of the stars in the different parts of the Hertzsprung–Russell diagram differed from one another, and how the position of a star changed as it evolved.

Four books have marked the gradual development of the physical theory necessary to tackle this problem. The first of these was the classical treatise of Emden, *Gaskugeln*, published in 1907. The theory of polytropic gas spheres was treated in this book and it became the basis for much of the discussion of stellar structure in the ensuing years. The fundamental point of view adopted at that time was that energy transport in the interior of the star would be by convective motions, and it was noted that a particular polytropic index corresponded to a gas sphere in convective equilibrium.

However, during the next two decades the importance of radiative energy transfer in the interior of the star was realized, and the theory of radiative transfer was developed in some detail. The situation was surveyed at the end of the third decade of the century by Eddington in his classic work, *The Internal Constitution of the Stars*, published in 1930.

During the following decade much work was done on the derivation of detailed stellar models, and the theory of degenerate matter and of white dwarf stars was developed. This led to the important monograph by Chandrasekhar, *An Introduction to the Study of Stellar Structure*, published in 1939. At this time the nuclear reactions responsible for stellar energy generation were being discovered, and Chandrasekhar's book contains an early account of these developments, with additional notes at the end of his book, added in proof stage, discussing the search for the nuclear reactions important during hydrogen-burning in a stellar interior.

The next two decades saw the working out of further details of the nuclear reactions and the determination of the reaction rates on the basis of laboratory measurements. Many detailed stellar models were constructed using a proper distribution of the nuclear energy sources in the interior. These developments are described in the book *Structure and Evolution of the Stars*, published by Schwarzschild in 1958, which he regards as an interim textbook.

In the last few years, the development of large fast computers has had a profound impact on the study of stellar structure. No longer is it necessary to use the laborious techniques of hand integration discussed at considerable length in Schwarzschild's book. Henyey has introduced a method of dividing the structure of a star into many concentric zones and solving the differential equations of stellar structure in difference form at the boundaries of these zones. This technique is especially suited for use with computers, although it does contain some tricky snares.

Although stellar evolution calculations have not progressed much beyond the stage of helium-burning, much work has been done on the physics which will be appropriate for inclusion in the study of the later stages when that becomes possible. This physics includes fairly exotic sets of nuclear reactions and a wide range of new processes in which neutrino-antineutrino pairs may be emitted. Considerable progress has also been made in the study of the theory of stellar pulsation, and a beginning has been made to the study of the hydrodynamics of the supernova collapse. The study of stellar evolution is being tied more and more to the requirements of the theory of nucleosynthesis. Thus, a conference on stellar evolution is likely to range widely across a great many of these topics, and the Goddard Institute conference of 1963 was no exception.

This proceedings volume is divided into seven sections. The first of these, the introduction, was written especially for this volume by one of us (R. F. Stein). In this introduction an attempt is made to account for as much of the theory of stellar evolution as is possible in simple analytical form. All models of stellar structure which are usually discussed in treatises on this subject, including polytropic gas spheres, have interior characteristics determined by numerical integration. Furthermore, the variables used in the presentation of such models are usually fairly far removed from the directly observable quantities such as the mass and radius of a star. This is something of a handicap when one attempts to teach the subject, particularly to students who may not care about the intricate details of the model construction. It is something of a challenge to see how the essential characteristics of stellar structure and evolution may be calculated with the use of crude analytical approximations. It turns out to be possible to make a surprising amount of progress in this direction, and the resulting analytical models are capable of giving new kinds of insight into the nature of stellar evolution. It is hoped that the techniques introduced in this introductory chapter may be of some use in courses on stellar evolution.

The second section contains some survey papers on the state of our knowledge of physics which must be used in the calculation of stellar structure. The discussions of energy generation and of neutrinos by Reeves and Chiu place emphasis upon the physics which will be required in the study of the later stages of evolution. The discussion of opacity by Cox lays emphasis upon the work on line absorption which he and his group have been doing lately at Los Alamos. The discussion of turbulent convection by Spiegel emphasizes our continuing and embarrassing inability to make fundamental improvements in the mixing length theory of convection.

The third section dealing with stellar evolution calculations has been sub-divided into four subsections. The first of these deals with pre-main sequence contraction. It opens with a paper by Hayashi, who discusses the important role played by convection in the structure of pre-main sequence stars, a topic to which he has made a fundamental contribution within the last few years. Two other papers in this section extend his considerations, and there is also a paper by Schatzman who discusses the role of magnetic activity during this stage of stellar evolution. The second subsection relates mainly to the hydrogen-burning phase of stellar evolution. It contains several papers which report the details of model evolutionary sequences and the effect of variations in abundances in stellar models, and there are also two papers relating to the important question of the detection of neutrinos from the sun. The third subsection contains papers dealing with more advanced stages of stellar evolution following the hydrogen-burning phase. This subsection is concluded with a paper on white dwarfs and a paper discussing the relation between our knowledge of stellar evolution and the problem of forming large amounts of helium in the interstellar medium. The fourth subsection contains one paper on stellar evolution with a time-varying gravitational "constant," by Dicke. It appears that if there is a scalar gravitational field present in space, its effect on the details of stellar evolution at different stages in galactic history may be more profound than any other physical process so far considered.

The fourth section contains papers reviewing the good progress made in theories of stellar variability by Baker, Cox, and Christy.

The fifth section deals with stellar mass loss. The theory of this mass loss is in an extremely incomplete state. One suspects that stellar mass loss results from a generalization of the process which leads to the solar wind, which in turn appears to be a consequence of the sun's possession of a hot corona heated by some form of wave motion: acoustic, gravitational, or hydromagnetic waves. However, the rate of mass loss is so sensitive to the coronal temperature that the uncertainties in the theories of coronal heating preclude the possibility of predicting the mass loss rate. Consequently our present survey of the mass loss process must deal primarily with observational evidence. This section contains discussions of observed mass loss processes in T Tauri stars, red giants, and planetary nebulae, by Kuhi, Deutsch, and Osterbrock.

The sixth section contains papers covering a wide range of observations concerning stellar evolution. The evolutionary state of a star, its composition, and its position in the Hertzsprung–Russell diagram are intimately related. The observations reported deal with many aspects of these relationships and present challenges to theoretical astrophysicists for the future.

The seventh section contains the summarizing lecture by Burbidge. After such a long and varied conference, it is very useful to have this assessment of the state of the subject and of the current avenues of research which seem most promising.

This conference was tape-recorded and a transcript made. From this transcript first drafts of the papers were prepared for the assistance of the authors in readying their papers for publication. We wish to thank the authors for their cooperation in this enterprise. We also wish to thank Mrs. M. McCorkle for her assistance in the preparations of the first drafts. The conference secretary was Mrs. E. Silva, who handled very capably the arrangements for the conference and subsequently made important contributions to the preparation and typing of the manuscripts. We are very grateful. Thanks are also due to Mr. B. Kramer, who typed the final manuscripts, and Mrs. L. C. Stein, who assisted in proofreading. Mr. G. Goodstadt,

Mr. N. Panagakos, and Mr. L. J. Stein, and several other staff members of the Goddard Institute for Space Studies gave valuable assistance in the organization of the conference, and we wish to thank them for their help in making it a success.

R. F. STEIN

A. G. W. CAMERON

CONTENTS

ix

C. Advanced Stages of Evolution

D. Stellar Evolution with Varying G

Part IV. Stellar Variability

Part V. Stellar Mass Loss

Part VI. Observations Concerning Stellar Evolution

Part VII. Summary

CONTRIBUTORS

L. H. ALLER*
University of California
Los Angeles, California

JOHN BAHCALL
California Institute of Technology
Pasadena, California

NORMAN BAKER
Columbia University
New York, New York

W. BIDELMAN
University of Michigan
Ann Arbor, Michigan

G. R. BURBIDGE
University of California, San Diego
La Jolla, California

A. G. W. CAMERON
Institute for Space Studies
Goddard Space Flight Center, NASA
New York, New York

H. Y. CHIU
Institute for Space Studies
Goddard Space Flight Center, NASA
New York, New York

R. F. CHRISTY
California Institute of Technology
Pasadena, California

STIRLING A. COLGATE
New Mexico Institute of Mining and
 Technology
Secorro, New Mexico

ARTHUR N. COX
Los Alamos Scientific Laboratory
Los Alamos, New Mexico

JOHN P. COX
Joint Institute for Laboratory Astro-
 physics
Boulder, Colorado

D. L. CRAWFORD
Kitt Peak National Observatory
Tucson, Arizona

RAYMOND DAVIS
Brookhaven National Laboratory
Upton, New York

PIERRE DEMARQUE
David Dunlap Observatory
University of Toronto
Ontario, Canada

A. J. DEUTSCH
Mt. Wilson and Palomar Observatories
Pasadena, California

R. H. DICKE
Princeton University
Princeton, New Jersey

O. J. EGGEN
Royal Greenwich Observatory
Herstmonceaux Castle, Sussex, England

DILHAN EZER
Institute for Space Studies
Goddard Space Flight Center, NASA
New York, New York

JESSE L. GREENSTEIN
California Institute of Technology
Pasadena, California

J. GUNN
Mt. Wilson and Palomar Observatories
Pasadena, California

C. J. HANSEN
Yale University
New Haven, Connecticut

C. HAYASHI
Kyoto University
Kyoto, Japan

* Senior National Science Foundation Postdoctoral Fellow at Australian National University 1960–61.

H. L. HELFER
University of Rochester
Rochester, New York

G. H. HERBIG
Lick Observatory
Mt. Hamilton, California

E. HOFMEISTER
Max-Planck-Institut
München, Germany

I. IBEN
Massachusetts Institute of Technology
Cambridge, Massachusetts

R. KIPPENHAHN
Max-Planck-Institut
München, Germany

L. V. KUHI
Mt. Wilson and Palomar Observatories
Pasadena, California

J. B. OKE
California Institute of Technology
Pasadena, California

D. E. OSTERBROCK
University of Wisconsin
Madison, Wisconsin

B. J. E. PAGEL
Royal Greenwich Observatory
Herstmonceaux Castle, Sussex, England

JØRGEN O. PETERSEN
University Observatory
Copenhagen, Denmark

VERN L. PETERSON
National Bureau of Standards
Boulder, Colorado

H. REEVES
University of Montreal
Montreal, Quebec, Canada

A. REIZ
University Observatory
Copenhagen, Denmark

JORGE SAHADE*
Observatorio Astronomico
La Plata, Argentina

M. P. SAVEDOFF
University of Rochester
Rochester, New York

E. SCHATZMAN
Institute d'Astrophysique
Paris, France

R. L. SEARS
W. K. Kellogg Radiation Laboratory
Pasadena, California

K. VON SENGBUSCH
Max-Planck-Institut
München, Germany

E. A. SPIEGEL
New York University
New York, New York

R. F. STEIN
Institute for Space Studies
Goddard Space Flight Center, NASA
New York, New York

B. G. STRÖMGREN
Institute for Advanced Study
Princeton, New Jersey

S. TEMESVÁRY
Institute for Advanced Study
Princeton, New Jersey

J. W. TRURAN
Yale University
New Haven, Connecticut

M. WALKER
Lick Observatory
Mt. Hamilton, California

G. WALLERSTEIN
University of California
Berkeley, California

A. WEIGERT
Max-Planck-Institut
München, Germany

RICHARD H. WHITE
Lawrence Radiation Laboratory
Livermore, California

M. H. WRUBEL
Indiana University
Bloomington, Indiana

* Member of the Carrera del Investigador Científico, Consejo Nacional de Investigaciones Científicas y Técnicas, Argentina.

Part I. Introduction

STELLAR EVOLUTION: A SURVEY WITH ANALYTIC MODELS

Robert F. Stein

INTRODUCTION

To increase our physical insight into the evolution of stars, we shall, in this introductory paper, reproduce the basic features of stellar structure (as found from accurate calculations), by purely analytic considerations. We will, therefore, not attempt accurate calculations of structures and evolutionary tracks. Rather, we shall first discuss the general properties of stellar structure and evolution, and then construct analytic models for the early homogeneous and advanced inhomogeneous stages of evolution.

EQUATIONS OF STELLAR STRUCTURE

The structure of stars is determined by the conditions of mass conservation, momentum conservation, energy conservation, and the mode of energy transport (Schwarzschild, 1958, and Wrubel, 1958). Rotation and magnetic fields will be neglected so that a star will be spherically symmetric.

Hydrostatic Equilibrium

A star changes very slowly during most of its life and so may be considered in hydrostatic equilibrium. Two forces balance to keep a nonrotating star in hydrostatic equilibrium: the gravitational force directed inward and the gas and radiation pressure force directed outward. The equation of hydrostatic equilibrium is

$$\frac{dP}{dr} = -\frac{GM(r)\rho}{r^2} \tag{1.1}$$

The total pressure is the sum of gas and radiation pressure

$$P = P_{gas} + P_{rad}$$

For an ideal gas

$$P_{gas} = \frac{k}{\mu H}\rho T \tag{1.2}$$

where $H = 1.67 \times 10^{-24}$ g is the mass of a proton, $k = 1.38 \times 10^{-16}$ ergs/deg, and μ is the mean molecular weight.

$$P_{rad} = \tfrac{1}{3}aT^4 \tag{1.3}$$

where $a = 7.57 \times 10^{-15}$ ergs/cm^3-deg^4. $M(r)$ is the mass inside a sphere of radius

r. The equation of mass conservation is

$$\frac{dM(r)}{dr} = 4\pi r^2 \rho \qquad (1.4a)$$

so that

$$M(r) = \int_0^r 4\pi r^2 \rho \, dr \qquad (1.4b)$$

G is the gravitational constant, $G = 6.67 \times 10^{-8}$ dyn-cm^2/g^2.

Energy Conservation

The total energy of an element of material is

$$E = U + \Omega + K \qquad (1.5)$$

where U is the internal energy of the gas, Ω is gravitational potential energy, and K is the kinetic energy of large-scale mass motion, which we are neglecting here. The internal energy of a gas plus radiation is

$$U = \frac{1}{\gamma - 1} NkT + \frac{1}{\rho} aT^4$$

where γ is the ratio of specific heats ($\gamma = \frac{5}{3}$ for a monatomic ideal gas). The sources and sinks of energy are (1) energy release or absorption by nuclear reactions, and (2) energy transport into and out of the element of material.

Let $\mathscr{E}$ be the net release of energy per gram per second, and F be the energy flux. The equation of conservation of energy is then

$$\frac{dE}{dt} = \frac{dU}{dt} + \frac{d\Omega}{dt} = \mathscr{E} - \frac{1}{\rho} \operatorname{div} F \text{ ergs/g-sec}$$

The change of gravitational potential energy is

$$d\Omega = -dW = P \, dv = -\frac{P}{\rho^2} d\rho$$

Define the luminosity L_r as the total net energy flux through a spherical shell of radius r, so that

$$L_r = 4\pi r^2 F$$

Then the equation of energy conservation is

$$\frac{dL_r}{dr} = 4\pi r^2 \rho \left[\mathscr{E} + \frac{P}{\rho^2} \frac{d\rho}{dt} - \frac{dU}{dt} \right] \qquad (1.6)$$

Energy Transport

Energy is transported by radiation and convection, and by conduction when the electrons are degenerate:

$$\frac{L_r}{4\pi r^2} = F_{\text{rad}} + F_{\text{conv}} \qquad (1.7)$$

In the interior of a star, where the radiation is almost isotropic, the force due to the gradient of the radiation pressure is equal to the momentum absorbed from the radiation beam in passing through matter:

$$\frac{dP_R}{dr} = -\frac{\varkappa\rho}{c}\frac{L_r}{4\pi r^2}$$

where $P_R = \frac{1}{3}aT^4$ is the radiation pressure, $(\varkappa\rho)^{-1}$ is the photon mean free path, and c is the velocity of light. Thus, in the interior of the star, the radiative energy flux is

$$F_{\text{rad}} = -\frac{4acT^3}{3\varkappa\rho}\frac{dT}{dr} \tag{1.8}$$

and the temperature gradient necessary to drive the radiation flux is

$$\frac{dT}{dr} = -\frac{3}{4ac}\frac{\varkappa\rho}{T^3}\frac{L_r}{4\pi r^2} \tag{1.9}$$

Convection occurs in those regions of a star where the temperature gradient is steeper than the adiabatic gradient. The convective flux is (Spiegel, 1965), crudely, the energy fluctuation (excess or deficiency) of an element of gas times its velocity, averaged over horizontal directions,

$$F_{\text{conv}} = \overline{\rho c_p w\theta} \tag{1.10}$$

where w is the radial velocity fluctuation and θ the temperature fluctuations in the matter. The convective flux depends on the superadiabatic gradient

$$\beta = -\left[\frac{dT}{dr} - \left(\frac{dT}{dr}\right)_{\text{ad}}\right] \tag{1.11}$$

Because convection is an extremely efficient energy-transport mechanism in the interior of a star, the superadiabatic gradient is very small and the temperature gradient will be very nearly equal to the adiabatic gradient:

$$\frac{dT}{dr} = \left(\frac{dT}{dr}\right)_{\text{ad}} = \frac{\Gamma - 1}{\Gamma}\frac{T}{P}\frac{dP}{dr} \tag{1.12}$$

where Γ is the effective ratio of specific heats, including ionization, dissociation, and radiation. Near the surface, where the photon mean free path is long, there is a leakage of heat by radiation from the convective elements and the convective temperature gradient is greater than the adiabatic gradient.

Stellar Structure

Order of magnitude estimates of the density, pressure, and temperature of a star can easily be made from the condition of hydrostatic equilibrium. The mean density of a star of mass M and radius R is

$$\bar{\rho} = \frac{M}{\frac{4}{3}\pi R^3} \tag{1.13}$$

In the equation of hydrostatic equilibrium (1.1), setting

$$\frac{dP}{dr} \approx \frac{P_c - P_0}{R} \approx \frac{P_c}{R}$$

where P_c is the central and P_0 the surface pressure, gives

$$P_c \approx \frac{GM\bar{\rho}}{R} \approx \frac{GM^2}{R^4} \tag{1.14}$$

Let $\beta = P_{gas}/P$ be the ratio of gas pressure to total pressure and assume that the material of the star is a perfect gas. Then the central temperature is obtained from the perfect-gas law (equation 1.2):

$$T_c \approx \frac{\mu\beta H}{k} \frac{P_c}{\bar{\rho}} \approx \frac{\mu\beta H}{k} \frac{GM}{R} \tag{1.15}$$

The mean energy-generation rate is

$$\bar{\mathscr{E}} = \frac{L}{M}$$

For the sun

$$L = 3.89 \times 10^{33} \text{ ergs/sec}$$
$$M = 1.99 \times 10^{33} \text{ g} \tag{1.16}$$
$$R = 6.95 \times 10^{10} \text{ cm}$$

Thus, the internal conditions of stars are of the order of magnitude

$$\bar{\rho} = 1.41 \left(\frac{M}{M_\odot}\right)\left(\frac{R_\odot}{R}\right)^3 \text{ g/cm}^3$$

$$P_c \approx 1.1 \times 10^{16} \left(\frac{M}{M_\odot}\right)^2 \left(\frac{R_\odot}{R}\right)^4 \text{ dyn/cm}^2$$

$$T_c = 2.3 \times 10^7 \mu\beta \left(\frac{M}{M_\odot}\right)\left(\frac{R_\odot}{R}\right) \text{°K} \tag{1.17}$$

$$\bar{\mathscr{E}} = 1.9 \left(\frac{L}{L_\odot}\right)\left(\frac{M_\odot}{M}\right) \text{ ergs/g-sec}$$

as functions of the stars' mass, radius, and luminosity given in solar units.

For a more detailed account of the restrictions imposed by hydrostatic equilibrium on stellar structure, see Chandrasekhar (1939).

This section is concluded by calculating the gravitational potential energy of a sphere of uniform density. The gravitational potential energy is

$$\Omega = -\int_0^R \frac{GM(r)}{r} dM(r) = \frac{1}{2}\int_0^R \Phi \, dM(r) \tag{1.18}$$

where Φ is the gravitational potential.

For a sphere of uniform density, the equation of hydrostatic equilibrium (1.1) is

$$\frac{1}{\rho}\frac{dP}{dr} = \frac{d}{dr}\left(\frac{P}{\rho}\right) = -\frac{d\Phi}{dr}$$

Upon integrating, using the boundary condition that $P/\rho \to 0$ at the surface, we get

$$-\Phi + \Phi_s = \frac{P}{\rho} \quad \text{and} \quad \Phi_s = \frac{GM}{R}$$

then

$$\Omega = -\tfrac{1}{2}\frac{GM}{R}\int_0^R dM(r) - \tfrac{1}{2}\int_0^R \frac{P}{\rho} dM(r)$$

$$= -\tfrac{1}{2}\frac{GM^2}{R} - \tfrac{1}{2}\int_0^R P4\pi r^2\, dr$$

$$= -\tfrac{1}{2}\frac{GM^2}{R} + \frac{4\pi}{6}\int \frac{dP}{dr}r^3\, dr$$

$$= -\tfrac{1}{2}\frac{GM^2}{R} + \tfrac{1}{6}\Omega$$

Thus, the gravitational potential energy of a sphere of uniform density is

$$\Omega = -\tfrac{3}{5}\frac{GM^2}{R} \tag{1.19}$$

The absolute value of the gravitational potential energy in an actual star will be somewhat larger, but of the same order of magnitude.

STELLAR EVOLUTION

A star is a self-gravitating mass of gas in space. The evolutionary trend of internal stellar conditions is determined by hydrostatic equilibrium and the loss of energy by a star due to radiation into space. The life history of a star is the progressive concentration of its mass toward its center, pulled by its own gravitational field, which releases gravitational energy, supplies the energy losses, and heats up the gas. As the gas becomes hotter, thermonuclear reactions among various nuclei become possible. At certain temperatures, the thermonuclear reactions can supply the energy losses, the gas and radiation pressure can support the star, and the gravitational contraction is temporarily halted.

A necessary condition for hydrostatic equilibrium is the virial theorem for a self-gravitating mass (Chandrasekhar, 1939):

$$2K + \Omega = 3(\gamma - 1)U + \Omega = 0 \tag{2.1}$$

Here, K is the total thermal energy of the mass, U is its internal energy, and Ω is its gravitational potential energy. The virial theorem requires that the thermal energy of a star equal half the absolute value of its gravitational potential energy (since Ω is intrinsically negative). As a star contracts and releases gravitational energy, Ω becomes more negative, and the thermal energy must increase. Half of the gravitational energy that is released is stored as thermal energy, increasing the temperature in the interior of the star, and half is radiated away.

The mean relation of temperature to density can be derived from the virial theorem. For a sphere of gas whose internal pressure is given by the perfect gas law with ratio of specific heats $\gamma = \tfrac{5}{3}$, the virial theorem (2.1) becomes

$$2U + \Omega = 0 \tag{2.2}$$

For a uniform density distribution, the gravitational energy (equation 1.19) is

$$\Omega = -\tfrac{3}{5}\frac{GM^2}{R}$$

and the internal energy is

$$U = \tfrac{3}{2}k\overline{T}\frac{M}{\mu H} \tag{2.3}$$

where $M/\mu H$ is the number of particles. Thus,

$$\overline{T} = \tfrac{1}{5}\frac{\mu H}{k}G\frac{M}{R} \tag{2.4}$$

The density (equation 1.13) is

$$\bar{\rho} = \frac{3}{4\pi}\frac{M}{R^3}$$

so that

$$\overline{T} = \tfrac{1}{5}\left(\frac{4\pi}{3}\right)^{\frac{1}{3}}\frac{G\mu H}{k}M^{\frac{2}{3}}\bar{\rho}^{\frac{1}{3}}$$

Thus, the relation between temperature and density for stars with negligible radiation pressure is

$$\overline{T} = 4.1 \times 10^6 \mu\left(\frac{M}{M_\odot}\right)^{\frac{2}{3}}\bar{\rho}^{\frac{1}{3}} \;^\circ K \tag{2.5}$$

For nonuniform density distributions, the same relation holds between the local temperature and density, but with a slightly different numerical coefficient.

The above temperature–density relation does not hold for those stars whose internal pressures are predominantly governed by radiation pressure. The gas and radiation pressures contribute equally to the total pressure when

$$\tfrac{1}{3}aT^4 = \frac{k}{\mu H}\rho T$$

or

$$T = 2.55 \times 10^7 \rho^{\frac{1}{3}} \tag{2.6}$$

This condition occurs at $5.5M_\odot$. For heavier stars, radiation pressure is predominant. In such cases, $\gamma = \tfrac{4}{3}$ and the virial theorem gives $U = -\Omega$. Thus,

$$U = \frac{1}{\rho}aT^4 = \tfrac{3}{5}\frac{GM^2}{R}$$

so

$$T = \left(\frac{9G}{20\pi a}\right)^{\frac{1}{4}}\frac{M^{\frac{1}{2}}}{R} \tag{2.7}$$

Expressing R in terms of the mean density (1.13), we obtain the temperature–density relation

$$T = \left(\frac{4\pi}{3}\right)^{\frac{1}{12}}\left(\frac{3G}{5a}\right)^{\frac{1}{4}} M^{\frac{1}{2}}\rho^{\frac{1}{3}}$$

$$= 1.92 \times 10^7 \left(\frac{M}{M_\odot}\right)^{\frac{1}{6}} \rho^{\frac{1}{3}}{}^\circ\text{K}$$

(2.8)

The temperature depends on density as before (to the $\frac{1}{3}$ power), but the effect of mass is less pronounced.

The temperature–density relations (2.5) and (2.8) describe the dependence of the temperature on the density inside a star. They also describe the evolution of stars, which consists of progressive gravitational contraction, increasing the central density and temperature according to

$$T \propto \rho^{\frac{1}{3}}$$

Some simplified evolutionary tracks for internal stellar conditions are shown in Figure 1.

When the central density of a star gets very great, the electrons may become degenerate and the equation of state thus changes. The boundary of degeneracy in terms of density and temperature has the asymptotic forms for low and high density (nonrelativistic and relativistic energies):

$$T = 1.2 \times 10^5 \left(\frac{\rho}{\mu_e}\right)^{\frac{2}{3}} \qquad \text{Low density}$$

$$T = 1.49 \times 10^7 \left(\frac{\rho}{\mu_e}\right)^{\frac{1}{3}} \qquad \text{High density}$$

(2.9)

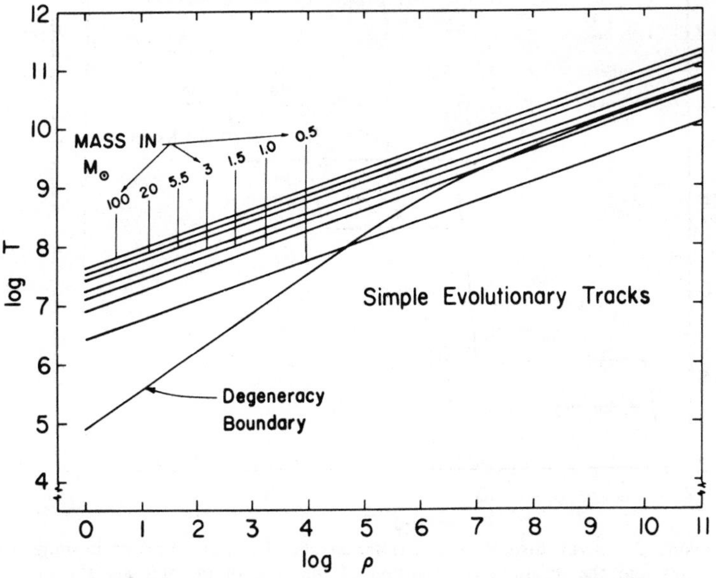

Figure 1. Simple evolutionary tracks for the internal conditions of stars of various masses.

The full boundary curve has been derived by Chandrasekhar (1939). This boundary is also plotted in Figure 1.

Stars of mass less than about $1.3M_\odot$ enter the degenerate region. For these stars, the pressure due to degenerate electrons is so high that further compression is no longer possible. This is essentially the end-point in the evolution of a star of small mass. The star becomes a white dwarf, achieving in this process some maximum temperature which depends specifically on its mass.

The general evolutionary trend of contraction, increasing the central density and temperature, is interrupted periodically by nuclear burning. The energy-generation history of a star is a succession of gravitational contractions which raise the central temperature of the star sufficiently to initiate thermonuclear reactions; the thermonuclear reactions transform a given type of fuel nuclei into heavier nuclei and release energy; the supply of the given fuel nuclei becomes exhausted and the core resumes its gravitational contraction. The order of thermonuclear reactions is determined by the nuclei present and their charges. The larger the nuclear charge, the higher its Coulomb barrier and the higher the kinetic energy (temperature) of the bombarding particles must be to penetrate the barrier and initiate nuclear reactions. A schematic sketch of the energy history of a star is shown in Figure 2. During nuclear burning, the temperature is almost constant. During gravitational contraction, the isotopic composition does not change.

The most abundant element is hydrogen, which also has the lowest charge, 1. It is transformed into He^4, releasing 6×10^{18} erg/g at temperatures above 10^7 °K (Reeves, 1965). Helium is transformed into C^{12} at temperatures above about 10^8 °K, and at slightly higher temperatures carbon reacts with helium to form O^{16}. The

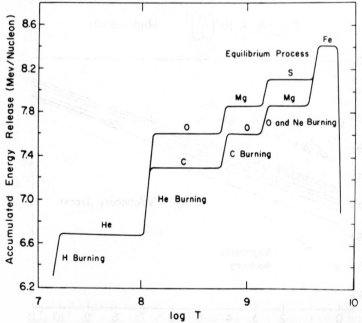

Figure 2. Energy history of a star (schematic diagram). Nuclear burning stages and the resulting composition of the core of the star are shown. Where two curves are drawn, they represent the lower and upper limits of the range of nuclei produced (Reeves, 1963).

amounts of carbon and oxygen produced in the core during helium-burning depend on the central temperature and therefore on the mass of the star. The C and O curves in Figure 2 are the lower and upper limits, respectively. Carbon reacts with itself at temperatures above about 7×10^8 °K; carbon-burning produces nuclei in the range O^{16} to Mg^{26}. The two curves again are the upper and lower limits. Neon photodisintegrates and oxygen reacts with itself at still higher temperatures, about 1.4×10^9 °K. Neon-burning produces predominantly O^{16} and Mg^{24}. Oxygen produces isotopes in the mass range $A = 25$ to 32, with a strong peak at Si^{28}. The two curves show the lower and upper limits.

The full chain of thermonuclear reactions does not occur in all stars. For a star of given mass, there is a maximum central temperature attainable in a non-degenerate core. The exclusion principle requires that the average separation of particles be greater than the electron wavelength:

$$\bar{r} = \left(\frac{m_p}{\rho}\right)^{\frac{1}{3}} > \lambda_e = \frac{\hbar}{(2m_e kT)^{\frac{1}{2}}} \tag{2.10}$$

where $\bar{r}$ is the size of a cube containing one proton and $\lambda_e = \hbar/P$ and $P = (2m_e kT)^{\frac{1}{2}}$. Using expressions (1.13) and (2.4) for $\bar{\rho}$ and T, we must have

$$1 > \frac{\hbar \bar{\rho}^{\frac{1}{3}}}{m_p^{\frac{1}{3}}(2m_e kT)^{\frac{1}{2}}} = 0.0914 \mu^{-\frac{1}{2}} \left(\frac{M_\odot}{M}\right)^{\frac{1}{6}} \left(\frac{R_\odot}{R}\right)^{\frac{1}{2}}$$

Thus, the condition for nondegeneracy requires

$$\left(\frac{R}{R_\odot}\right) > 8.36 \times 10^{-3} \mu^{-1} \left(\frac{M_\odot}{M}\right)^{\frac{1}{3}} \tag{2.11}$$

The necessary central temperature for hydrogen-burning is 10^7 °K, so that mass and radius must satisfy the condition, from (2.4),

$$\mu \left(\frac{M}{M_\odot}\right) \left(\frac{R_\odot}{R}\right) \geq \frac{10^7}{4.61 \times 10^6} = 2.16 \tag{2.12}$$

Combining these two requirements, equations (2.11) and (2.12), the minimum mass of a star that can burn hydrogen is

$$\mu^{\frac{3}{2}} \frac{M}{M_\odot} \geq 0.05 \tag{2.13}$$

For helium-burning, the central temperature must be 10^8 °K. The maximum central temperature occurs when the hydrogen-burning shell has burnt its way almost to the surface, so we can treat the core as a homogeneous star. The minimum mass for helium-burning is thus

$$\mu_c^{\frac{3}{2}} \frac{M}{M_\odot} \geq 0.28 \quad \text{or} \quad \frac{M}{M_\odot} \geq 0.18 \tag{2.14}$$

The necessary central temperature for carbon-burning is about 7×10^8 °K, so the minimum mass for carbon-burning is

$$\mu_c^{\frac{3}{2}} \frac{M}{M_\odot} \geq 1.2 \tag{2.15}$$

The necessary central temperature for neon- and oxygen-burning is 1.3×10^9 °K, so the minimum mass for neon- and oxygen-burning is

$$\mu_c^{\frac{3}{2}} \frac{M}{M_\odot} \geq 1.9 \tag{2.16}$$

Oxygen- and neon-burning are the end point of thermonuclear burning stages. Nuclear reactions among larger-mass nuclei (further photodisintegrations and recombinations) would occur in the temperature range of 2 to 4×10^9 °K. However, at these temperatures, the rate of energy dissipation by neutrinos (which are produced in the core and escape directly from the star) is so large that further nuclear reactions are unable to halt the gravitational contraction, but can merely slow it down. These reactions can, however, produce nuclei all the way up to Fe^{56}, and the temperature is high enough to produce statistical equilibrium among the various nuclei.

EARLY STAGES OF EVOLUTION—HOMOGENEOUS STARS

Hydrostatic equilibrium and overall energy conservation determine the evolution of central stellar conditions. For more of the details of evolution, including the star's radius and luminosity, the mode of energy transport from the interior to the surface must also be considered.

The equations of stellar structure—mass conservation, hydrostatic equilibrium, energy conservation, and energy transport—form a system of nonlinear differential equations which must be integrated numerically. It is possible, however, to obtain crude analytic stellar models by separating the condition of hydrostatic equilibrium from the energy transport. In the previous section, the condition of overall hydrostatic equilibrium was expressed by the virial theorem. Now, since a more detailed stellar model is desired, we assume an analytic density distribution, namely, that the density in a star varies linearly from the center to the surface (Cameron, 1963). It is then possible to integrate the equations of mass conservation, hydrostatic equilibrium, and energy generation through the star. Hence, together with the equation of state of an ideal gas, the run of density, mass, pressure, temperature, and luminosity through the star is determined. Also, the central density, pressure and temperature, and the total rate of energy generation are determined as a function of the star's mass and radius. Finally, the different modes of energy transport—radiative transport with Kramer's or electron-scattering opacity and convective transport—are considered. The energy-transport equation can be satisfied at only one typical point of the star because of the approximation made in assuming a given density distribution. This gives a mass–luminosity–radius relation which gives the evolutionary track of the star in the Hertzsprung–Russell diagram.

To summarize: Hydrostatic equilibrium and energy conservation determine the changes in the central stellar conditions, including the mode of energy transport, which gives the changes in the surface conditions—the track in the Hertzsprung–Russell diagram.

Linear Stellar Model

Assume that the density in a star varies linearly from the center to the surface:

$$\rho(r) = \rho_c \left(1 - \frac{r}{R}\right) \tag{3.1}$$

where R is the radius of the star. We call this a "linear star model." The equations of hydrostatic equilibrium and energy generation can now be integrated, but the energy-transport equation can only be satisfied at one point in the star. The mass distribution is [from equation (1.4)]

$$M(r) = \int_0^r 4\pi r^2 \rho(r)\, dr$$

$$= \frac{4\pi}{3}\rho_c r^3 \left(1 - \frac{3}{4}\frac{r}{R}\right) \tag{3.2}$$

Hence

$$M(R) = \tfrac{1}{3}\pi \rho_c R^3$$

Thus, the central density is

$$\rho_c = \frac{3M}{\pi R^3}$$

$$= 5.64\left(\frac{M}{M_\odot}\right)\left(\frac{R_\odot}{R}\right)^3 \text{g/cm}^3 \tag{3.3}$$

The pressure is obtained from the equation of hydrostatic equilibrium (1.1)

$$P = P_c - \int_0^r \frac{GM(r)\rho(r)\, dr}{r^2}$$

where P_c is the pressure at the center. Hence

$$P = P_c - \frac{2\pi}{3}G\rho_c^2 r^2 \left(1 - \frac{7}{6}\frac{r}{R} + \frac{3}{8}\frac{r^2}{R^2}\right)$$

Applying the boundary condition $P(R) = 0$, we get

$$P = \frac{\pi}{36}G\rho_c^2 R^2 \left(5 - \frac{24r^2}{R^2} + \frac{28r^3}{R^3} - \frac{9r^4}{R^4}\right) \tag{3.4a}$$

and the central pressure is

$$P_c = \frac{5\pi}{36}G\rho_c^2 R^2 = 4.44 \times 10^{15}\left(\frac{M}{M_\odot}\right)^2\left(\frac{R_\odot}{R}\right)^4 \text{dyn/cm}^2 \tag{3.4b}$$

Assume that the radiation pressure is negligible; the temperature is then given by the perfect gas law, equation (1.2),

$$T = \frac{\mu H}{k}\frac{P}{\rho} = \frac{\pi}{36}\frac{G\mu}{kN_0}\rho_c R^2 \left(5 + \frac{5r}{R} - \frac{19r^2}{R^2} + \frac{9r^3}{R^3}\right) \tag{3.5a}$$

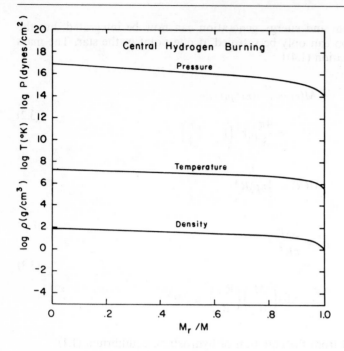

Figure 3.

and the central temperature is

$$T_c = \frac{5\pi}{36} \frac{G\mu H}{k} \rho_c R^2 = 9.62 \times 10^6 \mu \left(\frac{M}{M_\odot}\right)\left(\frac{R_\odot}{R}\right) °K \tag{3.5b}$$

We now know how the density, temperature, and pressure vary throughout the interior of this linear star model. We have satisfied the condition of hydrostatic equilibrium. The run of pressure, temperature, and density through the star is shown in Figure 3.

We must now consider the condition of energy conservation. The rate of thermonuclear energy generation can be expressed in the form (Reeves, 1965)

$$\mathscr{E} = \mathscr{E}_0 \rho^k \left(\frac{T}{T_0}\right)^n \text{ ergs/g-sec}$$

The total net rate of energy generation is equal to the luminosity of the star (equation 1.6)

$$L = \int_0^R 4\pi\rho(r)\mathscr{E}r^2 \, dr$$

For a linear density distribution,

$$L = 4\pi R^3 \mathscr{E}_0 \rho_c^2 \left(\frac{T_c}{T_0}\right)^n I_n$$

$$= \frac{36}{\pi} \mathscr{E}_0 \left(\frac{5}{12} \frac{G\mu H}{k} \frac{1}{T_0}\right)^n \frac{M^{1+k+n}}{R^{3k+n}} I_n \tag{3.6}$$

where

$$I_n = \int_0^1 x^2(1 - x)^{n+k+1}(1 + 2x - 1.8x^2)^n \, dx$$

is the integral of $[\rho(r)/\rho_c]^2[T(r)/T_c]^n$ over the star and has values of the order 10^{-1} or 10^{-2}. Thus,

$$\frac{L}{L_\odot} = 35.58\mathscr{E}_0 I_n \left[\frac{0.962}{T_{0(7)}}\right]^n \mu^n \left(\frac{M}{M_\odot}\right)^{n+k+1} \left(\frac{R_\odot}{R}\right)^{n+3k} \tag{3.7}$$

Here, $T_{0(7)}$ is the temperature in units of $10^7\,°K$. The energy generation and luminosity in a $1M_\odot$ star are shown in Figure 4.

Radiative Energy Transport

Finally, consider the equation that governs the flow of energy through the star. First consider radiative energy transport, equation (1.8):

$$L_r = -4\pi r^2 \frac{4ac}{3} \frac{T^3}{\varkappa\rho} \frac{dT}{dr}$$

so the temperature gradient necessary to drive the radiative flux through the star is

$$\frac{dT}{dr} = -\frac{3\varkappa}{4ac} \frac{\rho}{T^3} \frac{L_r}{4\pi r^2}$$

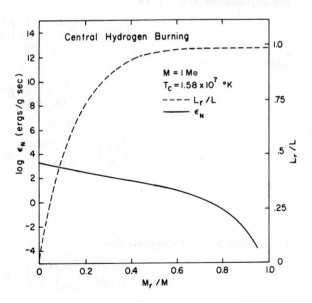

Figure 4.

We consider two types of opacity (Cameron, 1965, and Cox, 1965):
(1) Kramer's opacity

$$\varkappa = \varkappa_0 \rho T^{-3.5}$$

$$\varkappa_0 = 4.34 \times 10^{25}\left(\frac{\bar{g}_{bf}}{t}\right)Z(1 + X) + 3.68 \times 10^{22}\bar{g}_{ff}(1 + X)(X + Y)$$

(3.8)

where X, Y, Z are the mass fractions of hydrogen, helium, and all the heavier elements, respectively. The first term is the bound–free and the second the free–free absorption. We take $(t/\bar{g}_{bf}) = 3$ and $\bar{g}_{ff} = 1$. Kramer's opacity is a good approximation at intermediate internal temperatures.
(2) Electron-scattering opacity

$$\varkappa = \varkappa_e = 0.20(1 + X)$$

(3.9)

which is dominant at high internal temperatures. We also assume for convenience that all the energy is generated at the center of the star, so that

$$L_r = L = \text{constant}$$

The temperature gradient for Kramer's opacity is

$$\frac{dT}{dr} = \frac{3\varkappa_0}{4ac}\frac{\rho^2}{T^{6.5}}\frac{L}{4\pi r^2}$$

(3.10)

Compare this expression for dT/dr with the radial derivative of T from the linear model

$$\frac{dT}{dr} = \frac{\pi}{36}\frac{G\mu\rho_c}{kN_0}R\left(5 - 38\frac{r}{R} + 27\frac{r^2}{R^2}\right)$$

(3.11)

For our analytic model, these two expressions for the temperature gradient cannot be equal throughout the star. We determine the luminosity by equating the above two expressions at $r = 0.5R$:

$$L = -4\pi r_{\frac{1}{2}}^2\frac{4ac}{3\varkappa_0}\frac{T_{\frac{1}{2}}^{6.5}}{\rho_{\frac{1}{2}}^2}\left(\frac{dT}{dr}\right)_{\frac{1}{2}}$$

where $r_{\frac{1}{2}} = 0.5R$

$$T_{\frac{1}{2}} = \frac{31}{288}\pi\frac{G\mu\rho_c}{kN_0}R^2$$

$$\rho_{\frac{1}{2}} = 0.5\rho_c$$

$$\left(\frac{dT}{dr}\right)_{\frac{1}{2}} = \frac{-29\pi}{144}\frac{G\mu}{kN_0}\rho_c R$$

Now

$$\rho_c = \frac{3M}{\pi R^3}$$

Thus, when Kramer's opacity dominates,

$$L = \pi^3\frac{29}{81}\left(\frac{31}{96}\right)^{6.5}\frac{ac}{\varkappa_0}\left(\frac{GH}{k}\right)^{7.5}\mu^{7.5}\frac{M^{5.5}}{R^{0.5}}$$

(3.12)

or

$$\frac{L}{L_\odot} = \frac{0.988}{(1 + X)[Z + 2.54 \times 10^{-3}(1 - Z)]}\mu^{7.5}\left(\frac{M}{M_\odot}\right)^{5.5}\left(\frac{R}{R_\odot}\right)^{-0.5} \tag{3.13}$$

Solar matter is approximately two-thirds hydrogen and one-third helium by weight. The mean molecular weight for twelve nucleons, of which eight are hydrogen atoms and one a helium atom, is

$$\mu = \frac{\text{Mass}}{\text{Number of particles}} = \frac{8 \times 1 + 1 \times 4}{8 \times 2 + 1 \times 3}$$

$$= \frac{12}{19} = 0.632$$

$$\mu^{7.5} = 0.0320$$

Hence, for solar mass and radius,

$$\frac{L}{L_\odot} = 0.915$$

Thus, the linear star model gives a result which is within 10% of the observed value. The luminosity increases rapidly with the mass of the star and increases slightly with decreasing radius.

When electron scattering is the dominant opacity, the temperature gradient needed to transport the energy flux L is

$$\frac{dT}{dr} = -\frac{3\varkappa_e}{4ac}\frac{\rho}{T^3}\frac{L}{4\pi r^2} \tag{3.14}$$

Determining the luminosity by equating this temperature gradient with the expression for dT/dr obtained in the linear model (3.11) at the midpoint $r = 0.5R$ gives

$$L = -4\pi r_{\frac{1}{2}}^2 \frac{4ac}{3\varkappa_e}\frac{T_{\frac{1}{2}}^3}{\rho_{\frac{1}{2}}}\left(\frac{dT}{dr}\right)_{\frac{1}{2}}$$

which is

$$L = \frac{29}{2}\pi^2\left(\frac{31}{288}\right)^3\left(\frac{GH}{k}\right)^4\frac{ac}{\varkappa_e}\mu^4 M^3 \tag{3.15}$$

Hence, when electron scattering dominates,

$$\frac{L}{L_\odot} = \frac{179}{1 + X}\mu^4\left(\frac{M}{M_\odot}\right)^3 \tag{3.16}$$

The luminosity is independent of the radius and increases with mass, although less sensitively than for Kramer's opacity.

Equations (3.13) and (3.16) are the radiative mass–luminosity–radius relations for Kramer's and electron-scattering opacity. The effective surface temperature is defined by

$$\text{flux} = \sigma T_{\text{eff}}^4$$

or

$$T_e = \left(\frac{L}{4\pi\sigma R^2}\right)^{\frac{1}{4}}$$

$$= 5.76 \times 10^3 \left(\frac{L}{L_\odot}\right)^{\frac{1}{4}} \left(\frac{R_\odot}{R}\right)^{\frac{1}{2}} {}^\circ K$$

(3.17)

Convective Energy Transport

Convection is an extremely efficient mode of energy transport. Therefore, in a convectively unstable region the entire energy flux can be transferred with only negligible adjustment in the super-adiabatic gradient: $-[(dT/dr) - (dT/dr)_{ad}]$. The energy flux is thus determined by the boundary layer of the convective region (Spiegel, 1965).

If the star has a substantial region with radiative transport, that region will determine the energy flux. If, however, the stellar interior is completely convective, the boundary layer determining the flux is the thin radiative photosphere surrounding the convective zone, where the energy must be transported by radiation since the material is becoming optically thin. The luminosity of the star is then determined by the temperature of the gas at the point from which photons can escape from the star,

$$L = 4\pi R^2 \sigma T_e^4$$

where T_e is the effective surface temperature of the star.

The depth in the star from which photons can escape nearly coincides with the transition point between the radiative and convective regions and occurs at an optical depth of about $\frac{2}{3}$. The radiative temperature gradient drops rapidly as the density decreases, so the temperature is practically constant from this point outward. We thus assume an isothermal photosphere and take the effective temperature as the temperature at the transition point between the convective and radiative regions (Hoyle and Schwarzschild, 1955, and Hayashi, Hoshi, and Sugimoto, 1962).

We assume an opacity law of the form

$$\varkappa = \varkappa_0 P^a T^b$$

Then, since the bottom of the photosphere is at an optical depth $\frac{2}{3}$,

$$\tau = \int_{r_{ph}}^\infty \varkappa\rho \, dr = \frac{2}{3} = \varkappa_0 T_e^b \int_{r_{ph}}^\infty P^a \rho \, dr$$

and from equation (1.1)

$$\rho = -\frac{1}{g}\frac{dP}{dr}$$

so

$$\frac{2}{3} = -\varkappa_0 T_e^b \frac{1}{g} \int_{r_{ph}}^\infty P^a \, dP = \frac{\varkappa_0 T_e^b P_{ph}^{a+1}}{(a+1)g}$$

where P_{ph} is the pressure at the bottom of the photosphere. Thus, one relation between the temperature and pressure (or density) at the bottom of the photosphere is

$$T_e^b P_{ph}^{a+1} = \tfrac{2}{3}(a+1)\frac{GM}{\varkappa_0 R^2} \qquad (3.18)$$

This relation is the boundary condition for the star:

$$P \to P_{ph} = \tfrac{2}{3}(a+1)\frac{g}{\varkappa_{ph}} \quad \text{as} \quad T \to T_{eff}$$

This condition is just that the photon mean free path $(\varkappa\rho)^{-1}$ equals the scale height $P/\rho g$ at the boundary so that the radiation can escape from the star at the effective temperature.

A second condition on T_e and P_{ph} can be obtained from the condition for the boundary of the convective zone, namely,

$$F_C = F_R$$

In the expression for the convective flux (1.10), approximate the velocity w by half the sound velocity

$$c = \left(\frac{\gamma kT}{\mu H}\right)^{\frac{1}{2}}$$

since c is an upper limit to the velocity. Also, approximate $\rho c_p \theta$ by γ times the internal energy

$$U = \tfrac{3}{2}kT\frac{\rho}{\mu H} = \tfrac{3}{2}P$$

Then the convective flux is

$$F_C = \tfrac{1}{2}\rho c_p w\theta \approx \tfrac{1}{2}\gamma\frac{c}{2}U$$
$$= \frac{3\gamma}{8}\left(\gamma\frac{k}{\mu H}\right)^{\frac{1}{2}} PT^{\frac{1}{2}} \qquad (3.19)$$

The radiative flux is

$$F_R = \sigma T_e^4 \qquad (3.20)$$

Thus, equating (3.19) and (3.20), the transition point is given by

$$P_{ph} = \frac{8}{3\gamma}\left(\frac{\mu H}{\gamma k}\right)^{\frac{1}{2}} \sigma T_e^{3.5} \qquad (3.21)$$

The conditions (3.18) and (3.21) can be combined to determine the effective temperature, which is

$$T_e = \left[\tfrac{2}{3}(1+a)\left(\frac{3\gamma}{8}\right)^{1+a}\frac{G}{\sigma^{1+a}\varkappa_0}\left(\frac{\gamma k}{H}\right)^{(1+a)/2}\mu^{-(1+a)/2}\left(\frac{M}{R^2}\right)\right]^{1/[b+3.5(1+a)]} \qquad (3.22)$$

In the outer layers of stars, the opacity is due primarily to H^- and is an increasing function of pressure and temperature, so $a, b > 0$. The H^- opacity is very temperature sensitive, so b is large. Thus, T_{eff} is nearly constant; it increases slightly with increasing mass and decreases slightly with increasing radius.

The approximate power law form for the opacity obtained from the detailed opacity calculations in the region about 3500 °K is as follows:

For population I stars ($X = 0.6$, $Y = 0.38$, $Z = 0.02$)

$$\varkappa = 6.9 \times 10^{-26} P^{0.7} T^{5.3}$$

For population II stars ($X = 0.9$, $Y = 0.099$, $Z = 0.001$)

$$\varkappa = 6.1 \times 10^{-40} P^{0.6} T^{9.4}$$

where X, Y, Z are the mass fractions of hydrogen, helium, and all the heavier elements, respectively. The luminosity is found by inverting equation (3.17):

$$\frac{L}{L_\odot} = \frac{4\pi R^2 \sigma T_e^4}{L_\odot} = \left(\frac{T_e}{5.76 \times 10^3}\right)^4 \left(\frac{R}{R_\odot}\right)^2 \tag{3.23}$$

Then the effective temperature and luminosity are:

For population I

$$T_e = 7.5 \times 10^3 \left(\frac{M}{M_\odot}\right)^{0.089} \left(\frac{R}{R_\odot}\right)^{-0.178}$$

$$\frac{L}{L_\odot} = 2.86 \left(\frac{M}{M_\odot}\right)^{0.356} \left(\frac{R}{R_\odot}\right)^{1.288} \tag{3.24a}$$

For population II

$$T_e = 6.19 \times 10^3 \left(\frac{M}{M_\odot}\right)^{0.0666} \left(\frac{R}{R_\odot}\right)^{-0.133}$$

$$\frac{L}{L_\odot} = 1.34 \left(\frac{M}{M_\odot}\right)^{0.2665} \left(\frac{R}{R_\odot}\right)^{1.466} \tag{3.24b}$$

The effective temperature is less sensitive to the radius for population II than for population I stars because the opacity is more sensitive to temperature. In population II stars, there are fewer metals with low ionization potentials to provide electrons to form H^-. The electrons must now come partly from the ionization of hydrogen which has a high ionization potential, so the electron pressure will be very temperature sensitive.

In stars with high surface density, the relation (3.21) between the pressure and temperature at the bottom of the photosphere is not valid, because in deriving it from the boundary condition $F_C = F_R$ we evaluated the convective flux by assuming that the temperature fluctuation is of the order of magnitude of the temperature itself. This assumption is valid only in stars where convection is inefficient near the surface due to low density and large radiative losses from the convective elements. In stars with high surface density, convection is very efficient and the temperature gradient in the convective region is nearly adiabatic throughout. In this case, the temperature fluctuations are much smaller than the order of magnitude of the temperature itself.

For stars with high surface density, we therefore go to the opposite extreme from the low surface density case and assume that the temperature gradient is adiabatic throughout the convective zone. We may then use the adiabatic relation between pressure and temperature. In the interior

$$P = KT^{\gamma/(\gamma-1)} = KT^{2.5}$$

since (neglecting radiation pressure) $\gamma = \frac{5}{3}$, except in the hydrogen-ionization zone. For a fully convective star

$$K = \text{constant} = \frac{P_c}{T_c^{2.5}}$$

From the linear model (3.4) and (3.5)

$$P_c = \frac{5}{4\pi} \frac{GM^2}{R^4}$$

$$T_c = \frac{5}{12} \frac{G\mu H}{k} \frac{M}{R}$$

thus

$$K = \frac{5}{4\pi} \left(\frac{12}{5}\right)^{2.5} \left(\frac{k}{\mu H}\right)^{2.5} G^{-1.5} M^{-0.5} R^{-1.5}$$

$$= 1.53 \times 10^{-2} \mu^{-2.5} \left(\frac{M}{M_\odot}\right)^{-0.5} \left(\frac{R}{R_\odot}\right)^{-1.5}$$

In particular, the above relation holds at the bottom of the hydrogen-ionization zone.

If we neglect the effect of hydrogen ionization, which reduces γ, then at the boundary between the convective zone and the photosphere

$$P_{\text{ph}} = KT_e^{2.5}$$

with the same K as for the interior. This relation, combined with the optical depth condition from equation (3.18), gives the effective temperature

$$T_e = \left[\tfrac{2}{3}(1+a) \frac{GM}{\varkappa_0 R^2} K^{-(1+a)} \right]^{1/[b+2.5(1+a)]}$$

$$= \left\{ \tfrac{2}{3} \frac{1+a}{\varkappa_0} \left[\frac{4\pi}{5}\left(\frac{5}{12}\right)^{2.5}\left(\frac{H}{k}\right)^{2.5} \right]^{1+a} G^{2.5+1.5a} \right. \tag{3.25}$$

$$\left. \times \, \mu^{2.5(1+a)} M^{1.5+0.5a} R^{1.5a-0.5} \right\}^{1/[b+2.5(1+a)]}$$

For population I

$$T_e = 2.18 \times 10^3 \left(\frac{M}{M_\odot}\right)^{0.194} \left(\frac{R}{R_\odot}\right)^{0.0576}$$

$$\frac{L}{L_\odot} = 0.02 \left(\frac{M}{M_\odot}\right)^{0.776} \left(\frac{R}{R_\odot}\right)^{2.23}$$

For population II

$$T_e = 2.5 \times 10^3 \left(\frac{M}{M_\odot}\right)^{0.1715}\left(\frac{R}{R_\odot}\right)^{0.0298}$$

$$\frac{L}{L_\odot} = 0.035\left(\frac{M}{M_\odot}\right)^{0.686}\left(\frac{R}{R_\odot}\right)^{2.119}$$

The hydrogen ionization can, however, be treated exactly and we can relate $K_e = P_{\rm ph}/T_e^{2.5}$ at the top of the hydrogen-ionization zone to

$$K = P_b/T_b^{2.5} = P_c/T_c^{2.5}$$

at its bottom. The effect of the ionization zone is to decrease

$$\frac{d \ln T}{d \ln P} = \frac{\Gamma - 1}{\Gamma}$$

so that the temperature will decrease less than the pressure going outward through the ionization zone. Then $K_e < K$ and $T_{\rm eff}$ will be increased. Since the temperature varies adiabatically through the ionization zone, the entropy is constant across it. The entropy per unit mass is

$$s = \frac{Xk}{H}\left[\tfrac{5}{2}(1 + x + \delta) + \frac{\chi}{kT} + \ln\left(\frac{2\pi H}{h^2}\right)^{\frac{3}{2}} + \delta \ln\left(\frac{8\pi H}{h^2}\right)\right.$$

$$\left. + x \ln\left(\frac{2\pi m_e}{h^2}\right)^{\frac{3}{2}} + (1 + x + \delta) \ln\frac{(kT)^{\frac{5}{2}}(1 + x + \delta)}{P}\right] \tag{3.26}$$

where χ is the ionization energy of hydrogen, $\delta = Y/4X$, and x is the fraction of hydrogen ionized. Evaluating $s = $ constant above and below the hydrogen-ionization zone, that is, for $x = 0$ and $x = 1$, respectively, gives

$$\frac{P_{\rm ph}}{T_e^{2.5}} = K_e = (1 + \delta)\left[\left(\frac{2\pi m_e}{h^2}\right)^{\frac{3}{2}}(ke)^{\frac{5}{2}}\right]^{-1/(1+\delta)}\left(\frac{K}{2 + \delta}\right)^{(2+\delta)/(1+\delta)}$$

Thus, the effective temperature is

$$T_{\rm eff} = \left[\tfrac{2}{3}\frac{1 + a}{\varkappa_0}\left\{\frac{1}{1 + \delta}\left[\left(\frac{2\pi m_e^2}{h^2}\right)^{\frac{3}{2}}(ke)^{\frac{5}{2}}\right]^{1/(1+\delta)}\left[\frac{4\pi(2 + \delta)}{5}(\tfrac{5}{12})^{\frac{5}{2}}\left(\frac{H}{k}\right)^{\frac{5}{2}}\right]^{(2+\delta)/(1+\delta)}\right\}^{1+a}\right.$$

$$\times G^{1 - 1.5(1+a)[(2+\delta)/(1+\delta)]}\mu^{2.5(1+a)[(2+\delta)/(1+\delta)]}M^{1+0.5(1+a)[(2+\delta)/(1+\delta)]}$$

$$\left. \times R^{1.5(1+a)[(2+\delta)/(1+\delta)]-2}\right]^{1/[b+2.5(1+a)]} \tag{3.27}$$

Again, in the high surface density as in the low surface density case, the effective temperature is very insensitive to mass and radius.

For population I

$$T_e = 5.65 \times 10^3 \left(\frac{M}{M_\odot}\right)^{0.27} \left(\frac{R}{R_\odot}\right)^{0.288}$$

$$\frac{L}{L_\odot} = 0.93 \left(\frac{M}{M_\odot}\right)^{1.08} \left(\frac{R}{R_\odot}\right)^{3.15}$$

(3.28a)

For population II

$$T_e = 4.63 \times 10^3 \left(\frac{M}{M_\odot}\right)^{0.1925} \left(\frac{R}{R_\odot}\right)^{0.204}$$

$$\frac{L}{L_\odot} = 0.42 \left(\frac{M}{M_\odot}\right)^{0.77} \left(\frac{R}{R_\odot}\right)^{2.816}$$

(3.28b)

Evolutionary Tracks

The evolutionary tracks of stars in the Hertzsprung–Russell diagram depend on the mode of energy transport, which determines the mass–luminosity–radius relation. For fully convective stars, the luminosity is determined by the surface condition. Since the opacity is very temperature sensitive, the effective temperature is nearly constant, independent of the radius, and the track in the Hertzsprung–Russell diagram is a nearly vertical line. The mass–luminosity–radius relations for a fully convective star are given by equations (3.24) and (3.28), so the track in the H–R diagram will be:

For population I ($X = 0.6$, $Y = 0.38$, $Z = 0.02$)

$$\log\left(\frac{L}{L_\odot}\right) = -7.236 \log\left(\frac{T_e}{8.65 \times 10^3}\right) + \log\left(\frac{M}{M_\odot}\right) \quad \text{(Low surface density)}$$

(3.29a)

$$\log\left(\frac{L}{L_\odot}\right) = 10.94 \log\left(\frac{T_e}{5.69 \times 10^3}\right) - 1.874 \log\left(\frac{M}{M_\odot}\right) \quad \text{(High surface density)}$$

For population I ($X = 0.6$, $Y = 0.38$, $Z = 0.02$)

$$\log\left(\frac{L}{L_\odot}\right) = -11 \log\left(\frac{T_e}{6.19 \times 10^3}\right) + \log\left(\frac{M}{M_\odot}\right) \quad \text{(Low surface density)}$$

(3.29b)

$$\log\left(\frac{L}{L_\odot}\right) = 13.8 \log\left(\frac{T_e}{4.93 \times 10^3}\right) - 1.887 \log\left(\frac{M}{M_\odot}\right) \quad \text{(High surface density)}$$

For stars with radiative energy transport, the flux [equation (1.8)] is

$$L \propto r^2 \frac{T^3}{\varkappa\rho} \frac{dT}{dr} \propto \frac{R^2}{\bar{\varkappa}} \frac{T_c}{R}$$

since T^3/ρ is approximately constant. The opacity in the interior of a star is nearly constant, so the luminosity is approximately proportional to RT_c, which is constant, independent of the radius. Thus the track in the Hertzsprung–Russell diagram is a nearly horizontal line. As the temperature rises, Kramer's opacity

$$\varkappa \propto \rho T^{-3.5} \propto T^{-0.5}$$

decreases, and the luminosity increases slightly. The mass–luminosity–radius relation for a radiative star with Kramer's opacity is given by equation (3.13) and the path in the H–R diagram is

$$\log\left(\frac{L}{L_\odot}\right) = 0.8 \log\left(\frac{T_e}{5.83 \times 10^3}\right) + 4.4 \log\left(\frac{M}{M_\odot}\right) + 6 \log \mu$$
$$- 0.8 \log[(1 + X)\{Z + 2.54 \times 10^{-3}(1 - Z)\}] \qquad (3.30)$$

If the central temperature becomes very high and the central density is low, the dominant opacity is due to electron scattering. Electron-scattering opacity is independent of temperature and density, so the luminosity is constant. The mass–luminosity–radius relation is given by equation (3.16):

$$\frac{L}{L_\odot} = \frac{179}{1 + X} \mu^4 \left(\frac{M}{M_\odot}\right)^3$$

$$T_e = 2.14 \times 10^4 (1 \times X)^{-\frac{1}{4}} \mu \left(\frac{M}{M_\odot}\right)^{\frac{3}{4}} \left(\frac{R}{R_\odot}\right)^{-\frac{1}{2}} \qquad (3.31)$$

The changes in the stellar radius depend on the sources of energy and the internal structure of the star.

PRE-MAIN-SEQUENCE CONTRACTION PHASE

The linear stellar model is now applied to the pre-main-sequence contraction stage of evolution. A star is formed from a condensation of the interstellar gas that is dense enough to become opaque to its own radiation. Then, as the gas contracts, its temperature will rise. As the temperature rises, the gas, composed predominantly of hydrogen and helium, is ionized. Much energy is necessary to ionize the gas, which means that the temperature cannot rise much above 10^4 °K until the hydrogen is ionized. The ionization of the hydrogen and helium leads to gravitational instability, since the energy released by the contraction does not increase the kinetic energy per particle (the temperature) but goes into the ionization energy of the atoms. Hence, the contraction of the gas does not raise the pressure sufficiently to permit the gas to remain in hydrostatic equilibrium; the ratio of specific heats γ falls below $\frac{4}{3}$, and the collapse must continue.

A stable star is not formed until its major constituents (hydrogen and helium) are ionized throughout most of the gas fragment. The energy necessary for ionization comes from gravitational potential-energy release (Truran, 1964). Let I be the total dissociation plus ionization energy per gram of stellar material. The gravitational potential energy is $\Omega \approx -GM^2/R$. From the virial theorem, half the gravitational energy released goes into thermal energy. Thus,

$$IM = \frac{1}{2}\frac{GM^2}{R}$$

so

$$R_{\max} = \frac{1}{2}\frac{GM}{I}$$

The dissociation plus ionization energy is

$$I = N_0\left(\frac{X}{2}D_H + XE_H + \frac{Y}{4}E_{He}\right)$$

where $N_0 = 6.025 \times 10^{23}$ is the number of hydrogen atoms per gram; D_H is the dissociation energy of hydrogen molecules, $D_H = 4.476\,eV = 7.16 \times 10^{-22}$ ergs; E_H is the ionization energy of hydrogen, $E_H = 13.595\,eV = 21.75 \times 10^{-12}$ ergs; E_{He} is the ionization energy of helium, $E_{He} = 24.581 + 54.403\,eV = 78.984\,eV = 1.26 \times 10^{-10}$ ergs. Since $Y = 1 - X$

$$I = 1.9 \times 10^{13}(1 - 0.3X)\,\text{ergs/g}$$

The maximum radius of a stable star is thus

$$\left(\frac{R}{R_\odot}\right)_{max} = \frac{50.3}{1 - 0.3X}\left(\frac{M}{M_\odot}\right) \tag{3.32}$$

Such a marginally stable star is the starting point of stellar evolution.

When a star becomes stable, its internal temperatures are of the order of 10^5 °K, and the opacity is so high that the radiative transport of energy is impeded. Further, there are extensive ionization zones which increase the specific heat and reduce γ to less than $\frac{4}{3}$ throughout large regions of the star. Thus the adiabatic gradient

$$\left(\frac{dT}{dr}\right)_{ad} = -\frac{\gamma - 1}{\gamma}\frac{\mu H}{k}g$$

will be small and the star will be unstable to convection throughout most of its interior. Its luminosity will then be determined by the surface conditions. For a given star, the rate of contraction is limited by the rate at which energy can be radiated away

$$L = -\frac{1}{2}\frac{d\Omega}{dt} \approx -\frac{1}{2}\frac{GM^2}{R}\left(\frac{1}{R}\frac{dR}{dt}\right)$$

so

$$-\frac{1}{R}\frac{dR}{dt} = \alpha\frac{LR}{GM^2}$$

where $\alpha \approx 1$. The equilibrium stellar structure is that with the highest contraction rate, i.e., with the highest luminosity, so the condition for a fully convective star is that the convective luminosity exceed the radiative luminosity. Initially, stars are fully convective. The effective temperature and luminosity of a marginally stable star, as given by the fully convective linear model for low surface density (3.24), are

$$\left.\begin{aligned}T_e &= 3.6 \times 10^3\left(\frac{M}{M_\odot}\right)^{-0.089} \\ \frac{L}{L_\odot} &= 5.71 \times 10^2\left(\frac{M}{M_\odot}\right)^{1.644}\end{aligned}\right\} \text{population I}$$

$$\left.\begin{aligned}T_e &= 3.52 \times 10^3\left(\frac{M}{M_\odot}\right)^{-0.0667} \\ \frac{L}{L_\odot} &= 5.91 \times 10^2\left(\frac{M}{M_\odot}\right)^{1.733}\end{aligned}\right\} \text{population II}$$

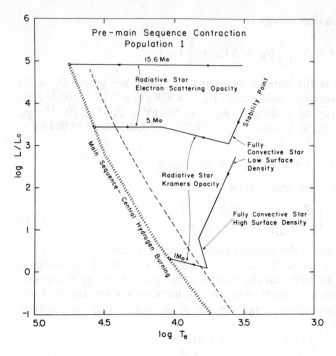

Figure 5. Hertzsprung–Russell diagram of the pre-main-sequence contraction evolutionary tracks and initial main sequence. The tracks are labeled with the type of energy transport determining the direction of that portion of the track. Dotted curve is observed main sequence (Hayashi, Hoshi, and Sugimoto, 1962, and Schwarzschild, 1958).

These relations give the starting point of a star's evolutionary track in the Hertzsprung–Russell diagram.

As a star contracts, when fully convective, the effective temperature is nearly constant. The tracks in the H–R diagram for fully convective stars follow equation (3.29). These tracks are shown in Figure 5. The tracks for population II stars are similar to, but slightly steeper than, those for population I stars.

As a star contracts, its central temperature increases according to (equation 3.5):

$$T_c \approx \frac{\mu H}{k} \frac{GM}{R}$$

The rising temperature increases the emission of radiation and reduces the opacity. A central core which is in radiative equilibrium will develop. When about half the star is in radiative equilibrium, it leaves the fully convective path. The luminosity of a radiative star is nearly constant, and the effective temperature varies as $T_e \sim L^{\frac{1}{4}} R^{-\frac{1}{2}}$. Thus, as the star contracts, the effective temperature rises and the star moves to the left in the H–R diagram at nearly constant luminosity. The track in the H–R diagram for a radiative star follows equation (3.30) for Kramer's opacity and equation (3.31) for electron-scattering opacity. Typical radiative tracks in the H–R diagram are shown in Figure 5.

Time Scale of Contraction

The luminosity of a star is the rate of change of total energy

$$\bar{L} = \frac{\Delta E}{\Delta t} = -\tfrac{1}{2} \frac{\Delta \Omega}{\Delta t}$$

The gravitational energy is [from equation (1.19)]

$$-\,\Omega \approx \frac{GM^2}{R}$$

so

$$\bar{L} = \tfrac{1}{2}\frac{GM^2}{R\Delta t}$$

Thus, the time scale of the contraction phase is

$$\Delta t = \tfrac{1}{2}\frac{GM^2}{\bar{L}R}$$

$$= 1.59 \times 10^7 \left(\frac{M}{M_\odot}\right)^2 \left(\frac{R_\odot}{R}\right)\left(\frac{\bar{L}}{L_\odot}\right)^{-1} \text{ years}$$

(3.33)

Pre-main-sequence contraction times are listed in Table I.

CENTRAL HYDROGEN-BURNING

As a star contracts, its central temperature rises until it is high enough for hydrogen thermonuclear reactions to produce the energy radiated away from the star. At this point, the contraction stops and the star spends most of its lifetime

Table I. Evolutionary Time Scales (Years)

Mass ($M_\odot$)	Population	Pre-main-sequence contraction	Central hydrogen-burning	Hydrogen shell-burning	Central helium-burning
0.7	I	7×10^7	5×10^{10}	8×10^8	1×10^8
	II	8×10^7	5×10^{10}	5×10^9	6×10^7
1	I	2×10^7	9×10^9	4×10^8	5×10^7
	II	3×10^7	9×10^9	2×10^9	4×10^7
2	I	2×10^6	5×10^8	6×10^7	2×10^7
	II	3×10^6	7×10^8	2×10^8	2×10^7
5	I	1×10^5	3×10^7	4×10^6	1×10^7
	II	4×10^5	8×10^7	1×10^7	2×10^7
7	I	6×10^4	1×10^7	2×10^6	7×10^6
	II	2×10^5	4×10^7	7×10^6	8×10^6
10	I	3×10^4	8×10^6	1×10^6	3×10^6
	II	1×10^5	2×10^7	3×10^6	3×10^6
15.6	I	1×10^4	3×10^6	5×10^5	1×10^6
	II	6×10^4	9×10^6	1×10^6	1×10^6

burning hydrogen into helium. The locus of luminosity versus effective surface temperature of such stars (burning hydrogen in their cores and still of nearly homogeneous composition) defines the main sequence in the Hertzsprung–Russell diagram.

The luminosity of a star is determined mainly by the thermal conductivity (radiative) of the stellar material. The central temperature is determined by the adjustment of the nuclear-energy generation to maintain mechanical and thermal equilibrium throughout the star. Nuclear-energy generation processes are very temperature sensitive, and thus nuclear-energy sources play the role of thermostats. The radius of the star depends on the temperature and the mass distribution.

The basic features of the structure of homogeneous stars can be determined by dimensional analysis. The dependence of the central temperature and density on chemical composition, mass, and radius is determined by the condition of hydrostatic equilibrium and the equation of state [from equations (1.15) and (1.13)]

$$T_c \propto \mu\beta \frac{M}{R}$$

$$\rho_c \propto \frac{M}{R^3}$$

(3.34)

The luminosity and radius are then determined by the energy balance. The equation for radiative energy transport (1.8) is

$$L = -4\pi r^2 \frac{16\sigma}{3} \frac{T^3}{\kappa\rho} \frac{dT}{dr}$$

Assuming an opacity law of the form

$$\kappa = \kappa_0 \rho^a T^b$$

the luminosity is

$$L \propto \kappa_0^{-1}(\mu\beta)^{4-b} M^{3-a-b} R^{3a+b}$$

(3.35)

The rate of nuclear-energy generation is [equation (1.6)]

$$L = 4\pi \int \mathscr{E}\rho r^2 \, dr$$

Assuming the rate of nuclear-energy generation per gram has the form

$$\mathscr{E} = \mathscr{E}_0 \rho^k T^n$$

the total rate of energy generation is

$$L \propto \mathscr{E}_0(\mu\beta)^n M^{1+k+n} R^{-3k-n}$$

(3.36)

When the rate of energy generation equals the rate of energy loss (luminosity), then the dependence of the radius, luminosity, and effective temperature on the mass and chemical composition is (Hayashi, Hoshi, and Sugimoto, 1962)

$$R \propto (\mathscr{E}_0\kappa_0)^{1/l}(\mu\beta)^{(n+b-4)/l} M^{(k+n+a+b-2)/l}$$

$$L \propto \kappa_0^{-(n+3k)/l} \mathscr{E}_0^{(3a+b)/l}(\mu\beta)^{[n(4+3a)+3k(4-b)]/l} M^{[n(3+2a)+k(9-2b)+3a+b]/l}$$

(3.37)

$$T_e^4 \propto \kappa_0^{-(n+3k-2)/l} \mathscr{E}_0^{(3a+b-2)/l}(\mu\beta)^{[n(2+3a)+3k(4-b)-2b+8]/l} M^{[n(1+2a)+k(7-2b)+a-b+4]/l}$$

where $l = n + 3k + 3a + b$ and $b \le 0$ in the interior. The central temperature and density are

$$T_c \propto (\mathscr{E}_0 \varkappa_0)^{-1/l} (\mu\beta)^{(4 + 3k + 3a)/l} M^{2(k + a + 1)/l}$$

$$\rho_c \propto (\mathscr{E}_0 \varkappa_0)^{-3/l} (\mu\beta)^{-3(n + b - 4)/l} M^{-2(n + b - 3)/l} \tag{3.38}$$

Thus the radius, luminosity, effective temperature, and central temperature increase with mass, and the central density increases with mass for the p-p chain, $n = 4$, but decreases with mass for the CNO cycle, $n \approx 18$.

The main sequence is the locus of points in the luminosity–effective temperature diagram

$$\log\left(\frac{L}{L_\odot}\right) = 4\frac{n(3 + 2a) + k(9 - 2b) + 3a + b}{n(1 + 2a) + k(7 - 2b) + a - b + 4}\log T_e + \text{constant}$$

$$= 4\frac{5n + 15.5}{3n + 15.5}\log T_e + \text{constant} \quad \text{(Kramer's)} \tag{3.39}$$

$$= 4\frac{3n + 9}{n + 11}\log T_e + \text{constant} \quad \text{(electron scattering)}$$

The central temperature of a contracting star is

$$T_c = 9.62 \times 10^7 \mu\left(\frac{M}{M_\odot}\right)\left(\frac{R_\odot}{R}\right)$$

Hydrogen burning starts at about $T_c = 8 \times 10^6$ °K. Thus, a star will start generating energy by nuclear reactions when its radius is

$$\frac{R}{R_\odot} = 1.2\mu\left(\frac{M}{M_\odot}\right) \tag{3.40}$$

Stars of small mass, $M \le 2M_\odot$, burn hydrogen by the p-p chain at a temperature around 1.5×10^7 °K. The rate of energy generation is approximately

$$\mathscr{E} = \mathscr{E}_0 \rho \left(\frac{T}{1.5 \times 10^7}\right)^4 \text{ergs/g-sec}$$

$$\mathscr{E}_0 = X_H^2$$

where X_H is the mass fraction of hydrogen. Massive stars, $M \gtrsim 2M_\odot$, burn hydrogen by the CNO cycle at a temperature of about 2×10^7 °K. The rate of energy generation is approximately

$$\mathscr{E} = \mathscr{E}_0 \rho \left(\frac{T}{2 \times 10^7}\right)^{18} \text{ergs/g-sec}$$

$$\mathscr{E}_0 = 451 X_H X_{CNO}$$

where X_{CNO} is the mass fraction of C + N + O. The energy-generation rates for the linear model are [from equation (3.7)]

$$\frac{L}{L_\odot} = 4.98 \times 10^{-3} \mu^4 \left(\frac{M}{M_\odot}\right)^6 \left(\frac{R_\odot}{R}\right)^7 \quad \text{(p-p chain)}$$

$$\frac{L}{L_\odot} = 1.157 \times 10^{-4} \mu^{14} \left(\frac{M}{M_\odot}\right)^{16} \left(\frac{R_\odot}{R}\right)^{17} \quad \text{(CNO cycle)}$$

The properties of population I stars on the main sequence—burning hydrogen in their cores—as given by the linear model, are:

For the p-p chain and Kramer's opacity

$$\frac{R}{R_\odot} = 0.312 \mu^{-0.538} \left(\frac{M}{M_\odot}\right)^{0.0769}$$

$$\frac{L}{L_\odot} = 49.1 \, \mu^{7.77} \left(\frac{M}{M_\odot}\right)^{5.46}$$

$$T_e = 2.18 \times 10^4 \, \mu^{2.21} \left(\frac{M}{M_\odot}\right)^{1.058}$$

$$\log\left(\frac{L}{L_\odot}\right) = 5.16 \log T_e - 0.74 \log \mu - 20.7$$ (3.41)

$$T_c = 3.05 \times 10^7 \, \mu^{1.54} \left(\frac{M}{M_\odot}\right)^{0.923}$$

$$\rho_c = 186 \, \mu^{1.615} \left(\frac{M}{M_\odot}\right)^{0.769}$$

For the CNO cycle and Kramer's opacity

$$\frac{R}{R_\odot} = 0.451 \, \mu^{0.395} \left(\frac{M}{M_\odot}\right)^{0.697}$$

$$\frac{L}{L_\odot} = 43.5 \, \mu^{7.3} \left(\frac{M}{M_\odot}\right)^{5.18}$$

$$T_e = 2.34 \times 10^4 \, \mu^{1.63} \left(\frac{M}{M_\odot}\right)^{0.871}$$ (3.42)

$$\log\left(\frac{L}{L_\odot}\right) = 5.948 \log T_e - 2.39 \log \mu - 24.36$$

$$T_c = 1.98 \times 10^7 \, \mu^{0.606} \left(\frac{M}{M_\odot}\right)^{0.364}$$

$$\rho_c = 65.8 \, \mu^{-0.455} \left(\frac{M}{M_\odot}\right)^{-0.909}$$

Stars switch over from the p-p chain to the CNO cycle at a central temperature of about $2 \times 10^7 \, °K$, which occurs at a mass of about $M = 2M_\odot$. For the CNO cycle and electron-scattering opacity,

$$\frac{R}{R_\odot} = 0.454\,\mu^{0.588}\left(\frac{M}{M_\odot}\right)^{0.765}$$

$$\frac{L}{L_\odot} = 112\,\mu^4\left(\frac{M}{M_\odot}\right)^3$$

$$T_e = 2.77 \times 10^4\,\mu^{0.706}\left(\frac{M}{M_\odot}\right)^{0.368}$$

$$\log\left(\frac{L}{L_\odot}\right) = 8.16 \log T_e - 1.76 \log \mu - 34.15$$

$$T_c = 2.12 \times 10^7\,\mu^{0.412}\left(\frac{M}{M_\odot}\right)^{0.235}$$

$$\rho_c = 60.3\,\mu^{-1.765}\left(\frac{M}{M_\odot}\right)^{-1.294}$$

(3.43)

Stars switch over from Kramer's to electron scattering as the dominant opacity in the deep interior, for mass $M > 3M_\odot$ for population I and $M > 2M_\odot$ for population II.

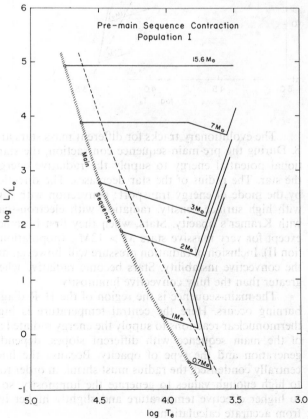

Figure 6. Evolutionary tracks of stars in H–R diagram during pre-main-sequence contraction. The main sequence is also shown. Dotted curve is observed main sequence.

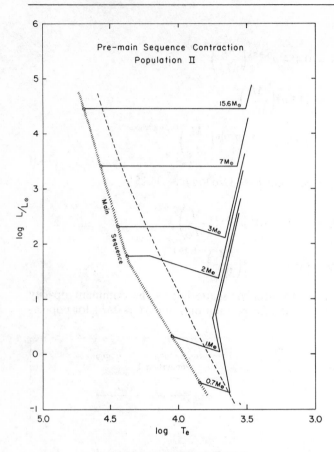

Figure 7. Evolutionary tracks of stars in H–R diagram during pre-main-sequence contraction. The main sequence is also shown. Dotted curve is observed main sequence.

The evolutionary tracks for different mass stars are shown in Figures 5 through 8. During the pre-main-sequence contraction, the stars contract to release gravitational potential energy to supply the radiative energy losses from the surface of the star. The radius of the star decreases. The direction of the track is determined by the mode of energy transport: convection with low surface density, convection with high surface density, radiation with electron-scattering opacity, or radiation with Kramer's opacity. Stars, when they first become stable, are fully convective, except for very massive stars $M > 12M_\odot$ (population I) and $M > 17M_\odot$ (population II). Inclusion of radiation pressure will, however, modify this result by increasing the convective instability. Stars become radiative when the radiative luminosity is greater than the fully convective luminosity.

The main-sequence is the region of the H–R diagram where central hydrogen-burning occurs. Here the central temperature is high enough for the hydrogen thermonuclear reactions to supply the energy radiated away. There are three sections of the main sequence with different slopes, depending on the mode of energy generation and the type of opacity. Because the linear model is not sufficiently centrally condensed, the radius must shrink in order to raise the central temperature to high enough values to generate the luminosity, so the main sequence is shifted to higher effective temperature and slightly higher luminosity than that obtained from accurate calculations.

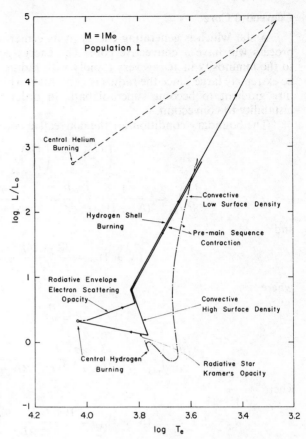

Figure 8. Evolutionary track in H–R diagram of star at one solar mass. Solid and dotted curves are from analytic models. Dashed-dot curve is results of calculations using Henyey method by Ezer and Cameron (1965).

The equation for the consumption of nuclear fuel is

$$\frac{dX}{dt} = -\frac{\mathscr{E}}{E} \quad \text{(radiative zone)}$$

$$\frac{dX}{dt} = -\frac{1}{E(M_2 - M_1)}\int_{M_1}^{M_2} \mathscr{E}\, dM(r) \quad \text{(convective zone)}$$

where X is the concentration of fuel nuclei and E is the energy released per gram of fuel consumed. This equation can be solved for the time scale of central nuclear-burning,

$$\Delta t \approx \frac{M_c}{L} E \Delta X \tag{3.44}$$

where $L/M_c \approx \bar{\mathscr{E}}$, the mean rate of energy generation; E is the energy release per gram; and $\Delta X \approx 1$. Lifetimes of stars near the main sequence are given in Table I.

Convective Core

A star which is generating energy at its center by a very temperature-sensitive process will have a convective core. The energy-generation region is very small, so the luminosity increases very rapidly with radius. The flux $F = L/4\pi r^2$ will then be extremely large, since the radius is very small, which forces the radiative temperature gradient to become superadiabatic in order to carry the flux. This causes instability to convection.

The boundary condition for the convective core is

$$\left(\frac{dT}{dr}\right)_{\mathrm{rad}} = \left(\frac{dT}{dr}\right)_{\mathrm{ad}} \tag{3.45}$$

$$\left(\frac{dT}{dr}\right)_{\mathrm{ad}} = \frac{1}{(N+1)_{\mathrm{ad}}}\frac{T}{P}\frac{dP}{dr} = -\frac{1}{(N+1)_{\mathrm{ad}}}\frac{T}{P}\frac{GM(r)\rho}{r^2} \tag{3.46}$$

and

$$(N+1)_{\mathrm{ad}} = \frac{32 - 24\beta - 3\beta^2}{8 - 6\beta} \tag{3.47}$$

where

$$\beta = \frac{P_g}{P}$$

$$\left(\frac{dT}{dr}\right)_{\mathrm{rad}} = -\frac{3}{16\sigma}\frac{\varkappa\sigma}{T^3}\frac{L}{4\pi r^2} = \frac{1}{(N+1)_{\mathrm{rad}}}\frac{T}{P}\frac{dP}{dr} \tag{3.48}$$

where

$$(N+1)_{\mathrm{rad}} = \frac{16\pi cG(1 - \beta)M(r)}{\varkappa L_r} \tag{3.49}$$

where

$$(1 - \beta) = \frac{P_{\mathrm{rad}}}{P} = \frac{1}{3}\frac{aT^4}{P}$$

Thus, the condition for convective instability is

$$(N+1)_{\mathrm{rad}} \leq (N+1)_{\mathrm{ad}} \tag{3.50}$$

Expressed in another form, the boundary of the convective core will be at

$$q = \frac{M(r)}{M} = (N+1)_{\mathrm{ad}}\frac{3}{16\pi acG}\frac{\varkappa P}{T^4}\frac{L_r}{M}$$

$$= (N+1)_{\mathrm{ad}}\frac{1}{16\pi cG}\frac{\varkappa L_r}{(1 - \beta)M} \tag{3.51}$$

For a convective core to exist, the effective polytropic index must be N_{ad} and decreasing inward at some point in the star (Naur and Osterbrock, 1953), i.e., at the core boundary

$$\frac{d\ln(N+1)_{\mathrm{rad}}}{d\ln r} \geq 0$$

Assuming $\varkappa = \varkappa_0 \rho^a T^{-b}$, then

$$\frac{d \ln (N + 1)_{\text{rad}}}{d \ln r} = 4\frac{d \ln T}{d \ln r} + \frac{d \ln M(r)}{d \ln r} - \frac{d \ln P}{d \ln r} - \frac{d \ln \varkappa}{d \ln r} - \frac{d \ln L_r}{d \ln r}$$

$$= \left[\frac{4 + b + a}{N + 1} - (1 + a)\right]\frac{d \ln P}{d \ln r} + \frac{d \ln M(r)}{d \ln r} - \frac{d \ln L_r}{d \ln r}$$

$$= -\left[\frac{4 + b + a}{N + 1} - (1 + a)\right]V + U - W \qquad (3.52)$$

where

$$U = \frac{d \ln M(r)}{d \ln r} = \frac{4\pi r^3 \rho}{M_r}$$

$$V = -\frac{d \ln P}{d \ln r} = \frac{GM(r)\rho}{rP}$$

$$W = \frac{d \ln L_r}{d \ln r}$$

Expand $M(r)$, P, L_r, and T about their central values:

$$M(r) = \tfrac{4}{3}\pi\rho_c r^3$$

$$P = P_c - \tfrac{2}{3}\pi G\rho_c^2 r^2$$

$$L_r = \tfrac{4}{3}\pi\rho_c \mathscr{E}_c r^3$$

$$T = T_c - \frac{1}{N + 1}\frac{T}{P}\Delta P = T_c - \frac{1}{N + 1}\tfrac{2}{3}\pi G\frac{T_c}{P_c}\rho_c^2 r^2$$

At the center, $U_c = 3$, $V_c = 0$, and $W_c = 3$, so $d \ln(N + 1)_{\text{rad}}/d \ln r = 0$ at the center. Thus, the condition for a convective core

$$D = \frac{d \ln(N + 1)_{\text{rad}}}{d \ln r} = -\left(\frac{4 + b + a}{N + 1} - a - 1\right)V + U - W \geq 0$$

becomes

$$\frac{dD}{dV} \geq 0$$

since $D_c = 0$ and V increases outward.

To evaluate dU/dV and dW/dV at the center, we must develop ρ and $M(r)$ to higher order, since in lowest order $dU = dW = 0$.

$$\rho = \frac{\mu H}{k}\frac{P}{T} = \rho_c\left(1 - \frac{N}{N + 1}\tfrac{2}{3}\pi G\frac{\rho_c^2}{P_c}r^2\right)$$

$$= \rho_c\left(1 - \frac{N}{N + 1}Cr^2\right)$$

where $C = \frac{2}{3}\pi G(\rho_c^2/P_c)$, so

$$M(r) = 4\pi \int_0^r \rho r^2 \, dr = \frac{4\pi}{3}\rho_c r^3 \left(1 - \frac{3}{5}\frac{N}{N+1}Cr^2\right)$$

Then

$$U = 3\left(1 - \frac{2}{5}\frac{N}{N+1}Cr^2\right)$$

$$V = 2Cr^2$$

thus

$$\frac{dU}{dV} = -\frac{3}{5}\frac{N}{N+1}$$

Now consider W.

$$L_r = 4\pi\mathscr{E}_0 \int_0^r \rho^{1+d}T^\nu r^2 \, dr$$

assuming an energy-generation rate of the form $\mathscr{E} = \mathscr{E}_0\rho^d T^\nu$, and

$$\rho = \rho_c\left(1 - \frac{N}{N+1}Cr^2\right) \quad \text{and} \quad T = T_c\left(1 - \frac{N}{N+1}Cr^2\right)$$

Thus

$$L_r = 4\pi\mathscr{E}_0\rho_c^{1+d}T_c^\nu \int_0^r \left[1 - \frac{N(1+d)}{N+1}Cr^2\right]\left(1 - \frac{\nu}{N+1}Cr^2\right)r^2 \, dr$$

$$= \frac{4\pi}{3}\mathscr{E}_0\rho_c^{1+d}T_c^\nu r^3\left[1 - \frac{3}{5}\frac{\nu + N(1+d)}{N+1}Cr^2\right]$$

$$= \frac{4\pi}{3}\mathscr{E}_c\rho_c r^3\left[1 - \frac{3}{5}\frac{\nu + N(1+d)}{N+1}Cr^2\right]$$

Then

$$W = 3 + \frac{d}{d\ln r}\left[\ln\left\{1 - \frac{3}{5}\frac{\nu + N(1+d)}{N+1}Cr^2\right\}\right]$$

$$= 3 - \frac{6}{5}\frac{\nu + N(1+d)}{N+1}Cr^2$$

so

$$dW = -\frac{12}{5}\frac{\nu + N(d+1)}{N+1}Cr \, dr$$

and

$$dV = 4Cr \, dr$$

Thus

$$\frac{dW}{dV} = -\tfrac{3}{5}\frac{v + N(1 + d)}{N + 1}$$

The criterion for the existence of a convective core is thus

$$\frac{dD}{dV} = \frac{1}{5(N + 1)}(3v + 3Nd + 5Na + 5N - 5b - 15) \geq 0 \qquad (3.53)$$

where

$$N + 1 = \frac{32 - 24\beta - 3\beta^2}{8 - 6\beta}$$

$$\mathscr{E} = \mathscr{E}_0 \rho^d T^v$$

$$\varkappa = \varkappa_0 \rho^a T^{-b}$$

Nuclear energy-generation processes have $d \geq 1$, so the condition for a convective core, assuming $\beta = 1$, is

$$v \geq 4.3 \qquad \text{(for Kramer's opacity)}$$

$$v \geq 1 \qquad \text{(for electron-scattering opacity)}$$

Thus, all central nuclear burning processes, except the equilibrium p-p chain in stars where Kramer's opacity dominates at the center, cause convective cores. The rate of gravitational energy release is $\mathscr{E}_{gr} = -T(\partial s/\partial t) \propto T$, so the condition for a convective core is

$$N \geq 2.4 \quad \text{or} \quad \beta \leq 0.75 \quad \text{(for electron-scattering opacity)}$$

$$N \geq 2.95 \quad \text{or} \quad \beta \leq 0.32 \quad \text{(for Kramer's opacity)}$$

Thus, a convective core can exist from gravitational contraction alone in a massive star. For $\beta = 0$, no temperature or density dependence of $\mathscr{E}$ is needed in order to have a convective core.

Assuming the existence of a convective core and $L_r = L$ at the core boundary, its size is given by

$$q_1 = \frac{(N + 1)_{\text{ad}}}{1 - \beta} \frac{\varkappa L}{16\pi cGM} \qquad (3.54)$$

For the linear model with electron-scattering opacity [equation (3.15)], but including radiation pressure

$$q_1 = \frac{(N + 1)_{\text{ad}}}{1 - \beta} \frac{29\pi}{32}\left(\frac{31}{288}\right)^3 \left(\frac{GH}{k}\right)^4 \frac{a}{G} M_\odot^2 \left(\frac{1 + X_c}{1 + X_e}\right)(\beta\mu_e)^4 \left(\frac{M}{M_\odot}\right)^2$$

$$= 6.8 \times 10^{-4} \frac{(N + 1)_{\text{ad}}}{1 - \beta}\left(\frac{1 + X_c}{1 + X_e}\right)(\beta\mu_e)^4 \left(\frac{M}{M_\odot}\right)^2$$

Assuming β is constant through the star,

$$\frac{1}{1 - \beta} = \frac{3P}{aT^4} = \frac{3k}{\beta a\mu H}\frac{\rho}{T^3} = \frac{3k}{\beta a\mu H}\frac{\rho_c}{T_c^3}$$

then the size of the convective core is

$$q_1 = 0.14(N + 1)_{ad}\left(\frac{1 + X_c}{1 + X_e}\right)\left(\frac{\mu_e}{\mu_c}\right)^4 \tag{3.55}$$

Note that the size of the convective core depends on the mass of the star only through the radiation pressure.

ADVANCED STAGES OF EVOLUTION—INHOMOGENEOUS STARS

A star spends most of its life burning hydrogen into helium in its core. The advanced stages of evolution comprise the star's life after central hydrogen-burning. When the hydrogen in the core is completely transformed into helium, the core of the star contracts and heats up. The rising temperature enables hydrogen thermonuclear reactions to occur in a hydrogen-burning shell source surrounding the core. A star in this stage is composed of a helium core, a hydrogen-burning shell source, and a hydrogen envelope. If the star is massive enough, the core continues to contract and heat up, until, at about 10^8 °K, helium-burning thermonuclear reactions occur in the core. Depending on its mass, a star may thereafter proceed to carbon, neon, and oxygen burning.

As a star evolves, each nuclear burning process starts first in the core, exhausts its fuel there, and then burns outward in a shell source as the star heats up. Thus, a star that has passed through several nuclear burning stages will be composed of concentric shells of the products of the different processes, with a hydrogen envelope on the outside and a core of the products of the last nuclear burning stage through which the star has passed. Figure 9 illustrates the shell structure of a star that has passed through all the nuclear burning stages.

We first consider some general properties of stars in advanced stages of evolution. The evolutionary trend of stars is toward greater central condensation. Stars contract and increase their central density and temperature. This contraction is occasionally interrupted (but the evolutionary trend is not altered) by nuclear burning in the core of the star.

The increasing central density as a star evolves, together with the existence of nuclear burning shell sources, causes the development of large radii and extended envelopes. The large radii are caused by increasing central condensation, that is, increasing central density but decreasing envelope density. The degree of central condensation is measured by

$$U = \frac{d \ln M(r)}{d \ln r} = \frac{4\pi r^3 \rho}{M(r)} = 3\frac{\rho(r)}{\bar{\rho}_r}$$

where $\bar{\rho}_r$ is the mean density interior to r. Since

$$d \ln r = \frac{1}{U} d \ln q$$

where $q = M(r)/M$, the radius is

$$\ln R = \int_{q_1}^{1} \frac{1}{U} d \ln q + \ln R_1 \tag{4.1}$$

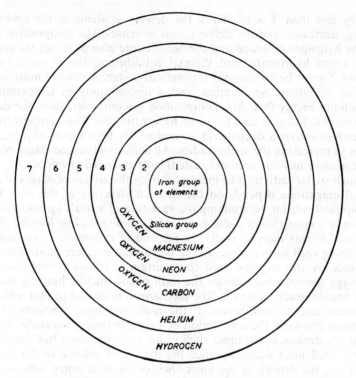

Figure 9. Schematic shell structure of a massive star at the end of nuclear burning. The star is assumed to have passed through all the nuclear burning stages and is approaching equilibrium among the nuclei in the core.

where q_1 and R_1 refer to the core–envelope interface. Now, from equation (1.15),

$$T_c \approx \frac{G\mu_c H}{k} \frac{M_1}{R_1}$$

so

$$R_1 \approx \frac{GHM_1}{k} \frac{\mu_c}{T_c} \qquad (4.2)$$

where M_1 is the mass of the core. Thus, the stellar radius is

$$\ln R = \ln\left(\frac{\mu_c}{T_c}\right) + \int_{q_1}^{1} \frac{1}{U} d\ln q + \ln\left(\frac{GHM_1}{k}\right) \qquad (4.3)$$

The larger the central condensation, the smaller the U near the shell source and the larger the stellar radius.

The central condensation develops as follows: As the central density increases, the pressure gradient $dP/dr = -\rho g$ increases. When a shell source contributes significantly to the energy output of a star, the core luminosity is less than the total luminosity, so the core tends toward an isothermal condition, that is, the temperature increases by less than $T \propto \rho^{\frac{1}{3}}$. Thus, the density gradient in the core increases, $U_1 = 3\rho_1/\bar{\rho}_c$ decreases, and the stellar radius increases. The composition difference between the hydrogen envelope and the helium core also increases the central condensation. From hydrostatic and thermal equilibrium, the physical variables r, $M(r)$, P, and T must be continuous throughout a star. A discontinuity in pressure would entail an infinite acceleration, and a discontinuity in temperature would entail an infinite energy flux. At a composition discontinuity, then, the density will be discontinuous, but ρ/μ and U/μ will be continuous. The composition discontinuity therefore causes a decrease in ρ_1 and so U_1 by a factor of μ_c/μ_e, and also contributes to increasing the stellar radius. All stars in advanced stages of evolution (until their nuclear fuel has been exhausted) have extended envelopes.

Although stellar radii tend to increase during the advanced stages of evolution, their actual magnitude depends on the detailed structure of the star. There is a general empirical rule for determining the variation of a star's radius: *The direction of expansion or contraction in a star is reversed at every nuclear burning shell source and unaffected by any inactive shell.* The reversal of expansion or contraction at a nuclear burning shell source is due to the thermostatic nature of an active nuclear-energy source. A star adjusts itself to maintain a constant temperature in the nuclear-energy source, which causes the radii of the nuclear burning shell sources to tend to remain nearly constant. The mechanism is similar to that which keeps a main-sequence star in equilibrium. If the radii and mass fractions of the shell sources remain constant, the contraction of a zone between two shells, for instance, means that the density at the inner shell of the zone increases but that the density at the outer shell must decrease, since the mass and volume of the zone remain constant. Thus, the density at the inner shell of the next outer zone is decreasing and that zone is expanding (see Figure 10).

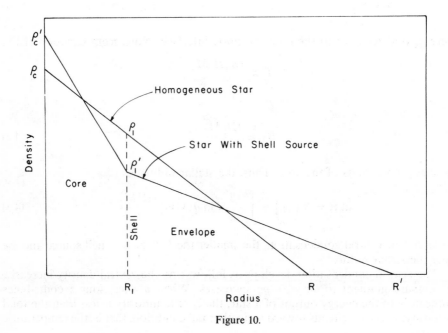

Figure 10.

Consider the zone between two shells of radii $R_0 < R_1$.

$$|\!\!\!\frac{}{R_0} \quad \rightarrow r \quad \frac{}{R_1}\!\!\!|$$

Let m be the mass of this zone and assume $R_1 \gg R_0$. The mean density of the zone is

$$\bar{\rho} = \frac{M_1 - M_0}{\frac{4}{3}\pi(R_1^3 - R_0^3)} \approx \frac{3m}{4\pi R_1^3} \tag{4.4}$$

Thus

$$\frac{\Delta R_1}{R_1} = -\frac{1}{3}\frac{\Delta\bar{\rho}}{\bar{\rho}} \tag{4.5}$$

We also assume the radiation pressure is negligible, so $\beta \approx 1$.

Consider what happens when the radius of the inner shell changes. Suppose R_0 changes by ΔR_0. If the shell at R_0 is not nuclear burning, its properties vary in a manner that preserves hydrostatic equilibrium, that is approximately homologously. Then, by equations (1.13) and (1.15),

$$T \propto \frac{1}{R} \qquad \rho \propto \frac{1}{R^3}$$

so

$$\frac{\Delta T_0}{T_0} = -\frac{\Delta R_0}{R_0}$$

$$\frac{\Delta\rho_0}{\rho_0} = -3\frac{\Delta R_0}{R_0} \tag{4.6}$$

$$\frac{\Delta P_0}{P_0} = -4\frac{\Delta R_0}{R_0}$$

Then

$$\frac{\Delta R_1}{R_1} = -\frac{1}{3}\frac{\Delta\rho_0}{\rho_0} = \frac{\Delta R_0}{R_0}$$

Thus, when the inner shell is not nuclear burning, the outer shell's radius changes in the same way as the inner shell's radius, and the shell has no effect on the expansion or contraction.

If a shell is nuclear burning, however, its structure initially changes homologously, but, due to the change in the rate of energy generation, there is an additional, nonhomologous change in the structure. The change in the rate of energy generation is

$$\mathscr{E}_N = \mathscr{E}_0(\rho + \Delta\rho)(T + \Delta T)^n = \mathscr{E}_{N_0}\left(1 + \frac{\Delta\rho}{\rho} + n\frac{\Delta T}{T}\right)$$

$$= \mathscr{E}_{N_0}\left(1 - (n + 3)\frac{\Delta R_0}{R_0}\right)$$

so that

$$\frac{\Delta\mathscr{E}_N}{\mathscr{E}_N} = -(n + 3)\left(\frac{\Delta R_0}{R_0}\right)_1$$

where $(\Delta R_0/R_0)_1$ is the initial change in R_0. Initially, this net change in energy is deposited (or removed) where it is generated and the material heats up (or cools down). The temperature changes until the fractional change in luminosity (rate of removal of energy from the region) is equal to the fractional change in the rate of energy generation. Since $L \propto T^4/\varkappa \propto T^{7.5}$, then

$$\left(\frac{\Delta T_0}{T_0}\right)_2 = \frac{1}{7.5}\frac{\Delta \mathscr{E}_N}{\mathscr{E}_N} = -\frac{n+3}{7.5}\left(\frac{\Delta R_0}{R_0}\right)_1$$

gives the additional, nonhomologous change in T_0. This additional temperature change produces a pressure change in addition to that produced by the initial homologous transformation:

$$\left(\frac{\Delta P_0}{P_0}\right)_2 = \left(\frac{\Delta T_0}{T_0}\right)_2 = -\frac{n+3}{7.5}\left(\frac{\Delta R_0}{R_0}\right)_1$$

An increase in pressure produced by a contraction of the shell will push the shell back out; a decrease in pressure produced by an expansion of the shell will allow the shell to fall back in. The pressure must return to its equilibrium homologous value and the shell must move back in the direction from which it came according to the homologous relation

$$\frac{\Delta R}{R} = -\frac{1}{4}\frac{\Delta P}{P}$$

The secondary correction to the radius of the shell is thus

$$\left(\frac{\Delta R_0}{R_0}\right)_2 = -\frac{1}{4}\left[-\left(\frac{\Delta P_0}{P_0}\right)_2\right] = -\frac{n+3}{4 \times 7.5}\left(\frac{\Delta R_0}{R_0}\right)_1 = -\frac{n+3}{30}\left(\frac{\Delta R_0}{R_0}\right)_1 \quad (4.7)$$

There is, therefore, a strong restoring force on the radii of nuclear burning shells, tending to keep them constant.

If the shell radii are precisely constant, the volume of the zone between them is constant, and if the mass fractions at the shells are constant, then the mean density (4.4) is constant. The changes in density can then be found from a linear model:

$$\rho(r) = \rho_0 - (\rho_0 - \rho_1)\frac{r - R_0}{R_1 - R_0}$$

and

$$M_1 - M_0 = \frac{4\pi}{3}R_1^3\left[\rho_0 - \frac{\rho_0 - \rho_1}{R_1 - R_0}(\tfrac{3}{4}R_1 - R_0)\right]$$

where R_0 is the radius of the inner shell and R_1 the radius of the outer shell. If $R_0 \ll R_1$, then

$$\Delta\rho_1 \approx -\frac{\Delta\rho_0}{3} \quad (4.8)$$

and ρ_1 changes in the opposite direction to ρ_0. Therefore, since the radii of the nuclear burning shells tend to remain constant, the sign of the change in the density will alternate from one shell to the next. Thus, the direction of expansion or contraction is reversed at a nuclear burning shell.

Apply the general rule for stellar radii changes to the various stages of evolution. During the pre-main sequence contraction stage, the core is contracting; there are no shells, so the whole star is contracting. During the hydrogen exhaustion phase, the core is contracting; there are no shells, so the whole star is contracting. Then a hydrogen-burning shell is ignited, and the helium core continues to contract; but now there is one shell, so the envelope expands. When the central helium-burning commences, the core expands; there is one shell, so the envelope contracts. These structural changes are illustrated in Figure 11. The structural changes during the stage of helium burning are illustrated in Hayashi (1965).

We now consider in some detail the evolution of stars from the depletion of hydrogen in the core to the onset of helium-burning in the core.

CENTRAL HYDROGEN DEPLETION

The depletion of a nuclear fuel in the core of a star and the ignition of a shell source is a process which changes the basic structure of a star. We therefore cannot construct an analytic model for this phase, but only give some of its general properties.

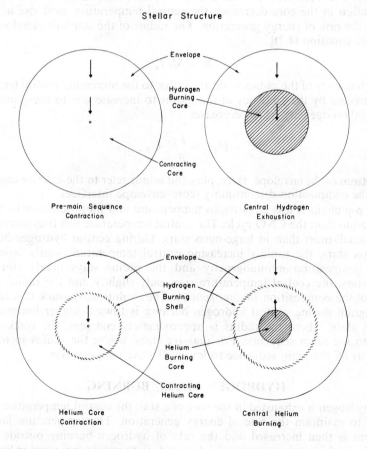

Figure 11. Schematic diagram of the changes in stellar structure from the pre-main-sequence contraction to the onset of central helium-burning.

During central hydrogen-burning, the luminosity of a star increases because of the increasing mean molecular weight as hydrogen is depleted in the core. Assuming that the homology relations for homogeneous stars are still valid, then in stars of small mass where Kramer's opacity is dominant from (3.13) and (3.35)

$$L \propto (\mu\beta)^{7.5} \tag{4.9}$$

while in massive stars where electron scattering is dominant from (3.16) and (3.35)

$$L \propto (\mu\beta)^4 \tag{4.10}$$

The molecular weight in the core increases by about a factor of 2 as hydrogen is consumed.

The energy-generation rate has the form

$$\mathscr{E} = \mathscr{E}_0 X_H X_2 \rho \left(\frac{T}{T_0}\right)^n$$

where X_2 is X_H for the p-p chain and is X_{CNO} for the CNO cycle. The temperature exponent is $n \approx 4$ for the p-p chain and $n \approx 18$ for the CNO cycle. As the hydrogen concentration in the core decreases, the central temperature must rise in order to maintain the rate of energy generation. The radius of the star will therefore tend to shrink [see equation (4.2)],

$$R \propto M/T_c \tag{4.11}$$

The tendency of the radius to decrease due to the increasing central temperature is counteracted by the tendency of the radius to increase due to the growing composition inhomogeneity which decreases

$$U_{1+} = \frac{\mu_{1+}}{\mu_{1-}} U_{1-} \tag{4.12}$$

at the bottom of the envelope. Here, plus and minus refer to the exterior and interior sides of the composition discontinuity (core–envelope interface).

The p-p chain is less sensitive to temperature and more sensitive to hydrogen concentration than the CNO cycle. The central temperature will thus increase much more in small-mass than in large-mass stars. During central hydrogen-burning in small-mass stars, the rapidly increasing central temperature nearly balances the growing composition inhomogeneity and the radius stays nearly constant. In massive stars, the central temperature rises only slightly and the radius increases because of the composition inhomogeneity. The evolutionary track of a star in the H–R diagram during central hydrogen-burning is toward higher luminosity. For low-mass stars, where the radius is approximately constant, the track is nearly parallel to the main sequence. For massive stars, where the radius increases, the track turns off the main sequence to lower effective temperatures.

HYDROGEN SHELL BURNING

As hydrogen is exhausted in the core of a star, the central temperature increases in order to maintain the rate of energy generation. The temperature farther out in the star is then increased and the rate of hydrogen-burning outside the core (where the hydrogen has not been exhausted) is therefore increased. Thus, a shell burning source is ignited.

When hydrogen becomes nearly exhausted in small-mass stars generating energy by the p-p chain, the central temperature has already increased and raised the temperature in the surrounding regions of higher hydrogen concentration sufficiently to produce hydrogen thermonuclear reactions there. When hydrogen becomes nearly exhausted in massive stars, the central temperature has not yet increased much due to the high temperature sensitivity of the CNO cycle. The energy requirements of the star must still be met by the core, so the central temperature must now increase greatly. This causes the radius of the star to contract, and its track in the H–R diagram swings to higher effective temperatures. Eventually, the decrease in X_c outruns the increase in T_c^n and the rate of nuclear-energy generation in the core decreases. The core then starts to contract rapidly and release gravitational energy to supplement the decreasing rate of central nuclear-energy generation. The gravitational contraction raises the central temperature $T_c \propto \rho_c^{\frac{1}{3}}$ and the shell temperature, and ignites the shell source. The more massive the star, the larger the size of the initial convective core and the farther out from the center the hydrogen-rich regions lie. Therefore, the temperature in the hydrogen-rich shell will be lower, the ignition of the shell source will be delayed, and gravitational energy release will supplant nuclear energy generation as the star's primary energy source. Eventually, the contraction will raise the temperature enough to ignite the shell source. To summarize, as hydrogen is exhausted in the core of a star, the temperature increases, nuclear-energy generation in the core decreases, and a hydrogen-burning shell source surrounding the core is ignited.

We now consider in greater detail the increase in central condensation that occurs when a shell source is set up.

It is convenient to discuss the structure of the star in terms of the nondimensional variables U, V, and $N + 1$:

$$U \equiv \frac{d \ln M(r)}{d \ln r} = \frac{4\pi r^3 \rho}{M(r)} = 3\frac{\rho(r)}{\bar{\rho}_r}$$

$$V \equiv -\frac{d \ln P}{d \ln r} = \frac{GM(r)\rho}{rP} = \frac{3}{2}\frac{GM(r)/r}{\frac{3}{2}P/\rho} \qquad (4.13)$$

$$N + 1 \equiv \frac{d \ln P}{d \ln T} = \frac{16\pi ac}{3}\frac{GM(r)T^4}{P \varkappa L(r)}$$

At the center of a star, $U \to 3$, $V \to 0$, and at the surface $U \to 0$, $V \to \infty$. The polytropic index N varies between 1.5 for a convective region and infinity for an isothermal region. Also

$$\frac{d \ln T}{d \ln r} = -\frac{V}{N + 1}$$

$$\qquad (4.14)$$

$$\frac{d \ln \rho}{d \ln r} = -\frac{NV}{N + 1}$$

Thus the r-dependence of the physical variables is given in terms of $U, V, N + 1$ by

$$M(r) \sim r^U$$

$$P \sim r^{-V}$$

$$T \sim r^{-V/(N+1)} \tag{4.15}$$

$$\rho \sim r^{-NV/(N+1)}$$

From the continuity of the physical variables r, $M(r)$, P, T, and ρ/μ, the continuity conditions on $U, V, N + 1$ are

$$\frac{U}{\mu} \qquad \frac{V}{\mu} \qquad \varkappa L_r(N+1) \qquad \text{(continuous)} \tag{4.16}$$

The $U - V$ locus of a star is given by

$$\frac{d \ln U}{d \ln r} = 3 - U - \frac{NV}{N+1}$$

$$\frac{d \ln V}{d \ln r} = U - 1 + \frac{V}{N+1} \tag{4.17}$$

or

$$\frac{d \ln V}{d \ln U} = \frac{U + V/(N+1) - 1}{3 - U - NV/(N+1)}$$

The points on the $V - U$ curve with horizontal or vertical tangent are given by

$$U + V/(N+1) - 1 = 0 \qquad \text{(horizontal)}$$

$$U + NV/(N+1) - 3 = 0 \qquad \text{(vertical)} \tag{4.18}$$

These two lines intersect at the point

$$U = \frac{N-3}{N-1} \qquad V = 2\frac{N+1}{N-1} \tag{4.19}$$

Thus, for $N > 3$, the intersection point is in the physical region and there is a loop point.

Typical $U - V$ curves for homogeneous and inhomogeneous stars are shown in Figure 12.

A star's central condensation is increased by the tendency toward isothermality in the core, when nuclear energy generation there ceases. For an isothermal core, $N = \infty$, so the $U - V$ curve has a loop point at $U = 1$, $V = 2$. The maximum V thus occurs for $U = 1$ and is somewhat larger than 2:

$$U_{1+} \approx \frac{\mu_e}{\mu_c} U \approx 0.5 \tag{4.20}$$

However, an isothermal core, if too large, cannot support the weight of the envelope. The critical size of an isothermal core can be found from the virial theorem (McCrea, 1957),

$$3(\gamma - 1)U + \Omega - 3PV = 0$$

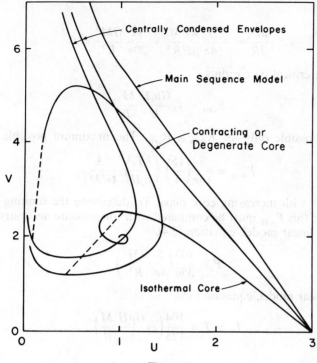

Figure 12.

so

$$P = (\gamma - 1)\frac{U}{V} + \tfrac{1}{3}\frac{\Omega}{V}$$

where U and V are now the internal energy and volume of the isothermal core, and P is the pressure on its surface. For an isothermal sphere the internal energy is, from equation (2.3)

$$U = \frac{1}{\gamma - 1}\frac{k}{\mu H}TM$$

and for a sphere of uniform density the gravitational energy is, from equation (1.19)

$$\Omega = -\tfrac{3}{5}\frac{GM^2}{R}$$

The pressure at the boundary of the isothermal core is therefore

$$P = \frac{3}{4\pi}\frac{k}{\mu H}\frac{TM}{R^3} - \tfrac{3}{5}\frac{1}{4\pi}\frac{GM^2}{R^4}$$

There is a maximum pressure consistent with the equilibrium virial theorem, which is given by

$$\frac{dP}{dR} = -\frac{9}{4\pi}\frac{kTM}{\mu HR^4} + \frac{12}{20\pi}\frac{GM^2}{R^5} = 0$$

Thus, there is a critical core radius

$$R_{crit} = \frac{4}{15}\frac{G\mu_c H}{k}\frac{M_1}{T_1} \tag{4.21}$$

with stability possible only for $R_{core} \geq R_{crit}$. The maximum possible pressure is

$$P_{max} = \frac{3}{16\pi}\left(\frac{15}{4}\right)^3\left(\frac{kT_1}{\mu_c H}\right)^4\frac{1}{G^3 M_1^2} \tag{4.22}$$

which decreases with increasing core mass. To determine the limiting mass of an isothermal core, this P_{max} must be compared with the pressure necessary to support a star. For the linear model, equation (3.4),

$$\bar{P} = \frac{103}{350}\left(\frac{5}{4\pi}\frac{GM^2}{R^4}\right)$$

Also for the linear model, equation (3.5),

$$T_1 \approx \bar{T} = \frac{104}{175}\left(\frac{5}{12}\frac{G\mu H}{k}\frac{M}{R}\right)$$

Thus, the condition for a stable star, that an isothermal nondegenerate core can support the surrounding envelope, is

$$P_{max} \gtrsim \bar{P}$$

or

$$\frac{M_1}{M} \lesssim 0.3\left(\frac{\mu_e}{\mu_c}\right)^2 \tag{4.23}$$

Accurate calculations (Schönberg and Chandrasekhar, 1942) give $q_1 < 0.37(\mu_e/\mu_c)^2$ with an accuracy of 8 %.

If the mass of the core is below the isothermal core limiting mass (Schönberg–Chandrasekhar limit), the core becomes isothermal.

In massive stars, the core exceeds the Schönberg–Chandrasekhar limit and gravitational contraction begins when nuclear-energy generation ceases to support the star. In small-mass stars, the core is initially below the limiting size, but shell burning adds material to the core until in this case too the core exceeds the Schönberg–Chandrasekhar limit, and it contracts. If the core mass is less than the white-dwarf limiting mass [equation (5.9)], the electrons become degenerate, while larger cores continue to contract. The pressure gradient needed to support the envelope is thus supplied by the electron-degeneracy pressure in small-mass stars and by a combination of the density and temperature gradients in large-mass stars. The development of a degenerate core greatly increases the central condensation.

We now consider the structure of a stellar envelope with extreme central condensation, characterized by $U \rightarrow 0$ as $r \rightarrow r_{shell}$. Since r_{shell} is very small, the structure at the base of the envelope will not be too different from that in the limit $r \rightarrow 0$.

Since $U \to 0$, $M(r) \approx M_1$ near the shell [equation (4.15)], that is, as $r \to 0$, $\rho \to \infty$ and $M(r) \to$ finite value. Then

$$\frac{dP}{dr} = \frac{k}{\mu_e H} \frac{d}{dr}(\rho T) = -\frac{GM_1}{r^2}\rho$$

and assuming some polytropic law, $\rho \propto T^N$, gives

$$\frac{dT}{dr} = -\frac{G\mu_e H}{(N+1)k}\frac{M_1}{r^2}$$

Thus

$$T = \frac{G\mu_e H}{(N+1)k}\left(\frac{M_1}{r}\right) + \text{constant} \propto \frac{1}{r} \tag{4.24}$$

Thus, for an extremely centrally condensed envelope

$$V \to N + 1 \tag{4.25}$$

in the limit as $r \to 0$ [from equations (4.15) and (4.24)]. The loop condition [equation (4.19)] shows that the limit of extreme central condensation, $U \to 0$ as $r \to 0$, requires $N \leqslant 3$. The radial dependence of the physical variables is

$$P \propto r^{-(N+1)} \qquad T \propto r^{-1} \qquad \rho \propto r^{-N} \tag{4.26}$$

For $N > 3$, the limit as $r \to 0$ is a loop point given by equation (4.19). In this case, $U > 0$ and the envelope is not so centrally condensed. The radial dependence of the physical variables is

$$P \sim r^{-2(N+1)/(N-1)} \qquad T \sim r^{-2/(N-1)} \qquad \rho \sim r^{-2N/(N-1)} \tag{4.27}$$

We now determine the effective polytropic index at the base of a centrally condensed envelope. In terms of nondimensional variables,

$$P = p\frac{GM^2}{4\pi R^4}$$

$$T = t\frac{\mu_e H}{k}\frac{GM}{R}$$

$$M(r) = qM$$

$$r = xR$$

the hydrostatic equilibrium equations are

$$\frac{dp}{dx} = -\frac{pq}{tx^2}\beta l$$

$$\frac{dq}{dx} = \frac{x^2 p}{t}\beta l \tag{4.28}$$

where $l = \mu/\mu_e$, and the flux equations are

$$\frac{dt}{dx} = -C_K \frac{p^2}{x^2 t^{8.5}} \qquad \text{(Kramer's opacity)}$$

$$\frac{dt}{dx} = -C_E \frac{p}{x^2 t^4} \qquad \text{(Electron scattering)}$$

(4.29)

For Kramer's opacity, combining equations (4.28) and (4.29),

$$\frac{dp^2}{dt^{8.5}} = \frac{q}{4.25 C_K}$$

so, near the surface and near the shell at the base of the envelope, where U is very small and the mass fraction q is nearly constant, the polytropic index is

$$N = 3.25$$

Thus, at the shell

$$\rho \sim r^{-2.89} \qquad \text{(Kramer's)} \tag{4.30}$$

Similarly, for electron scattering,

$$\frac{dp}{dt^4} = \frac{q}{4 C_E}$$

so, near the surface and near the shell, the polytropic index is

$$N = 3$$

Thus, the density distribution at the shell is

$$\rho \sim r^{-3} \qquad \text{(Electron scattering)} \tag{4.31}$$

The only envelope model which can be readily solved analytically is $\rho(r) \sim r^{-3}$. This, as was just shown, corresponds to the limiting case of extreme central condensation for both Kramer's and electron-scattering opacity. The internal structure will be well represented by such a model, but, because it is too centrally condensed, the stellar radii will be much too large. To calculate the radii, account must be taken of the fact that the mean polytropic index of the envelope is less than 3.

Inhomogeneous Analytic Stellar Model

We now construct an analytic model of a star with one shell using a linear density distribution in the core and an r^{-3} density distribution in the envelope.

1. *Core.* In the core assume a linear density distribution:

$$\rho(r) = \rho_c - (\rho_c - \rho_1)\frac{r}{R_1} \tag{4.32}$$

where ρ_c is the central density and ρ_1 the density at the shell. Then the mass distribution in the core is, by equation (1.4),

$$M(r) = \int_0^r 4\pi\rho(r)r^2\,dr$$

$$= \frac{4\pi}{3}r^3\left[\rho_c - \tfrac{3}{4}(\rho_c - \rho_1)\frac{r}{R_1}\right] \tag{4.33}$$

and the mass of the core is

$$M_1 = \frac{\pi}{3} R_1^3 (\rho_c + 3\rho_1) \qquad (4.34)$$

This relation can be turned around to give the radius of the core:

$$\frac{R_1}{R_\odot} = \left(\frac{3M_\odot}{\pi R_\odot^3}\right)^{\frac{1}{3}} \left(\frac{M_1}{M_\odot}\right)^{\frac{1}{3}} \left(\rho_c + 3\frac{\mu_c}{\mu_e}\rho_{1-}\right)^{-\frac{1}{3}}$$

$$= 1.78 \left(\frac{M_1}{M_\odot}\right)^{\frac{1}{3}} \left(\rho_c + 3\frac{\mu_c}{\mu_e}\rho_{1-}\right)^{-\frac{1}{3}} \qquad (4.35)$$

The pressure in the core is determined by hydrostatic equilibrium, equation (1.1),

$$P(r) = P_c - G \int_0^r \frac{M(r)\rho(r)}{r^2} dr$$

$$= P_c - \frac{2\pi}{3} G\rho_c^2 r^2 \left[1 - \frac{7}{6}\left(1 - \frac{\rho_{1-}}{\rho_c}\right)\frac{r}{R_1} + \frac{3}{8}\left(1 - 2\frac{\rho_{1-}}{\rho_c} + \frac{\rho_{1-}^2}{\rho_c^2}\right)\frac{r^2}{R_1^2}\right]$$

Applying the boundary condition that $P = P_1$ at the core boundary $r = R_1$, we find that the pressure in the core is

$$P(r) = P_1 + \frac{\pi}{36} G\rho_c^2 R_1^2 \left[5 + 10\frac{\rho_{1-}}{\rho_c} + 9\frac{\rho_{1-}^2}{\rho_c^2}\right.$$

$$\left. - 24\frac{r^2}{R_1^2} + 28\left(1 - \frac{\rho_{1-}}{\rho_c}\right)\frac{r^3}{R_1^3} - 9\left(1 - 2\frac{\rho_{1-}}{\rho_c} + \frac{\rho_{1-}^2}{\rho_c^2}\right)\frac{r^4}{R_1^4}\right] \qquad (4.36)$$

and the central pressure is

$$P_c = P_1 + \frac{5\pi}{36} G\rho_c^2 R_1^2 \left(1 + 2\frac{\rho_{1-}}{\rho_c} + 1.8\frac{\rho_{1-}^2}{\rho_c^2}\right) \qquad (4.37)$$

For a perfect gas, with negligible radiation pressure, the temperature is, by equation (1.2),

$$T(r) = \frac{\mu H}{k} \frac{P(r)}{\rho(r)}$$

Thus, the temperature in the core is

$$T(r) = \left[1 - \left(1 - \frac{\rho_{1-}}{\rho_c}\right)\frac{r}{R_1}\right]^{-1}\left[\frac{\rho_{1-}}{\rho_c}T_1 + \frac{\pi}{36}\frac{G\mu_c H}{k}\rho_c R_1^2\left\{5 + 10\frac{\rho_{1-}}{\rho_c}\right.\right.$$
$$\left.\left. + 9\frac{\rho_{1-}^2}{\rho_c^2} - 24\frac{r^2}{R_1^2} + 28\left(1 - \frac{\rho_{1-}}{\rho_c}\right)\frac{r^3}{R_1^3} - 9\left(1 - 2\frac{\rho_{1-}}{\rho_c} + \frac{\rho_{1-}^2}{\rho_c^2}\right)\frac{r^4}{R_1^4}\right\}\right] \tag{4.38}$$

and the central temperature is

$$T_c = \frac{\mu_c}{\mu_e}\frac{\rho_1}{\rho_c}T_1 + \frac{5\pi}{36}\frac{G\mu_c H}{k}R_1^2\rho_c\left(1 + 2\frac{\rho_{1-}}{\rho_c} + 1.8\frac{\rho_{1-}^2}{\rho_c^2}\right)$$

$$= \frac{\mu_c}{\mu_e}\frac{\rho_1}{\rho_c}T_1 + 0.17 \times 10^7\mu_c\left(\frac{R_1}{R_\odot}\right)^2\rho_c\left(1 + 2\frac{\rho_1}{\rho_c}\frac{\mu_c}{\mu_e} + 1.8\frac{\mu_c^2}{\mu_e^2}\frac{\rho_1^2}{\rho_c^2}\right) \tag{4.39}$$

However, when degenerate, the core is assumed to be isothermal, so $T_c = T_1$.

2. *Envelope.* In the envelope assume an r^{-3} density distribution:

$$\rho(r) = \rho_1\left(\frac{R_1}{r}\right)^3 \tag{4.40}$$

Notice that in this model the density does not go to zero at the surface of the star. The mass distribution is, from equation (1.4),

$$M(r) = M_1 + 4\pi\rho_1 R_1^3\int_{R_1}^r \frac{dr}{r}$$

$$= M_1 + 4\pi\rho_1 R_1^3 \ln\frac{r}{R_1}$$

where M_1 is the mass inside the shell. The mass of the envelope is

$$M - M_1 = 4\pi\rho_1 R_1^3 \ln\frac{R}{R_1} \tag{4.41}$$

The pressure is determined by hydrostatic equilibrium, equation (1.1),

$$P(r) = P_1 - G\int_{R_1}^r \frac{M(r)\rho(r)}{r^2}dr$$

$$= P_1 - G\rho_1 R_1^3\int_{R_1}^r \left(M_1 + 4\pi\rho_1 R_1^3 \ln\frac{r}{R_1}\right)\frac{1}{r^5}dr$$

so

$$P(r) = P_1 - \tfrac{1}{4}G\rho_1\frac{M_1}{R_1}\left[1 - \left(\frac{R_1}{r}\right)^4\right] - \frac{\pi}{4}G\rho_1^2 R_1^2 + \pi G\rho_1^2 R_1^6\frac{1}{r^4}\left(\tfrac{1}{4} + \ln\frac{r}{R_1}\right)$$

The boundary condition $P(R) = 0$ determines P_1, and we get

$$P(r) = \tfrac{1}{4}G\rho_1 \frac{M_1}{R_1}\left[\left(\frac{R_1}{r}\right)^4 - \left(\frac{R_1}{R}\right)^4\right]$$

$$+ \frac{\pi}{4}G\rho_1^2 R_1^2\left[\left(\frac{R_1}{r}\right)^4 - \left(\frac{R_1}{R}\right)^4 + 4\left(\frac{R_1}{r}\right)^4\ln\frac{r}{R_1} - 4\left(\frac{R_1}{R}\right)^4\ln\frac{R}{R_1}\right] \quad (4.42)$$

and the pressure at the shell is

$$P_1 = \tfrac{1}{4}G\rho_1\frac{M_1}{R_1}\left[1 - \left(\frac{R_1}{R}\right)^4\right] + \frac{\pi}{4}G\rho_1^2 R_1^2\left[1 - \left(\frac{R_1}{R}\right)^4\right] - \pi G\rho_1^2 R_1^2\left(\frac{R_1}{R}\right)^4\ln\frac{R}{R_1} \quad (4.43)$$

The temperature in the envelope, given by the equation of state with negligible radiation pressure (1.2), is

$$T(r) = \frac{G\mu_e H}{4k}\left[\frac{M_1}{R_1}\left\{\frac{R_1}{r} - \frac{R_1}{R}\left(\frac{r}{R}\right)^3\right\} + \pi\rho_1 R_1^2\left\{\frac{R_1}{r} - \frac{R_1}{R}\left(\frac{r}{R}\right)^3\right.\right.$$

$$\left.\left. + 4\frac{R_1}{r}\ln\frac{r}{R_1} - 4\frac{R_1}{R}\left(\frac{r}{R}\right)^3\ln\frac{R}{R_1}\right\}\right] \quad (4.44)$$

Note that the temperature is proportional to r^{-1}, except near the surface. The temperature at the shell is

$$T_1 = \frac{G\mu_e H}{4k}\left[\frac{M_1}{R_1}\left\{1 - \left(\frac{R_1}{R}\right)^4\right\} + \pi R_1^2\rho_1\left\{1 - \left(\frac{R_1}{R}\right)^4\left(1 + 4\ln\frac{R}{R_1}\right)\right\}\right]$$

$$\approx 0.577 \times 10^7\mu_e\left(\frac{M_1}{M_\odot}\right)\left(\frac{R_\odot}{R_1}\right) + 0.3066 \times 10^7\mu_e\left(\frac{R_1}{R_\odot}\right)^2\rho_1 \quad (4.45)$$

The stellar radius is extremely sensitive to the degree of central condensation. For our envelope density distribution $\rho \sim r^{-3}$, the radius is obtained from the mass relation (4.41):

$$R = R_1\exp\left(\frac{M - M_1}{4\pi\rho_1}R_1^{-3}\right)$$

$$= R_1\exp\left[\frac{1 - q_1}{12q_1}\left(\frac{\rho_c}{\rho_1} + 3\frac{\mu_c}{\mu_e}\right)\right] \quad (4.46)$$

Thus, the radius depends exponentially on ρ_c/ρ_1. This leads to extremely large radii, much larger than are observed. This is to be expected, since this envelope corresponds to the maximum degree of central condensation.

Consider now less centrally condensed envelopes with density distributions

$$\rho(r) = \rho_1 \left(\frac{R_1}{r}\right)^n \qquad (1.5 < n < 3) \tag{4.47}$$

The mass distribution in the envelope is then

$$M(r) = M_1 + 4\pi\rho_1 R_1^n \int_{R_1}^r r^{2-n}\, dr$$

$$= M_1 + \frac{4\pi}{3-n}\rho_1 R_1^n (r^{3-n} - R_1^{3-n})$$

and

$$M = M_1 + \frac{4\pi}{3-n}\rho_1 R_1^n (R^{3-n} - R_1^{3-n})$$

Thus, the radius is

$$R \approx \frac{R_1}{3-n} + \left[\frac{(3-n)(M-M_1)}{4\pi\rho_1 R_1^n}\right]^{1/(3-n)} \tag{4.48}$$

The dependence of the radius on ρ_c/ρ_1 varies from exponential for $n = 3$ to almost constant for $n = 1.5$. Thus, the radius of the star depends sensitively on the precise degree of central condensation. It is reasonable to treat the internal envelope structure and the radius using different effective polytropic indices n, because the internal properties of the envelope are determined by the behavior near the shell where $n \approx 3$, while the radius is determined by the entire envelope in which the polytropic index is less than 3 over large regions. In the radiative region near the shell $n = 3$ and the mass

$$M(r) \propto \ln \frac{r}{R_1}$$

increases very slowly with radius, while in a convective region $n = 1.5$ and the mass

$$M(r) \propto r^{\frac{3}{2}}$$

increases much faster with radius. As a rough approximation for all stars, we choose $\bar{n} = 2.4$. Then

$$\frac{R}{R_\odot} = \frac{5}{3}\frac{R_1}{R_\odot} + \left(\frac{3M_\odot}{20\pi R_\odot^3}\right)^{\frac{5}{3}}\left(\frac{M}{M_\odot}\right)^{\frac{5}{3}}(1-q_1)^{\frac{5}{3}}\left(\frac{R_1}{R_\odot}\right)^{-4}\rho_1^{-\frac{5}{3}}$$

$$= \frac{5}{3}\frac{R_1}{R_\odot} + 0.122\left(\frac{M}{M_\odot}\right)^{\frac{5}{3}}(1-q_1)^{\frac{5}{3}}\left(\frac{R_1}{R_\odot}\right)^{-4}\rho_1^{-\frac{5}{3}} \tag{4.49}$$

The internal structure of a typical inhomogeneous model is shown in Figure 13. Figure 14 shows the tremendous degree of central condensation of the mass as compared with the homogeneous model.

We now turn from the hydrostatics to the energy balance in the envelope. The rate of thermonuclear-energy generation in the shell is, from equation (1.6),

$$L = 4\pi\mathscr{E}_0 \int_{R_1}^R \rho^2 \left(\frac{T}{T_0}\right)^n r^2\, dr$$

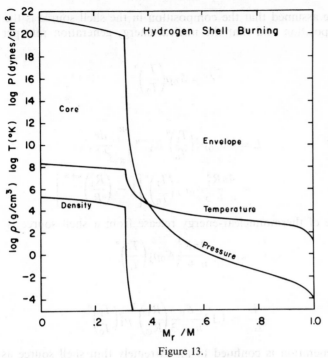

Figure 13.

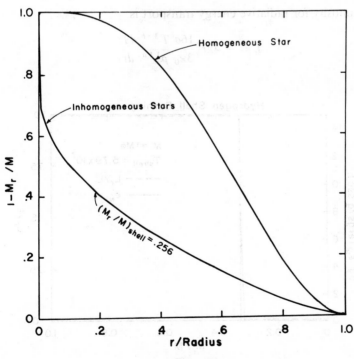

Figure 14.

where we have assumed that the composition in the shell source is the unevolved envelope composition and that the nuclear-energy-generation rate per gram has the form

$$\mathscr{E} = \mathscr{E}_0 \rho \left(\frac{T}{T_0}\right)^n$$

Then

$$L = 4\pi \mathscr{E}_0 \rho_1^2 \left(\frac{T_1}{T_0}\right)^n R_1^{6+n} \int_{R_1}^{R} \frac{dr}{r^{n+4}}$$

$$= \frac{4\pi R_1^3}{n+3} \mathscr{E}_0 \rho_1^2 \left(\frac{T_1}{T_0}\right)^n \left[1 - \left(\frac{R_1}{R}\right)^{n+3}\right]$$

Thus, the rate of thermonuclear-energy release from a shell source is

$$L = \frac{4\pi R_1^3}{n+3} \mathscr{E}_0 \rho_1^2 \left(\frac{T_1}{T_0}\right)^n$$

or

$$\frac{L}{L_\odot} = 1.12 \frac{\mathscr{E}_0}{n+3} \left(\frac{R_1}{R_\odot}\right)^3 \rho_1^2 \left(\frac{T_1}{T_0}\right)^n \tag{4.50}$$

The energy generation is confined to an extremely thin shell source as shown in Figure 15.

The luminosity for radiative energy transport is

$$L = -4\pi r^2 \frac{16\sigma}{3\varkappa_0} \frac{T^{3-b}}{\rho^{1+a}} \frac{dT}{dr}$$

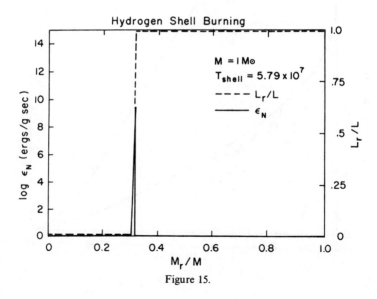

Figure 15.

assuming an opacity law of the form

$$\varkappa = \varkappa_0 \rho^a T^b$$

The temperature gradient given by the analytic model [equation (4.44)] is

$$\frac{dT}{dr} = - T_1 \frac{R_1}{r^2}$$

Thus, the radiative luminosity is

$$L = \frac{64}{3} \frac{\pi\sigma}{\varkappa_0} R_1 T_1 \frac{T^{3-b}}{\rho^{1+a}}$$

Evaluating it at the shell gives

$$L = \frac{64}{3} \frac{\pi\sigma}{\varkappa_0} R_1 \frac{T_1^{4-b}}{\rho_1^{1+a}}$$

$$= \frac{64}{3} \frac{\pi\sigma}{\varkappa_0} R_1 \frac{T_1^{7.5}}{\rho_1^2} \qquad \text{(Kramer's)} \qquad (4.51)$$

$$= \frac{64}{3} \frac{\pi\sigma}{\varkappa_0} R_1 \frac{T_1^4}{\rho_1} \qquad \text{(Electron scattering)}$$

Thus, the radiative luminosity of the envelope is, for population I ($X = 0.6$, $Y = 0.38$, $Z = 0.02$),

$$\frac{L}{L_\odot} = \begin{cases} 4.25 \times 10^3 \left(\dfrac{R_1}{R_\odot}\right) \dfrac{T_{1(7)}^{7.5}}{\rho_1^2} & \text{(Kramer's)} \\[2ex] 2.18 \times 10^3 \left(\dfrac{R_1}{R_\odot}\right) \dfrac{T_{1(7)}^4}{\rho_1} & \text{(Electron scattering)} \end{cases} \qquad (4.52)$$

and for population II ($X = 0.9$, $Y = 0.099$, $Z = 0.001$),

$$\frac{L}{L_\odot} = \begin{cases} 2.27 \times 10^4 \left(\dfrac{R_1}{R_\odot}\right) \dfrac{T_{1(7)}^{7.5}}{\rho_1^2} & \text{(Kramer's)} \\[2ex] 1.84 \times 10^3 \left(\dfrac{R_1}{R_\odot}\right) \dfrac{T_{1(7)}^4}{\rho_1} & \text{(Electron scattering)} \end{cases} \qquad (4.53)$$

The effective temperature is given by equation (3.17).

For a fully convective envelope, the luminosity and effective temperature are determined by the surface condition, equations (3.24) and (3.28). The track is the same as for pre-main sequence fully convective contraction, but traversed in the opposite direction.

The time scale of evolution is determined by the rate of release of energy:

$$L = \frac{dE}{dt} \qquad \text{so} \qquad \Delta t = \frac{\Delta E}{L} \qquad (4.54)$$

The time scale during stages of core contraction is determined by the gravitational energy release:

$$\Delta E = \tfrac{1}{2}|\Delta\Omega| = \tfrac{1}{2}\tfrac{6}{7}\frac{G\overline{M}_c^2}{\overline{R}_c}\left(-\frac{\Delta R_c}{R_c} + 2\frac{\Delta M_c}{M_c}\right)$$

where the factor $\tfrac{6}{7}$ arises in evaluating Ω using the linear model. The luminosity of the contracting core is determined by the rate at which material is added to the core by the hydrogen-burning shell source:

$$L_c = \Phi_c\frac{dM_c}{dt}$$

where Φ_c is the gravitational potential at the core surface

$$\Phi_c = \frac{GM_c}{R_c}$$

and

$$\frac{dM_c}{dt} = \frac{L_{\text{shell}}}{E_H X}$$

where E_H is the energy released per gram of hydrogen consumed. Thus the contracting core's luminosity is

$$L_c = \frac{GM_c}{R_c}\frac{L}{E_H X}$$

or

$$\frac{L_c}{L_\odot} = \frac{GM_\odot}{E_H R_\odot}\frac{1}{X}\left(\frac{M_c}{M_\odot}\right)\left(\frac{R_\odot}{R_c}\right)\left(\frac{L}{L_\odot}\right)$$

$$= 3.18 \times 10^{-4}\frac{1}{X}\left(\frac{M_c}{M_\odot}\right)\left(\frac{R_\odot}{R_c}\right)\left(\frac{L}{L_\odot}\right)$$

The time scale for contraction of the core is

$$\Delta t = \tfrac{6}{7}\frac{E_H X M_c}{2L}\left(2\frac{\Delta M_c}{M_c} - \frac{\Delta R_c}{R_c}\right)$$

$$= 4.31 \times 10^{10} X_e \left(\frac{M_c}{M_\odot}\right)\left(\frac{L}{L_\odot}\right)^{-1}\left(-\frac{\Delta R_c}{R_c} + 2\frac{\Delta M_c}{M_c}\right) \text{ years}$$

(4.55)

The amount of material added to the core by the hydrogen burning shell during this time is

$$\Delta M_c = \frac{\Delta t L}{E_H X_e}$$

so

$$\Delta q_1 = \frac{\Delta M_1}{M} = \left(\frac{L_\odot}{M_\odot E_H X_e}\right)\frac{M_\odot}{M}\frac{L}{L_\odot}\Delta t$$

(4.56)

Evolution During the Hydrogen Shell-Burning Phase

Evolution during the hydrogen shell-burning phase is toward greater central density and temperature and larger central condensation. The evolution of the shell structure can be expressed in terms of the central condensation U_1, the core mass M_1 and the ratio $\beta_1 = P_{g1}/P_1$ (Hayashi, Hoshi, and Sugimoto, 1962). From the definitions (4.13) of U, V, and $(N + 1)$

$$R_1 = \frac{G\mu\beta_1 H}{V_1 k}\frac{M_1}{T_1}$$

$$\rho_1 = \frac{1}{4\pi}\frac{M_1 U_1}{R_1^3}$$

$$= \frac{1}{4\pi}\left(\frac{k}{G\mu\beta_1 H}\right)^3 \frac{T_1^3 V_1^3 U_1}{M_1^2}$$

$$L = \frac{16\pi cG(1 - \beta_1)M_1}{(N + 1)_1 \varkappa_1}$$

The rate of energy generation in the shell, equation (4.50), is

$$L_{\text{shell}} = \frac{4\pi R_1^3}{n + 3}\mathscr{E}_0\rho_1^2 T_1^n$$

Thus, if the energy release in the core is negligible, the shell temperature is

$$T_1^{n+3} = 64\pi^2 cG\left(\frac{G\mu\beta_1 H}{k}\right)^3\left(\frac{n + 3}{\mathscr{E}_0\varkappa_1}\right)\frac{(1 - \beta_1)M_1^2}{V_1^3 U_1^2(N + 1)_1}$$

We have seen that for a centrally condensed envelope with electron-scattering opacity dominant $V_1 = (N + 1)_1 = 4$. Thus, the shell structure is

$$R_1 = \frac{G\mu_e H\beta_1 M_1}{4kT_1}$$

$$\rho_1 = \frac{16}{\pi}\left(\frac{k}{G\mu_e H}\right)^3\frac{T_1^3 U_1}{\beta_1^3 M_1^2}$$

$$T_1^{n+3} = \frac{\pi^2}{4}cG\left(\frac{G\mu_e H}{k}\right)^3\frac{n + 3}{\mathscr{E}_0\varkappa_1}\frac{\beta_1^3(1 - \beta_1)M_1^2}{U_1^2} \tag{4.57}$$

$$L = 4\pi cG\frac{(1 - \beta_1)M_1}{\varkappa_1}$$

As hydrogen shell-burning proceeds, the core contracts, U_1 decreases, and M_1 grows. The shell temperature T_1 increases, but only slightly, due to the high temperature sensitivity of nuclear reactions. The shell radius R_1 is approximately constant; it may increase slightly in the late red-giant phase, where the core mass increases rapidly, and it may decrease slightly in massive stars. The shell density ρ_1 decreases. The luminosity L is approximately constant, except in low-mass stars where $\beta_1 \approx 1$, so

$$(1 - \beta_1) = \frac{\pi}{48}\frac{a}{G}\left(\frac{G\mu H}{k}\right)^4\frac{\beta_1^4 M_1^2}{U_1}$$

increases substantially, and

$$L = \frac{\pi^2}{3}\sigma\left(\frac{G\mu H}{k}\right)^4 \frac{\beta_1^4 M_1^3}{\varkappa_1 U_1}$$

The stellar structure can be expressed as an explicit function of ρ_c, q_1, and M, using our analytic core and envelope models when $\rho_c \gg \rho_1$ and $R_1 \ll R$. The central density ρ_c is chosen as the parameter labeling the course of evolution, since ρ_c increases monotonically during the contraction of the helium core. A sequence of models with increasing ρ_c describes the course of evolution. It is necessary to choose an initial core size to start the sequence, since the details of the setting up of a shell source cannot be followed analytically.

The core radius is, from equation (4.35),

$$\frac{R_1}{R_\odot} = 0.178\left(\frac{M_1}{M_\odot}\right)^{\frac{1}{3}}\left(\frac{\rho_c}{10^3}\right)^{-\frac{1}{3}} \tag{4.58}$$

so R_1 shrinks with increasing ρ_c.

The central temperature for a nondegenerate core is found by substituting this expression for R_1 in the equation for T_c (4.39) and omitting the first term, which is negligible,

$$T_c = 5.4 \times 10^7 \mu_c\left(\frac{M_1}{M_\odot}\right)^{\frac{2}{3}}\left(\frac{\rho_c}{10^3}\right)^{\frac{1}{3}} \tag{4.59}$$

The shell temperature is found by substituting equation (4.58) for R_1 in the equation for T_1 (4.45) and neglecting the second term, which is small,

$$T_1 = 3.24 \times 10^7 \mu_e\left(\frac{M_1}{M_\odot}\right)^{\frac{2}{3}}\left(\frac{\rho_c}{10^3}\right)^{\frac{1}{3}} \tag{4.60}$$

For a nondegenerate core

$$\frac{T_c}{T_1} = 1.67\frac{\mu_c}{\mu_e}$$

In small-mass stars, $M \lesssim 3M_\odot$, the core is degenerate and isothermal, so instead of equation (4.59),

$$T_c = T_1$$

Thus, the core and shell temperatures increase during hydrogen shell-burning as $\rho_c^{\frac{1}{3}}$.

The shell density is determined by the energy balance: luminosity = energy-generation rate. The energy-generation rate per gram is assumed to be of the form

$$\mathscr{E} = \mathscr{E}_0\rho\left(\frac{T}{T_0}\right)^n$$

and all constants are evaluated for the CNO cycle at $T_0 = 3 \times 10^7$ °K, so $\mathscr{E}_0 = 3.87 \times 10^5 X_H X_{CNO}$ and $n = 16$. The total energy generation rate is given by equation (4.50). The luminosity depends on the opacity and the energy-transport mechanism, and two cases are considered: Radiative transfer with electron-scattering opacity, equation (4.51), and convective transport with low surface density, equation (3.23) with equation (3.22). For electron scattering

$$\rho_1 = C_1 \left(\frac{M_1}{M_\odot}\right)^{-2(n-3)/9} \left(\frac{\rho_c}{10^3}\right)^{-(n-6)/9} \tag{4.61a}$$

where

$$C_1 = \begin{cases} 217 & \text{(population I)} \\ 583 & \text{(population II)} \end{cases}$$

For a convective envelope

$$\rho_1 = C_2 \left(\frac{M}{M_\odot}\right)^{-\frac{1}{8}\{[(2n+1)A-4]/(2A-5)\}} (1-q_1)^{\frac{3}{4}[(A-4)/(2A-5)]} q_1^{-\frac{1}{8}\{[(2n+11)A-32]/(2A-5)\}} \tag{4.61b}$$
$$\times \left(\frac{\rho_c}{10^3}\right)^{-\frac{1}{8}\{[(n-11)A+32]/(2A-5)\}}$$

where $A = b + 3.5(a + 1)$ for an H^- opacity law of the form $\varkappa = \varkappa_0 P^a T^b$,

$$A = \begin{cases} 11 & \text{(population I)} \\ 15 & \text{(population II)} \end{cases}$$

and

$$C_2 = \begin{cases} 31.6 & \text{(population I)} \\ 63.5 & \text{(population II)} \end{cases}$$

The radius of the star depends on the mean polytropic index of the envelope. It is given by substituting equation (4.58) for R_1 and equation (4.61) for ρ_1 in the equation for the radius (4.49) and omitting the small first term. For electron scattering

$$\frac{R}{R_\odot} = C_3 \left(\frac{M}{M_\odot}\right)^{(10n-21)/27} (1-q_1)^{\frac{5}{3}} q_1^{(10n-66)/27} \left(\frac{\rho_c}{10^3}\right)^{(5n+6)/27} \tag{4.62a}$$

where

$$C_3 = \begin{cases} 1.54 \times 10^{-2} & \text{(population I)} \\ 2.97 \times 10^{-3} & \text{(population II)} \end{cases}$$

For a convective envelope

$$\frac{R}{R_\odot} = C_4 \left(\frac{M}{M_\odot}\right)^{\{[(10n+21)A-60]/24(2A-5)\}} (1-q_1)^{\frac{5}{4}[A/(2A-5)]}$$
$$\times q_1^{\{[(10n-9)A]/24(2A-5)\}} \left(\frac{\rho_c}{10^3}\right)^{\{[(5n+9)A]/24(2A-5)\}} \tag{4.62b}$$

where

$$C_4 = \begin{cases} 0.383 & \text{(population I)} \\ 0.120 & \text{(population II)} \end{cases}$$

Thus, the stellar radius increases with increasing central density.

The luminosity of the star is given by substituting equation (4.58) for R_1, equation (4.61) for ρ_1, and equation (4.60) for T_1 in the equation for the energy generation (4.50). For electron scattering

$$\frac{L}{L_\odot} = C_5 \left(\frac{M_1}{M_\odot}\right)^{(2n+21)/9} \left(\frac{\rho_c}{10^3}\right)^{(n+3)/9} \tag{4.63a}$$

where

$$C_5 = \begin{cases} 39.5 & \text{(population I)} \\ 5.01 & \text{(population II)} \end{cases}$$

For a convective envelope

$$\frac{L}{L_\odot} = C_6 \left(\frac{M}{M_\odot}\right)^{\{[(10n+21)A-40n-48]/12(2A-5)\}} (1-q_1)^{\frac{5}{2}[(A-4)/(2A-5)]}$$

$$\times q_1^{\{[(10n-9)(A-4)]/12(2A-5)\}} \left(\frac{\rho_c}{10^3}\right)^{\{[(5n+9)(A-4)]/12(2A-5)\}} \tag{4.63b}$$

where

$$C_6 = \begin{cases} 0.835 & \text{(population I)} \\ 0.0594 & \text{(population II)} \end{cases}$$

The effective temperature of the star is found from equation (3.17) with equations (4.63) and (4.62). For electron scattering

$$T_e = C_7 \left(\frac{M}{M_\odot}\right)^{-7(2n-15)/108} (1-q_1)^{-\frac{5}{6}} q_1^{-[(14n-195)/108]} \left(\frac{\rho_c}{10^3}\right)^{-(7n+3)/108} \tag{4.64a}$$

where

$$C_7 = \begin{cases} 1.16 \times 10^5 & \text{(population I)} \\ 1.58 \times 10^5 & \text{(population II)} \end{cases}$$

For a convective envelope

$$T_e = C_8 \left(\frac{M}{M_\odot}\right)^{-[(10n-3)/12(2A-5)]} (1-q_1)^{-\frac{5}{2}[1/(2A-5)]}$$

$$\times q_1^{-[(10n-9)/12(2A-5)]} \left(\frac{\rho_c}{10^3}\right)^{-[(5n+9)/12(2A-5)]} \tag{4.64b}$$

where

$$C_8 = \begin{cases} 8.89 \times 10^3 & \text{(population I)} \\ 8.22 \times 10^3 & \text{(population II)} \end{cases}$$

For radiative envelopes, the luminosity increases ($\sim \rho_c^2$), and the effective temperature decreases ($\sim \rho_c^{-1}$), with increasing central density. For convective envelopes, the luminosity increases rapidly ($\sim \rho_c^3$), and the effective temperature decreases slowly ($\sim \rho_c^{-0.4}$), with increasing central density. The mode of energy transport in the envelope switches from radiative to convective when the convective flux becomes larger than the radiative flux.

All of these relations simplify when the core is degenerate, for then the central density is a function of the core mass.

The tip of the red-giant sequence occurring in small-mass stars is determined by the onset of helium-burning in the center of a star when $T_c \approx 10^8$ °K. In small-mass stars with isothermal degenerate cores,

$$T_c = T_1 = 3.24 \times 10^7 \mu_e \left(\frac{M_1}{M_\odot}\right)^{\frac{4}{3}} \left(\frac{\rho_c}{10^3}\right)^{\frac{1}{3}} \tag{4.60}$$

so the central density at the start of helium-burning is

$$\frac{\rho_c}{10^3} = 29.4 \mu_e^{-3} \left(\frac{M_1}{M_\odot}\right)^{-2} \tag{}$$

Thus, the maximum luminosity at the tip of the red-giant branch, where the envelope is convective, is, from equation (4.63b),

$$\left(\frac{L}{L_\odot}\right)_{max} = C_9 \left(\frac{M}{M_\odot}\right)^{[(A+8)/4(2A-5)]} \left(\frac{1-q_1}{q_1^{0.9}}\right)^{\frac{2}{3}[(A-4)/(2A-5)]} \tag{4.65a}$$

At $T_0 = 9.5 \times 10^7$

$$\left(\frac{L}{L_\odot}\right)_{max} = \begin{cases} 1.485 \times 10^5 \left(\frac{M}{M_\odot}\right)^{0.276} \left[\dfrac{(1-q_1)^{1.033}}{q_1^{0.931}}\right] & \text{(population I)} \\[4mm] 1.95 \times 10^5 \left(\frac{M}{M_\odot}\right)^{0.23} \left[\dfrac{(1-q_1)^{1.1}}{q_1^{0.99}}\right] & \text{(population II)} \end{cases} \tag{4.65b}$$

The maximum luminosity is insensitive to the mass of the star and is much higher than obtained from accurate calculations (see Figure 20). The mean envelope index $\bar{n} = 2.4$ is much greater than the value $n = 1.5$ appropriate for a convective envelope. Therefore, in this model, the radius and luminosity are much too large.

The evolutionary changes in the central conditions are shown in Figure 16. For low-mass stars ($M \lesssim 3\,M_\odot$), the core is isothermal during hydrogen shell-burning; the central density increases with nearly constant central temperature and the electrons in the core become degenerate. The increasing luminosity along the red-giant branch eventually causes the shell temperature, and therefore the central temperature, to rise much more rapidly than $\rho_c^{\frac{1}{3}}$ until it approaches 10^8 °K and helium is ignited. In massive stars, an isothermal condition does not develop. The core contraction provides an appreciable part of the star's luminosity from the beginning of hydrogen shell-burning. The central temperature and density increase, with T_c increasing only slightly less rapidly than $\rho_c^{\frac{1}{3}}$.

The evolutionary tracks of stars in the H–R diagram during hydrogen shell-burning are shown in Figure 17 to 21. At the beginning of hydrogen shell-burning, the luminosity, except for very small mass population I stars, is much too low because we have taken the most centrally condensed model throughout and have not allowed the degree of central condensation of the envelope solutions to increase gradually. The temperature falls off extremely rapidly outside the shell and the hydrogen shell-burning region is therefore very thin, covering about 1% instead of an initial 10% of the mass, as found in accurate calculations. The total amount of energy generated is therefore too small. Since the thickness of the shell is constant, the rising shell-temperature during hydrogen shell-burning raises the luminosity in our models. Accurate calculations show, however, that the shells are originally much thicker than ours and the narrowing of the shells, due to the steepening temperature gradient on their outside and the exhaustion of fuel on their inside, counteracts the rising shell-temperature and the luminosity stays fairly constant unless a fully convective envelope is developed.

As stars evolve, their radii increase and they move to the right in the H–R diagram. The tracks depend in their grossest features on whether or not the star is small enough to develop a degenerate core. Those stars that develop isothermal degenerate cores must evolve to much higher central densities and much greater central condensation than those that do not. Thus, very small mass stars develop quite extensive envelopes, which are fully convective and very luminous. These form the red-giant branch (Figure 20). Intermediate mass stars develop fully convective but not so extensive envelopes, and their luminosity does not greatly increase (Figure 21). The very massive stars do not develop a very great central condensation before their central temperature has reached 10^8 °K, so they do not develop

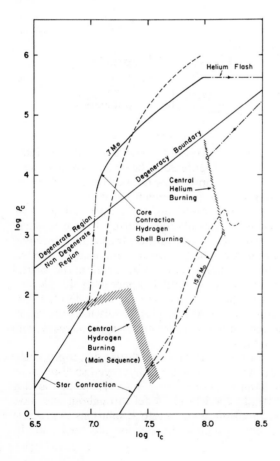

Figure 16. Evolution of central conditions during pre-main sequence contraction, central hydrogen-burning, helium core contraction, and central helium-burning. The solid lines and shaded regions are from the analytic models, the dashed-dot lines are interpolations. The dashed lines are from Hayashi, Hoshi, and Sugimoto (1962).

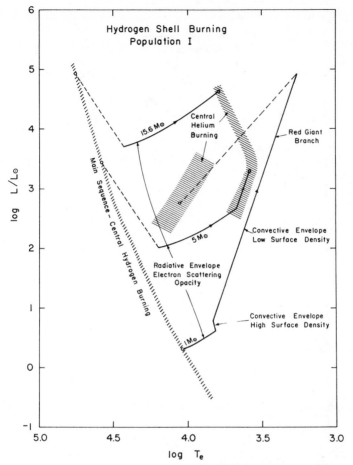

Figure 17. Evolutionary tracks of stars in H–R diagram during hydrogen-shell burning with helium core contraction. The nature of the energy-transport mechanism in the envelope, which determines the slopes of the tracks, is shown. The shaded area is the region where stars have just started to burn helium in their cores. The large extent of the central helium-burning region for low-mass stars that have passed through the red-giant phase is due to uncertainty in the mass of the helium core. The uncertainty in position of a given star is roughly parallel to the shading.

convective envelopes before helium-burning, and their luminosity increases only slightly.

The time scales for evolution during the hydrogen shell-burning–contracting helium-core stage are given in Table I.

To summarize: The cause of the extended envelopes of hydrogen-shell burning stars is their central condensation; the cause of their central condensation is as follows: When the core hydrogen is exhausted and the thermonuclear energy generation occurs in the shell, the core tends toward an isothermal state, that is,

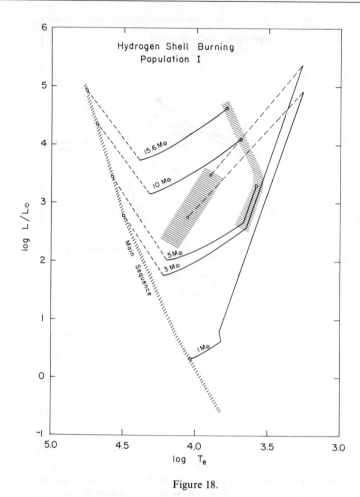

Figure 18.

as the core contracts the density rises by more than T^3. Thus, the density gradient in the core increases, which increases the star's central condensation. As a star's central density rises, its shell temperature tends to rise, since $T_1 \propto \rho_c^{\frac{1}{3}}$ for homologous contraction (equation 4.60). If the shell were not nuclear burning, its density would also rise. If, however, the shell is nuclear burning, the shell density must be low in order to keep down the rate of thermonuclear energy generation and increase the envelope's transparency, i.e., to bring the rate of energy generation and luminosity into balance. That is, the temperature sensitivity of the nuclear burning process keeps the shell temperature and radius nearly constant, so the shell density falls as the central density rises, in order to conserve mass. Thus, the star's central condensation increases. The greater the central density, the smaller the core radius; the higher the shell temperature, the lower the shell density [for nuclear burning processes with temperature sensitivity $n = d(\ln \mathscr{E})/d(\ln T) > 9$], and the larger the envelope.

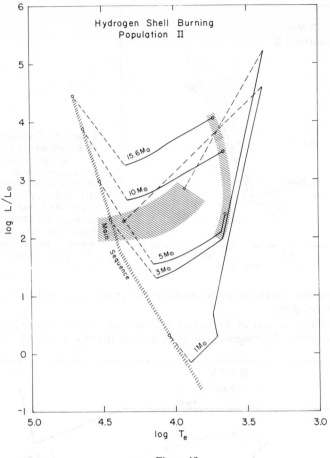

Figure 19.

CENTRAL HELIUM-BURNING

When the central temperature of the helium core is raised to about 10^8 °K, helium will begin to burn at the center of the star. In small-mass stars with degenerate cores, a helium flash occurs. The pressure of degenerate matter depends only on the density, not the temperature, so that the energy released by the onset of helium-burning increases the temperature without a corresponding increase in pressure. The increased temperature speeds up the helium reactions, which further increases the temperature, until the temperature is high enough (kT in the central degenerate region rises above the Fermi level) for the matter to become nondegenerate and the perfect gas law again holds. In nondegenerate material, increasing the temperature increases the pressure, which causes the core to expand, thereby reducing the density and temperature and damping the reaction. The core settles down to burning helium at a much lower density and slightly higher temperature than at the onset of the flash. In large-mass stars with nondegenerate cores, the

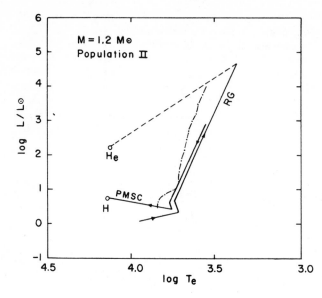

Figure 20. Evolutionary track in H–R diagram of 1.2 $M_\odot$ star. Solid lines are from analytic models for pre-main-sequence contraction (PMSC), central hydrogen-burning (H), hydrogen shell-burning red giants (RG), and central helium-burning (He). Dashed line is from Hoyle and Schwarzschild (1955), Schwarzschild and Selberg (1962), and Härm and Schwarzschild (1964).

pressure adjusts the structure to the increase in temperature and there is a slight decrease in central density.

A star burning helium at its center will be much more centrally condensed than a main sequence hydrogen-burning star. The density at the shell where the

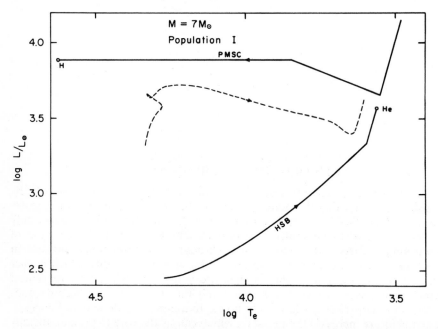

Figure 21. Evolutionary track in H–R diagram of $7M_\odot$ star. Solid lines are from analytic model for pre-main-sequence contraction (PMSC), central hydrogen-burning (H), hydrogen shell burning (HSB), and central helium-burning (He). The dashed curve is from Hofmeister, Kippenhahn, and Weigert (1965).

composition discontinuity, and possibly hydrogen-burning, occurs is much less than the central density. Thus the core by itself may be treated as a star with the density, but not the temperature, going to zero at its surface. The luminosity of the core is determined by the balance between the radiative energy transport and the helium energy-generation rate. This energy balance determines the central temperature, which together with hydrostatic equilibrium determines the central density. The radius of the core is determined by the density distribution, which we assume is linear. Thus, for the model of a star burning helium at its center, assume a linear density distribution in its core, a $\rho \sim r^{-3}$ density distribution in its envelope, and treat the core by itself as a star.

The helium energy-generation rate is

$$\mathscr{E}_{3\alpha \rightarrow C^{12}} = \mathscr{E}_0 \rho^2 \left(\frac{T}{T_0}\right)^n \tag{4.66}$$

for

$$T_8 \sim 1 \qquad (n = 41, \mathscr{E}_0 = 4.4 \times 10^{-8} X_{He}^2)$$

$$T_8 \sim 2 \qquad (n = 19, \mathscr{E}_0 = 15 X_{He}^2)$$

The total helium-burning energy-generation rate is

$$L = 4\pi R_1^3 \mathscr{E}_0 \rho_c^3 T_c^n J_n \tag{4.67}$$

where

$$J_n = \int_0^1 x^2 (1-x)^{n+3} (1 + 2x - 1.8x^2)^n \, dx$$

is the integral of $[\rho(r)/\rho_c]^3 [T(r)/T_c]^n$ over the star. Then

$$\frac{L}{L_\odot} = 201 \mathscr{E}_0 J_n \mu_c^n \left(\frac{M_1}{M_\odot}\right)^{n+3} \left(\frac{R_\odot}{R_1}\right)^{n+6} \left(\frac{9.6 \times 10^6}{T_0}\right)^n \tag{4.68}$$

The luminosity with electron-scattering opacity is given by equation (3.16). The central temperature and density are, from equations (3.6) and (3.3),

$$T_c = 9.62 \times 10^6 \mu_c \left(\frac{M_1}{M_\odot}\right) \left(\frac{R_\odot}{R_1}\right)$$

$$\rho_c = 5.65 \left(\frac{M_1}{M_\odot}\right) \left(\frac{R_\odot}{R_1}\right)^3$$

The core structure, for $T_c \sim 1 \times 10^8 \, °K$, is

$$T_c = 1.16 \times 10^8 \mu_c^{0.213} \left(\frac{M_1}{M_\odot}\right)^{0.128}$$

$$\rho_c = 9.9 \times 10^3 \mu_c^{-2.36} \left(\frac{M_1}{M_\odot}\right)^{-1.62}$$

$$\frac{R_1}{R_\odot} = 8.27 \times 10^{-2} \mu_c^{0.787} \left(\frac{M_1}{M_\odot}\right)^{0.872} \tag{4.69}$$

$$\frac{L_c}{L_\odot} = 179 \mu_c^4 \left(\frac{M_1}{M_\odot}\right)^3$$

The density and temperature at the shell are determined by the conditions of hydrostatic equilibrium and energy conservation. The shell temperature is given by equation (4.45) with the small second term neglected and equation (4.69) for R_1,

$$T_1 = 6.97 \times 10^7 \left(\frac{\mu_e}{\mu_c^{0.787}}\right) \left(\frac{M_1}{M_\odot}\right)^{0.128} \qquad (4.70)$$

The shell density is the solution of

$$\text{Luminosity} = L_{\text{core}} + L_{\text{shell}} \qquad (4.71)$$

where L_{core} is the core luminosity, equation (4.69); L_{shell} is the shell energy-generation rate, equation (4.50); and the luminosity is given by equations (4.51), (3.24), or (3.28). Once ρ_1 is known, the luminosity is found from equation (4.71). The effective temperature is given by equation (3.17), and the radius is given by equation (4.49) as in the case of hydrogen-shell burning.

The locus of points in the H–R diagram where initial central helium-burning occurs is shown as the shaded regions in Figures 17 to 19. The relative contributions of hydrogen- and helium-burning to the luminosity are found to be initially: $L_H > L_{He}$ for population I stars, while $L_H \ll L_{He}$ for population II stars. The contribution of the hydrogen shell source decreases in massive stars. We see that small-mass population II stars lie at the onset of central helium-burning in a strip of nearly constant luminosity, but with varying effective temperature depending on the mass; the smaller the mass, the higher the effective temperature. The time scales for evolution during the central helium-burning stage are given in Table I.

In the more advanced stages of evolution—helium-burning, carbon-burning, neon- and oxygen-burning—the core of the star continues to become denser and hotter, a complicated shell structure develops, with some shells active and others inactive, and the radius continues to grow. A schematic picture of the stages of central nuclear burning and shell formation is given in Figure 5 of Hayashi (1965). How far a star progresses through these stages of nuclear burning depends, as we have shown, on its mass.

FINAL STAGES OF EVOLUTION

After a star has exhausted all the nuclear fuels it is capable of burning, its only remaining sources of energy are its gravitational potential energy, which it can release by contracting, and its thermal energy, which it can release by cooling. Such a star will contract, increasing its central density and temperature. The core will, however, tend to be cooled off by energy losses from neutrino emission. The rate of emission of neutrinos increases with temperature, and since their mean free path is larger than the radius of the star they remove energy from the star. If neutrino emission is intense, all stars in the stage of gravitational contraction after the exhaustion of nuclear fuel will develop degenerate cores.

If the central density resulting from the gravitational contraction is low, only electrons, not nucleons, are degenerate and supply the pressure to support the star. There is a maximum density possible for a stable star supported by degenerate electron pressure. At higher densities, the electrons are forced onto the protons, creating neutrons. This process is a phase change and absorbs a great deal of energy, causing instability. The gravitational collapse of massive stars produces cores with densities above the critical density. The core of such a star will be composed of free degenerate neutrons and other baryons. If the mass of the remnant from the collapse is small enough, it can be supported by the pressure of the degenerate neutrons and a stable neutron star will be formed. If the mass is too large, the gravitational force, augmented by the relativistic effect that the pressure contributes to the effective mass, overwhelms the nuclear forces and the star collapses indefinitely. What happens to such core remnants remains to be discovered.

Structure of White Dwarfs

White dwarfs are stars whose support is provided by the pressure of degenerate electrons (only electrons, not nucleons, are degenerate) throughout most of the mass of the star. We assume that the electrons are completely degenerate throughout the star. This is, of course, not really possible, since the density in the surface layers is very low and the electrons are nondegenerate. However, the surface layers are extremely thin.

The equation of state of a degenerate gas is a complicated function

$$P = P(\rho)$$

approaching the limiting forms

$$P = K_1 \rho^{\frac{5}{3}} = 9.91 \times 10^{12} \left(\frac{\rho}{\mu_e}\right)^{\frac{5}{3}} \tag{5.1}$$

at low density, where the electrons are nonrelativistic ($p \ll m_e c$, where p is the electron Fermi momentum), and

$$P = K_2 \rho^{\frac{4}{3}} = 1.23 \times 10^{15} \left(\frac{\rho}{\mu_e}\right)^{\frac{4}{3}} \tag{5.2}$$

at high density where the electrons are relativistic, ($p \gg m_e c$). Thus, the temperature disappears from the equation of state, and the internal structure of a white dwarf is independent of the temperature.

Applying dimensional analysis to the equation of hydrostatic equilibrium gives, for a nonrelativistic degenerate gas (Osterbrock, 1963),

$$\rho_c \propto \frac{M}{R^3}$$

$$P_c \propto g\rho R \propto \frac{M^2}{R^4}$$

$$\propto K_1 \rho_c^{\frac{5}{3}} \propto \frac{M^{\frac{5}{3}}}{R^5}$$

Thus, for a nonrelativistic degenerate gas

$$R \propto M^{-\frac{1}{3}}$$

$$\rho_c \propto M^2 \tag{5.3}$$

There is thus a definite relation between the mass and radius of a white dwarf—the larger the mass, the smaller the radius. Since the central density increases with mass, for large-mass white dwarfs the electrons become relativistic. For an extreme relativistic degenerate gas, the pressure force is

$$\frac{dP}{dr} \propto \frac{\rho^{\frac{4}{3}}}{R} \propto \frac{M^{\frac{4}{3}}}{R^5}$$

and the gravitational force is

$$g\rho \propto \frac{GM(r)}{r^2}\rho \propto \frac{M^2}{R^5}$$

Thus, the pressure force and the gravitational force depend on the radius in the same way, but depend differently on the mass, so the two forces will be in balance for only one mass, the limiting mass of a white-dwarf star. For larger masses, the gravitational force always exceeds the pressure force.

The mass–radius relation for a white dwarf can be obtained from the virial theorem (Salpeter, 1964):

$$3(\gamma - 1)U + \Omega = 0 \tag{2.1'}$$

where Ω is the gravitational potential energy, given by equation (1.19),

$$\Omega \approx -\frac{GM^2}{R}$$

and the internal energy U is the electron kinetic energy

$$U = NK_e$$

Here N is the number of electrons and K_e is the kinetic energy per electron. The mass of the star is

$$M = N\mu_e m_p$$

where m_p is the proton mass and μ_e is the molecular weight per electron

$$\mu_e = \left[Xx_1 + \frac{Y}{4}(y_1 + 2y_2) + \frac{Z}{2} \right]^{-1}$$

where x_1, y_1 are the fractions of singly and y_2 of doubly ionized hydrogen and helium, respectively. $\mu_e = 2$ for a fully ionized gas if $X = 0$. Thus from the virial theorem

$$K_e = \frac{Gm_p^2 \mu_e^2 N}{3(\gamma - 1)R} \tag{5.4}$$

The electron kinetic energy is related to its momentum by

$$K_e = \frac{p_e^2}{2m_e} \qquad p_e \ll m_e c$$

$$K_e = p_e c \qquad p_e \gg m_e c$$

The average electron momentum p_e is related to the average interelectron spacing r_e by the uncertainty principle:

$$r_e p_e \geq \hbar$$

Using the equality sign gives for the kinetic energy

$$K_e = \frac{p_e^2}{2m_e} = \tfrac{1}{2}m_e c^2 \left(\frac{r_0}{r_e}\right)^2 \qquad [p_e \ll m_e c]$$

$$K_e = p_e c = m_e c^2 \left(\frac{r_0}{r_e}\right) \qquad [p_e \gg m_e c]$$

(5.5)

where $r_0 = \hbar/m_e c$ is the electron Compton wavelength.

These two limiting equations can be combined in the interpolation formula (Wheeler, 1964)

$$K_e = m_e c^2 \left(\frac{1}{s + 2s^2}\right)$$

(5.6)

where $s = r_e/r_0$. This formula is accurate to within 8%. The radius of the star is expressed in terms of r_e by

$$R = N^{\frac{1}{3}} r_e$$

(5.7)

Equating the expressions (5.4) and (5.6) for the electron kinetic energy gives the relation

$$1 + 2s = \left[\frac{3(\gamma - 1)r_0 m_e c^2}{G m_p^2 \mu_e^2}\right] N^{-\frac{2}{3}}$$

$$\equiv \frac{3(\gamma - 1)}{\mu_e^2} \left(\frac{N_0}{N}\right)^{\frac{2}{3}} \equiv \frac{3(\gamma - 1)}{\mu_e^{\frac{4}{3}}} \left(\frac{M_0}{M}\right)^{\frac{2}{3}}$$

(5.8)

where

$$N_0 \equiv \left[\frac{G m_p^2}{\hbar c}\right]^{-\frac{3}{2}} = 2.2 \times 10^{57}$$

$$M_0 \equiv \left(\frac{G}{\hbar c}\right)^{-\frac{3}{2}} m_p^{-2} = N_0 m_p = 3.7 \times 10^{33} \text{ g}$$

$$= 1.85 \, M_\odot$$

For a nonrelativistic electron gas $\gamma = \tfrac{5}{3}$, and for an extreme relativistic electron gas $\gamma = \tfrac{4}{3}$, so $3(\gamma - 1)$ varies between 1 and 2. We use the interpolation formula

$$3(\gamma - 1) = \frac{1 + 2s}{1 + s}$$

which has a maximum error of 30% (Schatzman, 1958). Thus

$$1 + s = \frac{1}{\mu_e^{\frac{4}{3}}} \left(\frac{M_0}{M}\right)^{\frac{2}{3}}$$

(5.8')

First note that the minimum value of the left-hand side of the above relation is 1, so there is a maximum mass for a white dwarf

$$M_{max} = \mu_e^{-2} M_0 = \frac{1.85 M_\odot}{\mu_e^2} \tag{5.9}$$

However, long before the density becomes infinite, inverse β-reactions will occur, and the above analysis will cease to apply. The increasing density causes instability of the white dwarf before the singularity is reached.

Second, the relation (5.8') can be written as a mass–radius relation

$$R = R_0 \mu_e^{-\frac{1}{3}} \left[\mu_e^{-\frac{4}{3}} \left(\frac{M}{M_0} \right)^{-\frac{1}{3}} - \left(\frac{M}{M_0} \right)^{\frac{1}{3}} \right] \tag{5.10}$$

where

$$R_0 = N_0^{\frac{1}{3}} r_0 = 5 \times 10^8 \text{ cm}$$

Thus, the radius of a white dwarf is very small, and it decreases as the mass increases.

The mean density of a white dwarf is

$$\bar{\rho} = \frac{M}{(4\pi/3)R^3}$$

$$= \frac{3M_0 \mu_e}{4\pi R_0^3} \left[\mu_e^{-\frac{4}{3}} \left(\frac{M}{M_0} \right)^{-\frac{2}{3}} - 1 \right]^{-3}$$

$$= 7.06 \times 10^6 \mu_e \left[\mu_e^{-\frac{4}{3}} \left(\frac{M}{M_0} \right)^{-\frac{2}{3}} - 1 \right]^{-3}$$

Since a white dwarf has a very thin nondegenerate surface layer, we may approximate such a star by a homogeneous model with a linear density distribution. Then the central density is

$$\rho_c = 4\bar{\rho} = 2.83 \times 10^7 \mu_e \left[\mu_e^{-\frac{4}{3}} \left(\frac{M}{M_0} \right)^{-\frac{2}{3}} - 1 \right]^{-3} \tag{5.11}$$

There is a maximum density possible for a stable white dwarf. As the density increases, the electron Fermi energy increases. An electron with energy greater than the β-decay energy for electron emission from a nucleus $(Z - 1, A)$ will produce inverse β-reactions

$$e^- + (Z, A) \rightarrow (Z - 1, A) + \nu$$

This process increases the value of μ_e in the interior, and thus the maximum stable mass is reduced. The predominant nuclei under white-dwarf conditions are elements in the range neon to iron, for which inverse β-decay will occur at densities of about 10^9 g/cm^3. Thus, the critical density for a white dwarf is about 10^9 g/cm^3. The relation between central density and mass for a white dwarf is shown in Figure 22, from Wheeler (1964). The stable configurations shown at higher densities are the neutron stars.

The degenerate interior of a white dwarf is practically isothermal because heat conduction by degenerate electrons is very efficient. This isothermal interior is blanketed by a nondegenerate surface layer, which is very thin and contains only

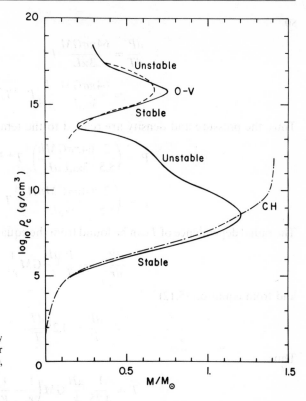

Figure 22. Schematic mass–density relation for white dwarfs and denser configurations from Wheeler (1964), calculated for cold catalyzed matter.

a minute fraction of the mass of the star. The small extent of the surface layer is easily seen by considering the scale height:

$$l = \frac{P}{\rho g} = \frac{RT}{\mu g}$$

The temperature at the transition layer is of the order of a million degrees, but $g = GM/R^2$ is extremely large because R is very small. Assuming $M \approx M_\odot$, $R \approx R_0$, and $T \approx 10^6$, then $g \approx 5 \times 10^8$ and $l \approx 10^6$ cm $= 10$ km. The density in the surface layer is less than about 10^3 g/cm^3, since it is nondegenerate. Again assuming $T \approx 10^6$, the mass of the surface layer will be

$$M_s = 4\pi R^2 \rho \Delta R \approx \pi 10^{27} \approx 10^{-6} M_\odot$$

Therefore, the equations for the surface layers may be integrated explicitly, since g, M, and L are practically constant.

The surface layer is in hydrostatic equilibrium, and energy transport is by radiation. We will assume Kramer's opacity (equation 3.8), with the quantum mechanical correction factor $(t/\bar{g}) \approx 10$. Then the equations for the structure of the envelope are (Schwarzschild, 1958, and Chandrasekhar, 1939)

$$\frac{dP}{dr} = -\frac{GM}{r^2}\rho$$

$$\frac{dT}{dr} = -\frac{3}{16\sigma}\frac{\varkappa\rho}{T^3}\frac{L}{4\pi r^2}$$

so

$$\frac{dP}{dT} = \frac{64\pi\sigma GM}{3\varkappa L}T^3$$

$$= \frac{64\pi\sigma GM}{3\varkappa_0 L}\frac{k}{\mu H}P^{-1}T^{7.5}$$

Thus, the pressure and density are related to the temperature by

$$P = \left(\frac{2}{8.5}\frac{64\pi\sigma GMk}{3\varkappa_0 L\mu H}\right)^{\frac{1}{2}}T^{4.25}$$

$$\rho = \left(\frac{2}{8.5}\frac{64\pi\sigma GM\mu H}{3\varkappa_0 Lk}\right)^{\frac{1}{2}}T^{3.25}$$

(5.12)

The radial dependence of T can be found from the equation of hydrostatic equilibrium

$$\frac{dP}{dr} = -\frac{P}{T}\frac{\mu H}{k}GM\frac{1}{r^2}$$

and from equation (5.12)

$$\frac{dP}{P} = 4.25\frac{dT}{T}$$

Thus

$$T = \frac{1}{4.25}\frac{\mu H}{k}GM\left(\frac{1}{r} - \frac{1}{R}\right)$$

(5.13)

These equations for T, P, and ρ can be used throughout the nondegenerate surface layer.

    The properties of the transition layer between the degenerate interior and the nondegenerate surface layer can be found as a function of the luminosity of the white dwarf (Schwarzschild, 1958). The isothermal nature of the interior gives a relation between interior temperature and the luminosity, which is constant through the surface layers, as follows:

$$L = \frac{2}{8.5}\frac{64\pi\sigma GM}{3\varkappa_0}\frac{\mu H}{k}\frac{T^{6.5}}{\rho^2}$$

(5.14)

Apply this to the transition layer. The boundary condition is the equality of the electron pressures in the two regions:

$$\frac{k}{\mu_e H}\rho T = K_1\left(\frac{\rho}{\mu_e}\right)^{\frac{5}{3}}$$

$$K_1 = \frac{h^2}{2m}\left(\frac{3}{\pi}\right)^{\frac{2}{3}}H^{-\frac{5}{3}} = 9.91 \times 10^{12}$$

so the boundary condition is

$$\rho_{tr} = \mu_e\left(\frac{kT_{tr}}{HK_1}\right)^{\frac{3}{2}} = 2.4 \times 10^{-8}\mu_e T_{tr}^{\frac{3}{2}}$$

(5.15)

Then the luminosity and internal temperature, $T_c = T_{tr}$, are related by

$$L = \frac{2}{8.5} \frac{64\pi\sigma GM}{3\varkappa_0 K_1^3} \left(\frac{H}{k}\right)^4 \frac{\mu}{\mu_e^2} T_c^{3.5}$$

$$= 5.7 \times 10^{25} \left(\frac{t/\bar{g}}{Z}\right) \frac{\mu}{\mu_e^2} \frac{M}{M_\odot} T_{c(6)}^{3.5} \tag{5.16}$$

The internal temperature, transition density, and extent of surface layer as a function of luminosity are shown in Table II for a 1 solar mass star with composition $X = 0$, $Y = 0.9$, $Z = 0.1$.

The source of energy for white dwarfs is the thermal energy of the nondegenerate nuclei. The energy source cannot be nuclear reactions. At the high densities found in white-dwarf interiors, the Coulomb barriers of nuclei are reduced. At densities greater than about 5×10^4 g/cm³, hydrogen reactions occur, and at densities greater than about 5×10^8 g/cm³, helium reactions occur. However, during a star's evolution before becoming a white dwarf, all the hydrogen in its core will have been exhausted, while white dwarfs with central densities great enough for helium reactions are massive enough to have exhausted the helium in their cores. In the surface layer, where hydrogen might be abundant, nuclear reactions would cause instability because of their temperature sensitivity. During a contraction, the rate of energy generation would increase above its equilibrium value, and during an expansion, it would decrease below its equilibrium value, thus feeding energy into the pulsations. The energy source cannot be gravitational, because a star's radius is fixed by the mass–radius relation after it has become almost completely degenerate, and no further contraction is possible. The energy source cannot be the thermal energy of the electrons, because they are degenerate and most are already in their lowest possible energy state.

The evolution of a white dwarf is a continual slow cooling at constant radius; its luminosity and effective temperature decrease in time. Evolutionary paths in the H–R diagram are shown for several masses in Figure 23.

The luminosity of a white dwarf is the rate of change of the thermal energy of the nondegenerate nuclei:

$$L = -\frac{d}{dt}\left(\tfrac{3}{2}kT\frac{M}{\mu_A H}\right) \tag{5.17}$$

where μ_A is the molecular weight of the nuclei, $\mu_A^{-1} = X + \tfrac{1}{4}Y$. This equation can be integrated to obtain the cooling time of a white dwarf (Schwarzschild, 1958,

Table II

$L/L_\odot$	$T_c(10^6 \,°K)$	$\log \rho_{tr}$	$\dfrac{R - r_{tr}}{R}$
10^{-2}	17	3.5	0.011
10^{-3}	9	3.1	0.006
10^{-4}	4	2.6	0.003

This table is taken from Schwarzschild (1958), *Structure and Evolution of Stars*, p. 238.

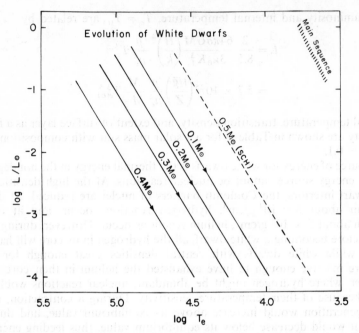

Figure 23. Evolutionary tracks of white dwarfs in the H–R diagram. Solid curves are from analytic expression (5.10); the dashed curve is from Schwarzschild (1958).

and Savedoff, 1965). Using the expression for the luminosity, equation (5.16),

$$L = K(\mu, M)T^{3.5}$$

gives

$$\frac{dT}{dt} = CT^{n}$$

where $n = 3.5$ and $C = -\frac{2}{3}(K\mu_A H/kM)$. Integration gives the cooling time from "infinite" temperature, setting the integration constant equal to zero, which is the time scale of evolution of white dwarfs

$$t = \frac{\left(\frac{3}{2}\frac{kT}{\mu_A H}M\right)}{(2.5L)}$$

$$= 1.73 \times 10^{11}\left(\frac{Z}{t/\bar{g}}\right)\left(\frac{\mu_e^2}{\mu\mu_A}\right)T_{c(7)}^{-2.5} \text{ years} \qquad (5.18)$$

$$= 8.92 \times 10^{7}\left(\frac{Z}{t/\bar{g}}\right)^{\frac{2}{7}}\left(\frac{\mu_e^2}{\mu}\right)^{\frac{2}{7}}\frac{1}{\mu_A}\left(\frac{M}{M_\odot}\right)^{\frac{5}{7}}\left(\frac{L}{L_\odot}\right)^{-\frac{5}{7}} \text{ years}$$

ACKNOWLEDGMENTS

I would like to thank A. G. W. Cameron, R. Stothers, E. Spiegel, and D. E. Osterbrock for stimulating discussions during the course of this work.

REFERENCES

A. G. W. Cameron (1963), *Space Physics*, lecture notes, Fairleigh Dickinson University, Rutherford, N.J.
A. G. W. Cameron (1965), *Galactic and Stellar Physics*, lecture notes, Yale University, New Haven, Conn.
S. Chandrasekhar (1939), *An Introduction to the Study of Stellar Structure*, Chicago University Press, Chicago, Ill.
A. N. Cox (1965), "Radiative Absorption and Opacity Calculations," this volume, p. 123.
D. Ezer and A. G. W. Cameron (1965), "The Contraction Phase of Solar Evolution," this volume, p. 203.
R. Härm and M. Schwarzschild (1964), *Astrophys. J.* **139**: 594.
C. Hayashi (1965), "Advanced Stages of Stellar Evolution," this volume p. 253.
C. Hayashi, R. Hoshi, and D. Sugimoto (1962), *Progr. Theoret. Phys. Suppl.* 22.
E. Hofmeister, R. Kippenhahn, and A. Weigert (1965), "Evolution of a Star of Seven Solar Masses," this volume, p. 263.
F. Hoyle and M. Schwarzschild (1955), *Astrophys. J. Suppl.* **2**: 1.
W. H. McCrea (1957), *Monthly Notices Roy. Astron. Soc.* **117**: 562.
P. Naur and D. E. Osterbrock (1953), *Astrophys. J.* **117**: 306.
D. E. Osterbrock (1963), Lecture notes, Summer Study Program in Geophysical Fluid Dynamics, Woods Hole Oceanographic Institution, Vol. II.
H. Reeves (1963), *Stellar Energy Sources*, Goddard Institute for Space Studies, NASA.
H. Reeves (1965), "Nuclear Energy Generation in Stars, and Some Aspects of Nucleosynthesis," this volume, p. 83.
E. E. Salpeter (1964) (see H. Y. Chiu, *Ann. Phys.* **26**: 364).
M. P. Savedoff (1965), "Cooling of White Dwarfs," this volume, p. 303.
E. Schatzman (1958), *White Dwarfs*, North Holland Publishing Co., Amsterdam.
M. Schönberg and S. Chandrasekhar (1942), *Astrophys. J.* **96**: 161.
M. Schwarzschild (1958), *Structure and Evolution of Stars*, Princeton University Press, Princeton, N.J.
M. Schwarzschild and H. Selberg (1962), *Astrophys. J.* **136**: 150.
E. A. Spiegel (1965), "Energy Transport by Turbulent Convection," this volume, p. 143.
J. W. Truran (1964), Private communication.
J. A. Wheeler (1964), in: H. Y. Chiu and W. F. Hoffman (eds.), *Gravitation and Relativity*, Benjamin, New York.
M. Wrubel (1958), *Handbuch der Physik*, *Vol. 51*, Springer-Verlag, Berlin.

Part II. Physics of Stellar Interiors

NUCLEAR ENERGY GENERATION IN STARS, AND SOME ASPECTS OF NUCLEOSYNTHESIS

Hubert Reeves

My aim in this paper will be threefold: (1) for each stage of stellar evolution, I will present the formula for the energy-generation rate*; (2) I will discuss the uncertainties attached to each formula—uncertainties issuing both from the nuclear-physics aspects of the problem and from possible variations in the chemical composition of the reacting material; (3) I will try to evaluate the size of the errors introduced in the analysis of stellar evolution when some minor nuclear reactions are neglected. (In other words, I shall try to find when these reactions can and cannot be neglected.) Because nuclear reactions produce new elements which may themselves become fuel at higher temperature, I shall consider some aspects of nucleosynthesis.

The basic equation describing the processes of energy generation and dissipation is

$$\frac{dL_r}{dM_r} = (\varepsilon_N - \varepsilon_v) + \frac{P}{\rho^2}\frac{d\rho}{dt} - \frac{du}{dt} \qquad (1)$$

This equation shows that the photon luminosity (per unit mass) dL_r/dM_r can be created by nuclear energy processes ε_N, by doing work on the gas or by decreasing its internal energy u. The term ε_v expresses the rate of energy dissipation by neutrinos. It may be worth mentioning that this formula implies the four basic interactions of modern physics: electromagnetic, nuclear, weak (neutrino-emitting), and gravitational. In this paper, I will consider only the nuclear and weak-interaction aspect of stellar energy generation.

In equation (1), I have grouped together the terms ε_N and ε_v. This way $(\varepsilon_N - \varepsilon_v)$ becomes a kind of "effective" nuclear energy rate and can be treated as such along the lines of previous analysis of stellar interiors (e.g., Schwarzschild, 1958).

This term may become locally negative in a given star. During a nuclear burning stage, this term must be mostly positive, or, more exactly,

$$\int_M (\varepsilon_N - \varepsilon_v)\, dM_r > 0 \qquad (2)$$

since we cannot otherwise expect the gravitational energy term to vanish, and hence the contraction to stop.

In cases where $(\varepsilon_N - \varepsilon_v)$ actually does become negative, the luminosity will assume the shape shown in Figure 1. In the zone $A \rightarrow B$, photons are effectively being converted into neutrinos.

---

* Much of the material presented here is discussed in Reeves (1964), where references are given. Only new material is given in any detail.

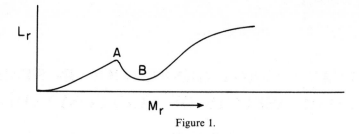

Figure 1.

To determine the presence of such zones, it is most useful to work with two parameters

$$n = \frac{d \ln \varepsilon}{d \ln T} \tag{3}$$

$$m = \frac{d \ln \varepsilon}{d \ln \rho} \tag{4}$$

n is called the temperature exponent and m the density exponent. Indeed, we can write

$$\varepsilon(T, \rho) = \varepsilon_0(T_0, \rho_0) \left(\frac{T}{T_0}\right)^n \left(\frac{\rho}{\rho_0}\right)^m \tag{5}$$

where T_0 and ρ_0 are chosen in the range of interest for a given model. In this work, I will quote the values of n and m for the various processes involved.

The behavior of the late stages of stellar evolution is ultimately related to the possibility (or impossibility) of creating neutrinos in processes involving a direct coupling between electrons and neutrinos, e.g., in processes such as $e^- + \gamma \rightarrow e^- + \bar{\nu}$ or $e^+ + e^- \rightarrow \nu + \bar{\nu}$. The physicists at CERN are now (1963) performing an experiment which may show the existence of the W^--particle, a charged vector boson expected to mediate the weak interactions. (I use "may show" because some experts in the field have emphasized the difficulty of drawing any definite conclusions from such involved and complicated experiments.) The existence of the W^- would most likely imply the reality of these neutrino processes. As of today, however, the subject is still unsettled.

The presence (or the absence) of this extra mode of energy dissipation would considerably alter the time scales associated with the late stages of stellar evolution. Time scales may be evaluated (e.g., from star counts in various regions of the Hertzsprung–Russell diagram). Hence, observational results may eventually be used as critical tests for (or against) the reality of the neutrino processes under consideration.

Here, following the more-or-less general consensus of opinion, I shall assume the existence of these processes. Occasionally, however, I shall quote results which were computed without them, and discuss the differences.

Three types of neutrino-producing mechanisms are suspected to be important during the course of stellar evolution: plasma neutrinos, photoneutrinos, and pair-annihilation neutrinos.* Their respective domains are mapped in Figure 2, together with the profile of isointensity. I shall discuss each of them in turn later.

---

* Sakashita and Nishida (to be published) have called attention to another process: neutrino pair-emission from excited nuclei. However, its emission rate never appears to become as high as the sum of the rates of the three processes mentioned here.

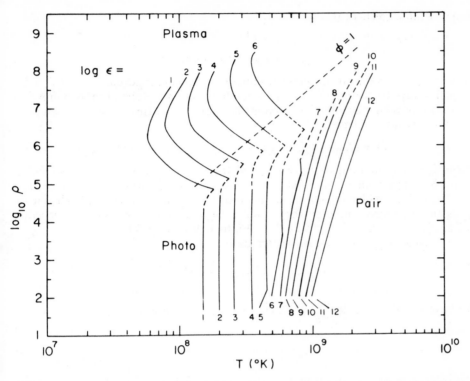

Figure 2. Isointensity curves for emission of neutrinos in the ρ-T plane. The label on each curve identifies the energy yield in ergs/g-sec along that line. Three processes are considered: plasma neutrinos, photoneutrinos, and pair-annihilation neutrinos. The dashed parts of the curves reflect the fact that photoneutrino rates in degenerate matter have not yet been properly computed.

The effect of these processes, together with the requirement stated in equation (2), seems to limit to four the maximum number of stellar nuclear-burning stages: hydrogen-, helium-, carbon-, and oxygen-burning. The time scale for neon photo-disintegration is so short that the energy generated will at best slow down the contraction. (Nuclear reactions subsequent to the fusion of oxygen will never be large enough to balance the tremendous output of neutrinos.)

Through the work of Hayashi *et al.* (1962) and Salpeter (to be published), we have learned that all stars will not pass through the four stages. Stars with masses smaller than approximately the Chandrasekhar limit (when due corrections are applied) will give up somewhere along the path to become white dwarfs. The limiting lower masses are about 0.1, 0.5, and $0.7 M_\odot$ to qualify for hydrogen-, helium-, and carbon-burning stages, respectively.

HYDROGEN-BURNING REACTIONS

Here, we consider a number of reactions burning four protons into one He^4 nucleus, yielding 6.68 MeV per nucleon (1 MeV/nucleon $\simeq 10^{18}$ ergs/g). The neutrino term ε_ν of equation (1) is, at best, a few percent of ε_N, so that $(\varepsilon_N - \varepsilon_\nu)$ never becomes negative.

Proton–Proton Cycle (Table I)

The energy-generation rate is given by

$$\varepsilon_{pp} = \varepsilon(X_4 = 0)\psi_{pp}(\alpha, W) \tag{6}$$

where $\varepsilon(X_4 = 0)$ is the rate of energy production if the core is devoid of He^4 ($X_4 = 0$). (X_i will always be the fractional mass abundance of an isotope of mass number i.) ψ_{pp} then represents the effect of reactions induced by the He^4 nuclei themselves.

Numerically, we have

$$\varepsilon(X_4 = 0) = (2.06 \pm 0.2) \times 10^{16} f_{1,1} g_{1,1} \, \rho X_1^2 \exp\left(-33.810 \, T_6^{-\frac{1}{3}}\right) T_6^{-\frac{2}{3}} \tag{7}$$

where $T_6 = T$ in millions of degrees K and the factor $f_{1,1}$ is the electron-screening factor. For stars with $M > 0.6 M_\odot$, a very good approximation is $f_{1,1} = (1 + 0.25 \times \rho^{\frac{1}{2}}/T_6^{\frac{3}{2}})$. For smaller stars, a more involved formalism must be used. The factor $g_{1,1}$ is another correction term whose value is close to unity. It grows from the value 1.02 at $T_6 = 1$ to the value 1.09 at $T_6 = 15$, to 1.19 at $T_6 = 50$.

The factor $\psi_{pp}(\alpha, W)$ can be written as

$$\psi_{p,p}(\alpha, W) = 1 + \gamma\left[0.96 - 0.49\left(\frac{W}{1 + W}\right)\right] \tag{8}$$

We have $1 < \psi_{pp} < 2$; for $T_6 > 30$, $\psi_{pp} \simeq 1.46$. The expression for γ is

$$\gamma = \left[\left(1 + \frac{2}{\alpha}\right)^{\frac{1}{2}} - 1\right]\alpha \tag{9}$$

The value of γ grows from 0 to 1 as the temperature increases. Then α, the term directly dependent on He^4 abundance, has the form

$$\alpha = 5.48 \times 10^{17}\left(\frac{X_4}{4X_1}\right)^2 \exp\left(-100 \, T_6^{\frac{1}{3}}\right) \tag{10}$$

Table I. The Proton–Proton Chain

$$H^1 + H^1 \rightarrow H^2 + e^+ + \nu$$

$$H^2 + H^1 \rightarrow He^3 + \gamma$$

$$He^3 + He^3 \rightarrow He^4 + 2H^1$$

or

$$He^3 + He^4 \rightarrow Be^7 + \gamma$$

$$Be^7 + e^- \rightarrow Li^7 + \nu$$

$$Li^7 + H^1 \rightarrow Be^8 + \gamma$$

$$Be^8 \rightarrow 2He^4$$

or

$$Be^7 + H^1 \rightarrow B^8 + \gamma$$

$$B^8 \rightarrow Be^{8*} + e^+ + \nu$$

$$Be^{8*} \rightarrow 2He^4$$

and, finally,

$$W = 1.22 \times 10^{16}\left(1 + \frac{\rho^{\frac{1}{2}}}{T_6^{\frac{3}{2}}}\right)T_6^{-\frac{1}{6}}\left(1 + \frac{1}{X_1}\right)^{-1}\exp\left(-102.6\,T_6^{-\frac{1}{3}}\right) \qquad (11)$$

The density exponent is $m = 1$ and the temperature exponent is given quite accurately by $n = (11.3/T_6^{\frac{1}{3}} - \frac{2}{3})$. At $T_6 = 15$, $n = 4$, while at $T_6 = 30$, $n \simeq 3$. In cases where electron screening becomes important (small masses), the exponents can be appreciably altered, and the above formula is no longer valid.

The governing reaction in the proton–proton cycle is the reaction $H^1 + H^1 \rightarrow D + e^+ + \nu$. This reaction can be thought of as occurring in two steps: (1) formation of an excited diproton, and (2) beta-decay of the diproton into a deuteron. The main uncertainties come from the second part, since both the beta-decay coupling constant and the matrix element for the transition are still somewhat in doubt. The uncertainty of the rate is about 10%.

We must now consider a complex of minor reactions involved in the proton–proton cycle. First, there is the reaction $He^3 + He^3$. Our estimate is based on an experiment performed at Oak Ridge in 1954. The people involved have assigned a 50% accuracy to their results; however, workers at the California Institute of Technology have expressed doubts about this, and plan to redo the experiments. Fortunately, because of the form of this cross-section in the energy-generation formalism, even a large change would not appreciably alter the total rate. In the absence of a better estimate, I will use the Oak Ridge value and uncertainty.

The rate of $He^3 + He^4$, through a recent experiment of Parker and Kavanagh (1963), is known to 15%, and the rate of $Be^7 + p$ is known to about 50%. These are the main uncertainties to worry about in the proton–proton cycle.

These effects on the total rate are roughly as follows: For $1 < T_6 < 10$, only the $p + p$ reaction matters ($\simeq 10\%$). In the region from $10 < T_6 < 20$, and most sensitively around $T_6 \simeq 14$, the reactions $He^3 + He^3$ and $He^3 + He^4$ bring in an extra source of uncertainty. The term α is known to about 60%. The term γ is relatively insensitive to α; in the worst case, $\Delta\gamma/\gamma \simeq 0.20$ (at $\alpha \simeq 3$). In this region, $W \ll 1$, so that $\psi_{pp} \simeq 1 + \gamma$ is known to better than 10%. Hence, in the range $10 < T_6 < 20$, the rate is known to better than 15%.

In the next region, $20 < T_6 < 30$, we have to consider the effect of the $Be^7 + p$ reaction. The expression W in equation (11) is known to about 50%. Its influence is most sensitively felt around $W \simeq 1$ ($T_6 \simeq 24$). It causes in ψ_{pp} an uncertainty of about 4% (the effect of α is nil in that region). Here again, the rate is known to better than 15%.

Toward higher temperatures, ψ_{pp} becomes constant, and only the $p + p$ uncertainty matters.

Errors in the evaluation of the chemical composition also influence our knowledge of the rates. The proton–proton reaction is proportional to X_1^2 (the hydrogen composition), but is also indirectly related to X_4 (the helium composition) through the term α in equation (9); changing X_4 by a factor of 2 never changes ψ_{pp} by more than 10%.

Carbon–Nitrogen–Oxygen Cycle (Table II)

Here, the energy-generation formula is

$$\varepsilon_{\text{CNO}} = (7.9 \pm 0.8) \times 10^{27} f_{14,1} g_{14,1} \rho X_{14} X_1 T_6^{-\frac{2}{3}} \exp(-152.31\,T_6^{-\frac{1}{3}})\ \text{ergs/g-sec} \qquad (12)$$

Table II. The CNO Bi-Cycle

$$C^{12} + H^1 \rightarrow N^{13} + \gamma$$
$$N^{13} \rightarrow C^{13} + e^+ + \nu$$
$$C^{13} + H^1 \rightarrow N^{14} + \gamma$$
$$N^{14} + H^1 \rightarrow O^{15} + \gamma$$
$$O^{15} \rightarrow N^{15} + e^+ + \nu$$
$$N^{15} + H^1 \rightarrow C^{12} + He^4$$
$$N^{15} + H^1 \rightarrow O^{16} + \gamma$$
$$O^{16} + H^1 \rightarrow F^{17} + \gamma$$
$$F^{17} \rightarrow O^{17} + e^+ + \nu$$
$$O^{17} + H^1 \rightarrow N^{14} + He^4$$

with $f_{14,1} = [1 + 1.75\,\rho^{\frac{1}{2}}/T_6^{\frac{3}{2}}]$, $g_{14,1} = [1 + 10^{-3}(3T_6^{\frac{1}{3}} - 4T_6^{\frac{2}{3}})]$; X_{14} is the fractional mass abundance of N^{14}.

The density exponent remains small ($m = 1$), while the temperature exponent ($n \simeq 50.8/T_6^{\frac{1}{3}} - \frac{2}{3}$) becomes larger than 10 in all cases of interest (for the sun $n \simeq 19.5$). For stars which draw most of their energy from this cycle (central temperatures $T_6 > 18$), such a high exponent drives the core into a convective state of energy transport. (For the sun $T_6 = 16$, and the energy contribution of this cycle is only 3%.)

At temperatures $T_6 > 16$, the cycle essentially transforms all the isotopes of carbon and nitrogen into N^{14}. Hence, the term X_{14} in equation (12) is very nearly equal to the initial abundance of these isotopes (Caughlan and Fowler, 1963). The same authors also show that at higher temperature the minor reactions $O^{16}(p, \gamma)\,F^{17}(e^+, \nu)\,O^{17}, O^{17}(p, \alpha)\,N^{14}$ will effectively include the O^{16} isotopes, thereby correspondingly increasing the value of X_{14}. Large uncertainties in the $O^{17}(p, \alpha)\,N^{14}$ rate make it difficult to evaluate with precision the onset of the transition.

As one approaches the very end of hydrogen-burning, a great number of minor reactions have to be taken into account; these have been considered by Parker et al. (1964). Although their energetic contribution is usually very small, it may become crucial in short periods. Great care must be taken in dealing with these last stages of nuclear burning.

After the exhaustion of hydrogen in the core, the nuclear burning of hydrogen takes place in a shell surrounding the core. The densities and temperature relevant to hydrogen-burning both in the core and in the shell are shown in Figure 3 (from Hayashi's work). The periods spent in each region of the graph are marked on the curves. Such graphs serve as a basis for the study of nucleosynthesis of minor products in stars.

We note that in a star of medium mass ($4M_\odot$), hydrogen-burning in the core and in the shell takes place at very similar values of ρ and T. In bigger stars, the shell burns at a higher temperature and slightly higher density than does the core.

Conditions in the contracting core are of interest in studying the pre-helium-burning stage.

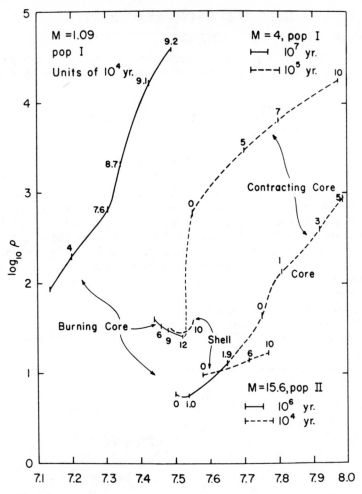

Figure 3. Densities and temperatures in hydrogen-burning core and shells for stars of 1.09, 4, and 15.6 solar masses. The numbers on the curves identify the length of time elapsed since the beginning of the hydrogen-burning phase and, further out on the curves, the length of time since the onset of the hydrogen-burning shell. The results are from Hayashi *et al.* (1962).

The nucleosynthetic effect of the hydrogen-burning stage is the transformation of H, He, Li, Be, and B into He^4 and the transformation of C, N, O, and F (and maybe Ne) mostly into N^{14} ($\sim 95\%$) but also into C^{12} ($\sim 4\%$) and C^{13} ($\sim 1\%$). In fact, at the end of hydrogen-burning, these ratios are not particularly temperature dependent (Caughlan and Fowler, 1963) and are characterized by the equalities

$$n_{12}P_{12,1} = n_{13}P_{13,1} = n_{14}P_{14,1} = n_{15}P_{15,1} \qquad (13)$$

where $P_{12,1}$ is the probability of the reaction $C^{12}(p,\gamma)N^{13}$, and n_{12} is the number of C^{12} atoms per unit volume. These equalities will be of importance to us later. In population II stars, the fractional mass of N^{14} created in this way is about 10^{-3} while it is about 10^{-2} in population I stars.

HELIUM-BURNING REACTIONS

C^{13} Substage $[C^{13}(\alpha, n)O^{16}, Q = 2.215 \text{ MeV}]$

During the contraction phase, the temperature becomes high enough for the reaction $C^{13}(\alpha, n)O^{16}$ to take place. The approximate rate is given by (Caughlan and Fowler, 1964)

$$\log \left(\frac{P_{13,4}}{\rho X_4 f_{14,4}} \right) = 17.1 - \tfrac{2}{3} \log T_8 - 30.2 T_8^{-\tfrac{1}{3}} \tag{14}$$

$$\log \left(\frac{\varepsilon_{13,4}}{X_{13}} \right) = \log P_{13,4} + 17.2 \tag{15}$$

The rate is plotted in Figure 4 (without electron screening, an important factor in small stars). The concentration of C^{13} varies from $X_{13} \simeq 10^{-5}$ (population II) to 10^{-4} (population I). The energy released in the entire fusion of C^{13} is $171 X_{13}$ keV per nucleon in the gas. Even in a population I star and even if the core material is highly degenerated (so that the nuclear-energy release will heat the core instead of expanding it), the rise in temperature would be less than one million degrees. Consequently, this reaction seems to have rather little influence on the course of stellar evolution.

Alan Liebman has used a sequence of models by Schwarzschild (1962) to investigate the onset and importance of this reaction (and of others to be described shortly) in a population II star of $1.3 M_\odot$ on its way to the helium-burning phase.

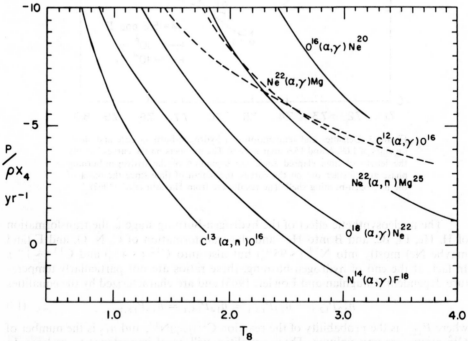

Figure 4. Rates of various nuclear reactions of importance during the helium-burning phase. The electron-screening factors are not included here. This effect will be important mostly during the period preceding the nitrogen–helium flash in small stars.

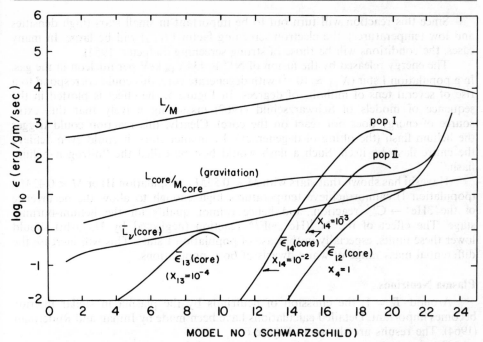

Figure 5. Behavior of various quantities during the pre-helium-flash period of a $1.2M_\odot$ star. The calculations have been made using a sequence of models by Schwarzschild et al. (1962). L/M is the mean energy-generation rate for the whole star (mostly from the hydrogen-burning shell). L_{core}/M_{core} (gravitation) is the energy-generation rate from the contraction of the helium core. $\bar{L}_\nu$(core) is the rate of neutrino dissipation by plasma neutrino processes. ε_{12} (core) is the nuclear energy released by the $3He^4 \to C^{12}$ burning. ε_{13} is the energy from the $C^{13}(He^4, n)O^{16}$ reaction, assuming $X_{13} = 10^{-4}$ (population I star). ε_{14} is the energy from the $N^{14}(He^4, \gamma)F^{18}(\beta^+, \nu)O^{18}$, assuming $X_{14} = 10^{-2}$ (population I star) or $X_{14} = 10^{-3}$ (population II star). The effects of these last two nuclear reactions on the sequence of models have been neglected (calculation by A. Liebman).

In Figure 5, the energetic evolution of this star is described. L/M is the overall energy-generation rate, while L_{core}/M_{core} follows the contracting helium core itself (here, most of the energy comes from hydrogen-burning in a shell). In the model, C^{13} (and N^{14}) burning are neglected. From the graph, it is clear that even the largest reasonable amount of C^{13} ($X_{13} = 10^{-4}$) cannot play a significant role.

N^{14} Substage $[N^{14}(\alpha, \gamma)F^{18}, Q = 4.404 \text{ MeV}, F^{18}(\beta^+, \nu)O^{18}, Q = 1.677 \text{ MeV}]$

Later in the contraction phase, N^{14} becomes combustible. New data on this rate are analyzed in Appendix A (Section A-3). In the range $0.4 < T_8 < 1.7$, the rate is given by

$$\log\left(\frac{P_{14,4}}{\rho X_4 f_{14,4}}\right) = -3.8 - \frac{12.45}{T_8} - \tfrac{3}{2}\log T_8 \tag{16}$$

$$\log\left(\frac{\varepsilon_{14,4}}{X_{14}}\right) = 17.7 + \log P_{14,4} \tag{17}$$

$$m = 1 \qquad n = \frac{28.7}{T_8} - 1.5 \quad \text{(unscreened)}$$

Since this reaction will turn out to be important in small stars (high densities and low temperatures), the electron-screening factor ($f_{14,4}$) will be large. In many cases, the conditions will be those of strong screening (Salpeter, 1954).

The energy released by the fusion of N^{14} is $434X_{14}$ keV per nucleon in the gas. In a population I star ($X_{14} \simeq 10^{-2}$) with degenerate core, this could correspond to a rise of several tens of millions of degrees. In Figure 5, the effect is plotted in the sequence of models of Schwarzschild (1962) (assuming naively that this extra source of energy does not react on the core). Clearly, this reaction could trigger the helium flash (the lifting of degeneracy). In smaller stars, it could even achieve the entire flash by itself. Such a flash would best be called the "nitrogen–helium flash."

Hayashi has shown that stars with $M < 0.53M_\odot$ (population II) or $M < 0.42M_\odot$ (population I) cannot achieve temperatures high enough to allow the occurrence of the $3He^4 \rightarrow C^{12}$ reaction, and hence cannot qualify for the helium-burning stage. The effect of the $N^{14}(He^4, \gamma)F^{18}$ reaction (neglected by Hayashi) should lower these limits, especially in the case of population I stars. This will increase the differential mass range of white dwarfs of both populations.

Plasma Neutrinos

Around $T_8 = 1$, the emission of neutrinos by the plasma in stellar interior becomes important. Detailed calculations have been made by Inman and Ruderman (1964). The results are shown in Figure 6.

One important parameter is the ratio (φ) of the plasma frequency ω_0 to the thermal energy

$$\varphi = \frac{\hbar\omega_0}{kT} = \frac{3.34 \times 10^{-4}}{T_9}\left(\frac{\rho}{\mu_e}\right)^{\frac{1}{2}}\left[1 + 1.0 \times 10^{-4}\left(\frac{\rho}{\mu_e}\right)^{\frac{2}{3}}\right]^{-\frac{1}{4}} \qquad (18)$$

Here, μ_e is the mean number of nucleons per electron ($\mu_e \simeq 2$). The line $\varphi = 1$ is plotted in Figure 2. In the region $\varphi > 1$, the density and temperature exponents are given by

$$m \simeq 2.75 - \frac{\varphi}{2} \qquad n \simeq 1.5 + \varphi \qquad (19)$$

The temperature exponent is always much smaller than the exponents associated with nuclear reactions.

In Figure 2, the plasma neutrinos are seen to be most intense around $\varphi \simeq 5.5$ (summit of the ridge). In Figure 5, the energy dissipation L_ν by plasma neutrinos in the core of a $1.3M_\odot$ star during the pre-helium phase is plotted.

During most of the contraction phase, L_ν is about ten times smaller than the gravitational-energy generation of the contracting core, but it reaches about one third of this value at the end of the contraction, after the nitrogen flash. It seems that the nitrogen flash comes just in time to prevent this process from influencing appreciably the course of the evolution.

Helium-Burning Stage ($He^4 \rightarrow C^{12}$) (Table III)

Around $T_8 = 1$, the reaction $3He^4 \rightarrow C^{12}$ becomes an important source of energy. The rate can be computed by methods of statistical mechanics or by ordinary reaction-rate theory. The latter method seems to give a rate twice as small as the former, if one overlooks the fact that the Breit–Wigner cross-section formula

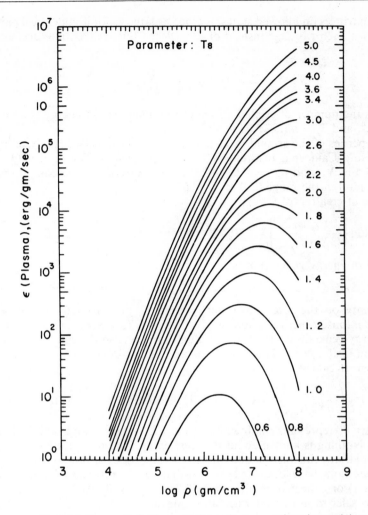

Figure 6. Energy-dissipation rate from plasma neutrinos in ergs/g/sec as a function of T_8 (as a parameter) and $\log \rho$ (as the abscissa).

Table III. Helium Reactions

$$He^4 + He^4 \rightleftarrows Be^8$$

$$Be^8 + He^4 \rightleftarrows C^{12*}$$

$$C^{12*} \rightarrow C^{12} + \gamma$$

$$C^{12} + He^4 \rightarrow O^{16} + \gamma$$

$$O^{16} + He^4 \rightarrow Ne^{20} + \gamma$$

$$Ne^{20} + He^4 \rightarrow Mg^{24} + \gamma$$

for identical particles (in allowed states) is twice as large as for nonidentical particles (the geometrical factor is $2\pi\lambda^2$ instead of $\pi\lambda^2$). When this is taken into account, the results agree:

$$\frac{n_{12*}}{n_4^3} = 3^{\frac{3}{2}}\lambda_\alpha^6 \exp\left(-\frac{Q}{kT}\right) \tag{20}$$

where n_{12*} is the number of excited C^{12} nuclei in a state at an energy Q above the mass of $3He^4$, and $\lambda_\alpha = h/(2\pi M_\alpha kT)^{\frac{1}{2}}$.

The properties of the second excited level in C^{12} have been reanalyzed by the Brookhaven and California Institute of Technology groups. The level is O^+ with $\Gamma_\gamma = 2.4 \pm 1.5$ meV, and $Q = 372 \pm 4$ keV. The reaction proceeds resonantly through this level in the range $0.80 < T_8 < 6$.

The rate of creation of C^{12} is

$$\log\left[\frac{P_{3\alpha\to C^{12}}}{(\rho X_4)^2 f_{3\alpha\to C}}\right] = -(6.72 \pm 0.25) - 18.9 T_8^{-1} - 3\log T_8 \tag{21}$$

and the rate of energy generation is

$$\log \varepsilon_{3\alpha\to C} = (11.52 \pm 0.25) + \log(f_{3\alpha\to C}\rho^2 X_4^3) - 3\log T_8 - 18.9 T_8^{-1} \tag{22}$$

The uncertainty on the rate comes partly from Γ_γ and from Q. The exponent $m = 2$, while n (unscreened) is given by $n = 3 + 44.5/T_8$, a very high value indeed.

The energy-generation rate during helium-burning also depends on the subsequent reaction $C^{12}(\alpha,\gamma)O^{16}$. This is still a weak point in our theory, since its rate is poorly known. It can be written as

$$\log\left(\frac{P_{12,4}}{\rho X_4 f_{12,4}}\right) = 10.7 + \log(\theta_\alpha^2) - 30.05 T_8^{-\frac{1}{3}} - 2\log T_8 \tag{23}$$

The parameter θ_α^2 represents the fraction of the $C^{12} + He^4$ state in the 7.112 MeV level in O^{16}. Nothing is known about its value except that $\theta_\alpha^2 < 1$, and most likely >0.01. It could, in principle, be determined by experiments, but so far it has not been. Some authors have discussed its possible value in terms of alpha-particle models ($\theta_\alpha^2 \simeq 1$) or collective models ($\theta_\alpha^2 \simeq 0.78$, or possibly 0.024). At the present time, it seems safer to use an intermediate value such as $\theta_\alpha^2 = 0.1$. The idea here is not to use an "average value," but rather to minimize possible damage to further theoretical development of an eventual experimental determination of this quantity.

On the $O^{16}(He^4,\gamma)Ne^{20}$ reaction ($Q = 4.730$ MeV), some very recent data are discussed in Appendix A (Section A-4). The rate becomes (in the range $2 < T_8 < 8$)

$$\log\left(\frac{P_{16,4}}{\rho X_4 f_{16,4}}\right) = (2.32 \pm 0.04) - \tfrac{3}{2}\log T_8 - 45.9 T_8^{-1} \tag{24}$$

It is three times smaller than the rate given, for example, in Reeves (1964). It is so much smaller than the rate of $C^{12} + He^4$ (for any reasonable value of θ_α^2) that, except for very heavy stars ($M \geq 30 M_\odot$), the helium process never goes beyond the former reaction. Consequently, we can write the total energy-generation rate as

$$\varepsilon = \varepsilon_{3He^4 \to C^{12}}\psi_\alpha \tag{25}$$

$$\psi_\alpha = 1 + \tfrac{1}{3}\frac{Q_{12,4}}{Q_{3He^4\to C^{12}}}\frac{X_{12}}{X_4^2}q_{16} = 1 + 0.33\frac{X_{12}}{X_4^2}q_{16} \tag{26}$$

$$\log(q_{16}\rho) = \log\left(\frac{P_{12,4}}{\rho X_4 f_{12,4}}\right) - \log\left(\frac{P_{3\alpha-C^{12}}}{(\rho X_4)^2 f_{3\alpha-C^{12}}}\right) \tag{27}$$

The term $P_{3\alpha \to C^{12}}$ is the *rate of formation of* C^{12} [equation (21)].

Here, $(q_{16}\rho)$ is independent of the density and of electron-screening effects. It contains the term θ_α^2 and hence is very poorly known (about a factor of 10). This usually has rather little effect on the total energy-generation rate, since ψ_α is less than 2 or 3 during most of the helium-burning phase. However, Deinzer and Salpeter (1964) have shown that at the end of helium-burning ($X_\alpha < 0.1$) ψ_α will become very large, and in this stage the uncertainty about θ_α^2 plays a major role.

To summarize, for stars with $M < 30M_\odot$, the energy-generation rate in the helium-burning phase is given by equations (22), (25), (26), and (27). The uncertainty in the rate is about a factor of 2 for $X_\alpha \geq 0.3$. For $X_\alpha \leq 0.1$, the uncertainty could reach a value of the order of 10. This uncertainty may in turn deeply effect the determination of the trajectory of the star in the Hertzsprung–Russell diagram.

Helium Reactions in Massive Stars. [$Ne^{20}(He^4,\gamma)Mg^{24}$ ($Q = 9.314$ MeV); $Mg^{24}(He^4,\gamma)Si^{28}$ ($Q = 9.986$ MeV); $Si^{28}(He^4,\gamma)S^{32}$ ($Q = 6.946$ MeV)]. In stars with $M \gtrsim 30M_\odot$, these reactions start contributing significantly to the energy generation rates.

Methods of evaluating these reactions and the uncertainties attached to them are described in Appendix A. We obtain

$$\log\left(\frac{P_{20,4}}{\rho X_4 f_{20,4}}\right) = 19.7 - 43.75 T_8^{-\frac{1}{3}} - 0.09 T_8^{\frac{2}{3}} - \frac{2}{3}\log T_8 \tag{28}$$

$$\log\left(\frac{P_{24,4}}{\rho X_4 f_{24,4}}\right) = 20.7 - 49.89 T_8^{-\frac{1}{3}} - 0.10 T_8^{\frac{2}{3}} - \frac{2}{3}\log T_8 \tag{29}$$

$$\log\left(\frac{P_{28,4}}{\rho X_4 f_{28,4}}\right) = 22.0 - \frac{2}{3}\log T_8 - 55.67 T_8^{-\frac{1}{3}} - 0.12 T_8^{\frac{2}{3}} \tag{30}$$

In this case, ψ_α in equation (26) becomes

$$\psi_\alpha = 1 + \sum_{i=3} \frac{X_{4i} X_4^{-2}}{i} \frac{Q_{4(i+1)}}{Q_{3He^4 \to C^{12}}} q_{4(i+1)} \tag{31}$$

where

$$\log(q_{4(i+1)}\rho) = \log\left(\frac{P_{4i,4}}{\rho X_4 f_{4i,4}}\right) - \log\left(\frac{P_{3\alpha \to C^{12}}}{(\rho X_4)^2 f_{3\alpha \to C^{12}}}\right) \tag{32}$$

Here, clearly, $i = 4$ represents O^{16}, etc. With the proper Q values, ψ_α is given by

$$\psi_\alpha = 1 + 0.328 X_4 X_{12} q_{16} + 0.162 X_4 X_{16} q_{20} + 0.256 X_4 X_{20} q_{24} + 0.229 X_4 X_{24} q_{28}$$
$$+ 0.136 X_4 X_{28} q_{32}$$

Nucleosynthesis from Helium-Burning Reactions

The regions of the ρ-T plane where helium-burning reactions are likely to be found (either in the core or in a shell during the carbon-burning stage) are shown in Figure 7 (again from Hayashi). (It should be remembered that neutrino emission has there been neglected. Its presence would affect the conditions of helium shell-burning but not the conditions of helium core-burning.)

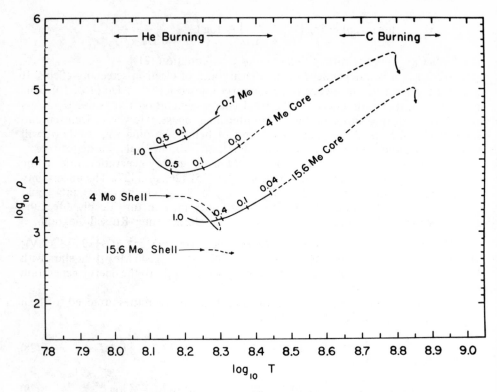

Figure 7. Densities and temperatures in helium-burning core and shells and carbon-burning core for stars of 0.7, 4, and 15.6$M_\odot$. The numbers on the curves identify the fractional weight of helium left in the core. The calculations pertaining to the carbon-burning phase do not take into account the effect of neutrino dissipation processes. The results are from the work of Hayashi et al. (1962).

To evaluate the nucleosynthetic effect of helium, the isotopic evolution has been followed until $X_\alpha \simeq 0.01$, assuming fixed temperature and density (Figure 8). The curves are isoabundance curves of various elements. For instance, $O^{16} > 50\%$ in a certain region means that in the corresponding range of density and temperature the final product is mostly O^{16}. Ne^{20} is almost bypassed; it appears only in a small region, where it reaches at best a value of 10%. (By using extreme values for the rates, a value of 20% could be obtained.)

Deinzer and Salpeter (1964) and Hayashi (1962) have followed in detail the isotopic evolution during helium-burning in the core for various masses. Deinzer and Salpeter start with initially homogeneous helium stars, a simplification which may be of importance in assessing the accuracy of the predicted isotopic results. For the $O^{16}(He^4, \gamma)Ne^{20}$ rate, they successively use $\theta_\alpha^2 = 0.1, 0.4, 1$. (Figure 8 is based on $\theta_\alpha^2 = 0.1$.) Their results are shown in Figure 9. The final abundance of C^{12} and O^{16} depends very much on θ_α^2, but the final abundance of heavier material does not. The isotopes Ne^{20} and Mg^{24} are essentially absent in stars with $M < 30M_\odot$. Even for larger masses, Ne^{20} reaches at best 20% [these authors used older data which gave an $O^{16}(He^4, \gamma)Ne^{20}$ three times larger than the rate quoted here].

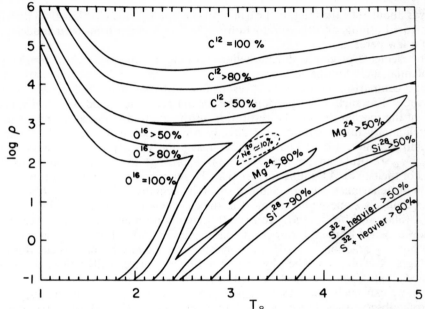

Figure 8. Isoabundance curves describing isotopic outcome of helium-burning processes as a function of (fixed) density and temperature. Except for neon, only the most important species is mentioned at any point of the ρ-T plane. The caption, e.g., $C^{12} > 50\%$, in a given region means that in the physical conditions pertaining to that region the fractional weight of C^{12} is more than 50% (but less than 80%). The rates used in this figure differed slightly from the rates quoted in the text. Present rates would allow Ne^{20} to go to about 20% in the region where $Ne^{20} \simeq 10\%$ is quoted. Everywhere else the abundance of Ne^{20} is at best a few percent.

Hayashi's results for C^{12} and O^{16} in the helium-burning core are quite similar. He does not consider the generation of Ne^{20}. Hayashi's group has also studied the formation of C^{12} and O^{16} in a helium-burning shell during carbon-burning. For stars of 4 and $15.6 M_\odot$, the temperature in the shell is very close to $T_8 = 2$ and the

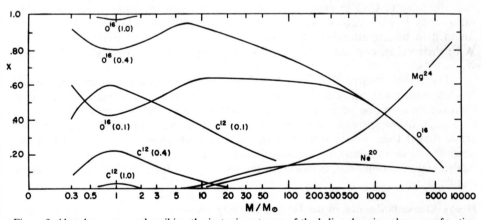

Figure 9. Abundance curves describing the isotopic outcome of the helium-burning phase as a function of stellar masses. [Salpeter and Deinzer (1964), pure helium stars.] Three curves are given for C^{12} and O^{16}, representing the choice of 0.1, 0.4, and 1.0 for the value of the parameter θ_α^2 (see text). The values of Ne^{20} and Mg^{24} are practically independent of the choice of θ_α^2.

density is about 800; the fractional carbon production is somewhat less than during helium-core-burning, and the production of Ne^{20} (evaluated from their models) is at best a few percent.

The elements produced during helium-burning are important in determining the potential fuels for the following stages of evolution. It is also of interest to attempt a comparison between these results and observational evidence on cosmic abundances. For such a comparison to be meaningful, one would need a detailed theory of the mechanism for the restitution of the evolved material to interstellar gas. Preliminary investigations have already been made on this matter (Salpeter, 1959, and Schmidt, 1959). These authors make the assumption that stars with $M > 0.7M_\odot$ restitute an amount $\Delta M = (M - 0.7M_\odot)$ to the interstellar gas at the end of their lives. Assuming further that the relative rate of star formation as a function of stellar mass has not changed since the beginning of the galaxy (the total rate *has* changed!), one finds that most of the gas came from stars with $2M_\odot < M < 5M_\odot$. (Stars with $M > 10M_\odot$ contribute virtually nothing.) The outer envelope, somehow expelled from such stars, will contain, among other things, the product of helium-burning reactions (mostly, presumably, from helium-burning shells during later stages of evolution). The physical conditions in these shells still have to be computed, taking into account the neutrino-emission processes. However, from Figure 8, we can safely guess that the neon production will always be small.

From observations based on the abundance of C, O, and Mg in the sun (Goldberg, Müller, and Aller, 1960) and the abundance of Ne^{20} from Suess–Urey "cosmic abundances" (Si^{28} is the standard), it appears that of the group C, O, Ne, Mg, Si, about 20% belongs to carbon, 50% to oxygen, 20% to neon, and a few percent to magnesium and silicon. In view of the previous discussion, the abundances of C and O are understandable, but the abundance of Ne is very high. Since neon is also a product of carbon-burning, most of it may come from that source. Magnesium and silicon most likely come from the carbon-burning stage, or from more advanced stages of stellar evolution.

Solar cosmic rays have recently been analyzed by Biswas and Fichtel (1963). The analysis gives strong support to the view that, except for protons, the accelerating mechanisms do not disturb the original relative abundances. They find, indeed, that the ratio of C/O in cosmic rays is very similar to the solar atmosphere C/O ratio. They have measured Ne^{20} (not spectroscopically detectable in the sun) and found it to be less abundant than the above-quoted value by a factor of about 3. With their value, one obtains $X_C = 0.26$, $X_O = 0.58$, $X_{Ne} = 0.09$, $X_{Mg} = 0.04$, and $X_{Si} = 0.04$.

The cosmic "normal" abundance of Ne^{20} has been evaluated from analyses of B stars, which according to Strömgren are very uncertain. The planetary nebulae typically have ratios of neon to oxygen which are about 10 times smaller than "normal" (Aller, private communication). It does not appear unreasonable to argue that if there is any setting in our universe where we may hope to find the products of helium-burning, unadulterated by nuclear processes associated with further stellar evolution, it is indeed in the gases which form the bulk of a planetary nebula.

Heavy Elements Buildup in the Helium-Burning Phase

The presence in nature of elements with $A > 56$ is attributed (Burbidge *et al.*, 1957) to the turning-on at a certain period (or periods) of stellar evolution of a rather large flux of neutrons. In particular, the existence of isotopes in the bottom of the valley of nuclear stability is proof of the action of neutron flux with characteristics such that the neutron-capture time was much longer than beta-decay lifetime of the unstable isotopes lining the bottom of the valley on both sides, and such that several tens of neutrons were absorbed for each seed nucleus (usually the metals Co-Fe and Ni*). This process is called the s-process, and the elements thus produced are called the s-elements. In this respect, the reaction $C^{13}(\alpha, n)O^{16}$ may be of importance, since it generates neutrons and hence may be a source of heavy element buildup. The C^{13} thus burned may have been left from hydrogen-burning (as discussed before, the fractional mass of the CNO group to be found in the form of C^{13} is about 1%), or it may have been formed by mixing some protons into the helium-burning core (Greenstein and Wallerstein, 1964). The effect of the originally present C^{13} is negligible for two reasons. First, even if every neutron were to be used for heavy element buildup, at best one neutron would be absorbed for each metal nucleus. Second, there is the question of the nuclear poisoning effect of N^{14}, a rather delicate question. From Figure 4, it is evident that the atoms of C^{13} will always burn before the atoms of N^{14}, and hence in the presence of N^{14} atoms. Through the $N^{14}(n, p)C^{14}$ reaction, which has a large cross-section, N^{14} will act as a neutron poison. But if the proton thus produced is then captured by a C^{12} atom, we shall have regeneration of the lost neutron by the reactions $C^{12}(p, \gamma)N^{13} \rightarrow C^{13}(He^4, n)O^{16}$. In other words, C^{12} will act as an antidote to N^{14} poison if it is abundant enough to capture a large fraction of the protons from the $N^{14}(n, p)$ reaction. Will that be our case? Clearly no, since in view of the hydrogen-burning equilibrium–abundance ratio prevailing in the core [see equation (13)], the proton has at best a 25% chance of being captured by a C^{12} nucleus. Hence, the formation of heavy elements from the original C^{13} must be ruled out.

After the onset of helium-burning, the abundance of C^{12} will increase sharply; the situation would be reversed if C^{13} and N^{14} had managed to survive (which will never be the case) or if they had been produced by admixture of some hydrogen from the envelope. In this case, C^{12} is a good antidote to N^{14} poisoning and heavy-element buildup is a possibility. Such mixing has been considered recently by Wallerstein and Greenstein (1964) in an attempt to explain the peculiar abundances found in CH stars.

In these stars, carbon and the products of the neutron-capture process on a slow time scale (Ba, La, Ce, Nd) are enhanced considerably, while oxygen is normal. The enhancement of C may imply that the neutron-capture processes took place after the onset of the $3He^4 \rightarrow C^{12}$ reaction. The absence of O^{16} enhancement may imply that these processes took place shortly after the onset of the $3He^4 \rightarrow C^{12}$, and most likely at rather low temperature.

Neutrons can also be generated by the sequence

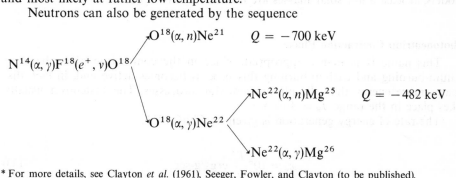

$$O^{18}(\alpha, n)Ne^{21} \qquad Q = -700 \text{ keV}$$

$$N^{14}(\alpha, \gamma)F^{18}(e^+, \nu)O^{18}$$

$$Ne^{22}(\alpha, n)Mg^{25} \qquad Q = -482 \text{ keV}$$

$$O^{18}(\alpha, \gamma)Ne^{22}$$

$$Ne^{22}(\alpha, \gamma)Mg^{26}$$

* For more details, see Clayton et al. (1961), Seeger, Fowler, and Clayton (to be published).

Because of its high threshold energy, the branching toward Ne^{21} is very difficult to achieve. It would occur if N^{14} atoms were to be suddenly brought to temperatures $T_8 \simeq 4$ even for only a matter of seconds. Such conditions may obtain during the helium flash. In Schwarzschild's star ($M = 1.3M_\odot$) $T_8 = 3.4$ is attained for a second or so. Evidently, more detailed models will be needed before we can evaluate the importance of this mechanism in stars of smaller masses.

For heavier stars, the lower branch is followed. There are two questions here: How far will the chain go before the supply of He^4 is exhausted? And, if the chain is carried to the end, will the final products be $Mg^{26} + \gamma$ or $Mg^{25} + n$? To answer these questions we must remember that at the end of helium-burning the main consumer of He^4 is C^{12}. From Figure 4, it appears that N^{14} and O^{18} will always burn before C^{12}, but not necessarily Ne^{22}. If the process takes place at $T_8 < 2$, Ne^{22} will be left intact; at higher temperature, Ne^{22} burns and, again from the graph, we get mostly $Mg^{25} + n$, with a small fraction of Mg^{26}. With the abundances of N^{14} discussed before, this chain of reaction can yield as much as 100 neutrons per seed nucleus, probably enough to meet the requirement of the s-process. At still higher temperature ($T_8 > 3$), Mg^{25} itself is reacting by $Mg^{25}(\alpha, n)Si^{28}$ to essentially double the output of neutrons. As a result, we do expect sufficient neutron generation from N^{14} isotopes if the helium-burning stage reaches a temperature of at least $T_8 = 2$ long enough to bring about fairly complete exhaustion of Ne^{22} isotopes. Such temperatures are reached during the nitrogen–helium flash, but for times which seem to be far too short (by factors of thousands, for instance in Schwarzschild's models). Studies of the flash for different masses may give a different answer, but at present this seems extremely unlikely.

The proper combination of time and temperature are reached at the end of helium-burning stages, at least for rather massive stars (e.g., $M = 7M_\odot$ as studied by Hofmeister et al., 1964). There, however, we should expect the presence of s-process elements such as Ba, La, Ce, and Nd to be accompanied by enhancement of C and O.

To summarize, neutrons will not be produced prior to the onset of helium-burning. Small stars may get N^{14}-generated neutrons during the helium flash; bigger stars do not get C^{13}- or N^{14}-generated neutrons in the early days of helium-burning. Neutrons may appear through admixture of hydrogen, producing (if the enriched material eventually reaches the stellar surface) enhancement of carbon and s-elements, but not necessarily of oxygen. Finally, N^{14}-generated neutrons will be produced at the very end of helium-burning if the mass is large enough to bring the central temperature above $T_8 = 2$ before the He^4 is exhausted. From existing models, at least a few solar masses are needed for this to occur.

Photoneutrino Contraction Phase

This name is probably appropriate since in the contraction phase between helium-burning and carbon-burning this process becomes active and, in fact, dissipates more energy than the photon emission processes. The switchover usually takes place in the range $T_8 = 3$ to 5.

The rate of energy generation is given by Chiu (1961)

$$\varepsilon_\nu = 10^8 \frac{T_9^8}{\mu_e} \text{ ergs/g-sec} \tag{33}$$

The term μ_e is the mean number of nucleons per electron in the gas. Usually, μ_e is equal to 2 (as for He^4, C^{12}, O^{16}, etc.). However, at higher temperature, electron–positron pairs will be produced, which must be taken into account in evaluating μ_e.

The density and temperature exponents are $m = 0$, $n = 8$. Again, this is a rather small temperature exponent compared to that for helium-burning ($n \simeq 30$) and the following carbon-burning ($n \simeq 30$ also). This contraction phase will end when the reaction $C^{12} + C^{12}$ becomes fast enough to stop the contraction. Shall we have a carbon flash? According to Hayashi, Hoshi, and Sugimoto (1962), no carbon flash will take place if the photoneutrino process occurs. The core will quietly warm up until the carbon-burning reaction equals the neutrino energy dissipation.

CARBON-BURNING REACTIONS

The Chalk River group has recently studied very thoroughly (experimentally and theoretically) the reactions between C^{12} and a few light nuclei (C^{12}, N^{14}, O^{16}). The elastic scattering data and the capture cross-section data have been improved to about 20% accuracy. Optical model parameters have been obtained which match the data all through the range of energy considered. In Appendix A (Section A-8) a method is described which allows determination of the low-energy part of the capture cross-section, and hence determination of the thermonuclear reaction rate, most likely accurate to better than a factor of 3. The probability of a reaction $C^{12} + C^{12}$ is given by

$$\log\left(\frac{P}{\rho X_c}\right) = (26.4 \pm 0.5) - \frac{36.55(1 + 0.070T_9)^{\frac{1}{3}}}{T_9^{\frac{1}{3}}} - \tfrac{2}{3}\log T_9 \qquad (34)$$

and the energy-generation rate by

$$\log \varepsilon_c = 17.7 + \log(P_{c+c}X_c) \text{ ergs/g-sec} \qquad (35)$$

The range of temperature and density at which a star will stop its contraction to burn its carbon can be investigated by means of the following crude model. We assume first that the energy is almost entirely dissipated by neutrino emission. Several detailed models have shown this assumption to be valid. Then we find in the ρ-T plane the locus of all the points representing conditions at the center of the star where carbon-burning energy generation (assuming pure carbon) equals the neutrino energy dissipation. For temperatures near 10^9, we should include pair-annihilation neutrinos (see Figure 2), which we shall discuss presently. This curve appears in Figure 10 under the label "first model." Correction has then been made to allow for the fact that neutrino emission takes place in a bigger volume than nuclear energy generation. This is the "corrected model" curve.

Hayashi has published models of stars in the carbon burning phase; however, he neglects the neutrino energy dissipation. He considers stars of $4M_\odot$ and $15.6M_\odot$, and sets at $0.7M_\odot$ the lower mass limit for a star to reach the carbon-burning stage. In Figure 7, the evolution of the central density and temperature (together with the helium-burning shells) are pictured. The Hertzsprung–Russell diagram of the cluster h and χ Persei can be used to test the validity of the model. There, two well-defined types of giant stars appear; a group called "early-type supergiants" (about 22 stars) and a group called "late-type supergiants" (about 15 stars). In between lies the Hertzsprung gap. After leaving the main sequence, Hayashi's

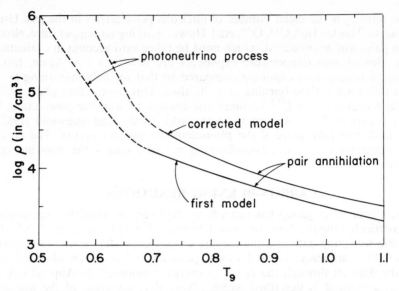

Figure 10. Carbon-burning stage curves. The "first-model" line describes the central stellar conditions where the nuclear energy-generation rate (from $C^{12} + C^{12}$) equals the neutrino energy-dissipation rate in the center of the star. The "corrected-model" curve describes the central stellar conditions when the total (volume integrated) nuclear energy-generation rate equals the total neutrino energy-dissipation rate (assuming an $n = 3$ polytropic model).

evolutionary track goes slowly through the first group, rapidly through the Hertzsprung gap, and then slowly again through the second group. We expect the ratio of the period spent in each group to be similar to the ratio of the number of stars in each group. The ratio of the periods is about one in Hayashi's model (neglecting neutrino emission). Hayashi has calculated (although crudely) the reduction in the carbon-stage period resulting from neutrino emission. He finds that the number of stars in the corresponding group should be down by a factor of about 10. Detailed models of carbon-burning stars with neutrino emission would be desirable. If similar results are found, such evidence would work against the assumption of direct coupling between electrons and neutrinos (although the statistics here may not be really relevant).

Nucleosynthesis During Carbon-Burning Phase

The reaction $C^{12} + C^{12}$ produces $Ne^{20} + He^4$ and $Na^{23} + H^1$ in roughly equal amounts. There is also a weak endothermic branching toward $Mg^{23} + n$.

The abundance equations of carbon-burning have been integrated for various carbon and oxygen initial abundances ($X_{12}^i + X_{16}^i = 1$) for different sets of densities and temperature by Cameron (1959), Reeves and Salpeter (1959), and Tsuda (1963). The results are rather insensitive to the choice of ρ_c and T_c and to the (slightly different) choice of rate made by each author. They can be qualitatively described in the following way (i denotes initial abundances and f denotes final abundances): (a) X_{16} remains the same ($X_{16}^i \simeq X_{16}^f$). (b) X_{20}^f is always less than $\simeq 0.35$; for $X_{12}^i < 0.40$, $X_{20}^f/X_{12}^i \simeq 0.6$; for $X_{12}^i > 0.40$, $X_{20}^f \simeq 0.30$. (c) $X_{24}^f/X_{12}^i \simeq 0.40$.

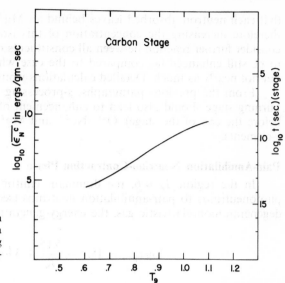

Figure 11. Lifetime and energy-generation rate of the carbon-burning stage as a function of the central temperature using the values given in Figure 10 and described in the text.

If the carbon reaction takes place at relative energy greater than 2.6 MeV, the branching $Mg^{23} + n$ occurs about 5% of the time (Bromley *et al.*, 1960). When the stellar central temperature is less than $T_9 = 0.8$, the neutron flux thus generated is negligible (see Appendix B).

On the other hand, at $T_9 < 0.75$ substantial fluxes of neutrons become available through the following set of reactions initiated by the protons and He^4 emitted in the main branching: $C^{12}(p, \gamma)N^{13}(e^+, \nu)C^{13}(\alpha, n)O^{16}$. Calculations have shown that the $C^{12}(p, \gamma)N^{13}$ reactions occur mostly in the early lower-temperature range of the carbon-burning stage. In this way, neutron fluxes capable of inducing about 50 neutron captures per metal nucleus could be released. This source of neutrons is quenched at $T_8 > 7.5$, because then the isotope N^{13} photodisintegrates into $C^{12} + p$ before it has time to beta-decay to C^{13}.

The introduction of pair-annihilation neutrino processes has recently called for a re-evaluation of the temperature at which the carbon burning stage will take place.

A combination of Hayashi's model (inhomogeneous models without neutrino emission) and Salpeter and Deinzer models (homogeneous models with neutrino emission) makes it clear that the carbon-burning temperature ranges from $T_9 = 0.8$ to about 1.1. In Table IV, the number of neutrons emitted per metal atom, assuming $X_{12} = \frac{1}{2} X_m = 5 \times 10^{-4}$, is given. Clearly, the $C^{12} + C^{12}$ burning stage provides an excellent source of neutrons and s-element processing. Here, the earmark would be probably the simultaneous enhancement of s-elements and of Na^{23} (especially when compared to Mg^{24}). This statement cannot be proven simply, but qualitative analyses have shown it to be true in most cases of interest. In a naive way, one sees

Table IV

	T_9							
	0.6	0.7	0.8	0.85	0.9	1.0	1.1	1.2
No. neutrons/metal	0.1	0.6	4	9	13	20	40	50

that each neutron absorbed leaves behind an Mg^{23} which quickly decays to Na^{23}, therefore increasing the concentration of this isotope. In real life, one must still consider further reaction between all constituents of the gas. Na^{23} usually turns out to be still enhanced (as compared to the case where the branching ratio is shut), but not nearly as much. Detailed calculations would be needed to settle this matter.

From the previous paragraphs, s-processing taking place during the carbon-burning stage should also lead to enhancement of C^{12} (if the material is extracted before the end of the stage), O^{16}, Ne^{20}, and Mg^{24}, together with enhancement of s-elements.

Pair-Annihilation–Neutrino-Contraction Phase

In the region $T_8 = 6$, the dominant emitting neutrino process passes from photoneutrinos to pair-annihilation neutrinos (see Figures 2 and 12). For a non-degenerate nonrelativistic gas, the energy-generation rate is given by

$$\log \varepsilon_{\text{pair}} = 18.7 - \frac{5.15}{T_9} + 3 \log T_9 - \log \rho \tag{36}$$

with exponents $n = 11.9/T_9 + 3$, $m = -1$.

Neon Photodisintegration (Flash?)

At $T_9 \simeq 1.25$, the contraction will be perturbed by the photodisintegration of Ne^{20}. The lifetime of an atom of Ne^{20} against photodisintegration is given by $\log \tau = -(12.15 \pm 0.05) + 28.4/T_9$. Computation (Tsuda, 1963) has shown that one half of the destroyed neon becomes O^{16} and the other half becomes Mg^{24} (through $Ne^{20}(\alpha, \gamma)Mg^{24}$). The energy yield is about 2.3 MeV per initial neon atom. Hence

$$\log \varepsilon = 29.2 - \frac{28.4}{T_9} \text{ ergs/g/sec} \tag{37}$$

As discussed in the previous section, the core at this moment contains about 25% Ne^{20} (in fractional mass). The photodisintegration will yield approximately 600 keV for each nucleus in the gas or about 30 keV for each particle in the gas. In a partially degenerated core, we may have a neon flash bringing us all the way up to the oxygen-burning stage similar to the N^{14} flash. For large stellar masses, the Ne^{20} photodisintegration will at best slow down the contraction toward the next stage.

At the end of this period, the isotopic abundance is roughly as follows ($X_{12}^i + X_{16}^i = 1$ describes the results of the previous helium stage): (a) $X_{16} \simeq X_{16}^i + 0.15$. (b) $X_{20} \simeq 0$. (c) $X_{24} \simeq 0.4X_{12}^i + 0.15$. (d) $X_{28} = 2$ to 4%.

OXYGEN-BURNING REACTIONS

As the core temperature approaches $T_9 = 1.3 \simeq 1.4$, reactions between oxygen nuclei start releasing energy. Here, there is a lack of experimental data. Only the $O^{16}(O^{16}, \alpha)Mg^{24}$ has been measured (Bromley, Kuehner, and Almqvist, 1960), and the data extend only slightly below the Coulomb barrier. More work is being done on that reaction but is not available as of now. To an accuracy no better than a

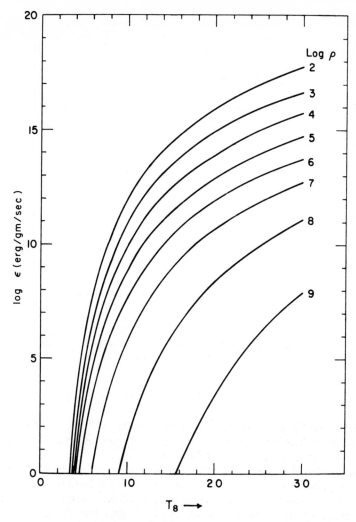

Figure 12. Energy dissipation by pair-annihilation neutrinos in erg/g/sec. The abscissa is T_8 and the parameter $\log \rho$ is in g/cm^3.

factor of 10, the following equation can be extracted from the data:

$$\log \frac{P}{\rho X_0} = 40.5 - 59.02\left(\frac{1 + 0.14T_9}{T_9}\right)^{\frac{1}{3}} - \tfrac{2}{3}\log T_9 \tag{38}$$

with

$$\log \varepsilon_0 = 17.8 + \log\left(\frac{P}{X_0}\right)$$

In Figure 13, the temperature and densities at which oxygen and neon are expected to burn are described by using the crude model described in the section on carbon-burning. The nucleo-synthetic effect of the oxygen burning stage has been studied

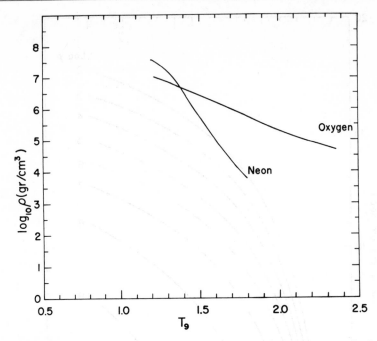

Figure 13. Oxygen- and neon-burning stage curves. The model is described
in the text and in the caption of Figure 10.

by Tsuda (1963) and Cameron (1959). The main results are the elements from Mg^{24}
to S^{32}. The elements Mg^{24}, Si^{28}, and S^{32} are not really preponderant over inbetween
isotopes (as is the case in cosmic abundances).

After the fusion of oxygen, the emission of neutrinos becomes so strong that
nuclear energy generation most likely never succeeds in halting the contraction.
Hence, we do not expect any other burning stage. Nuclear reactions leading to the
equilibrium process have been considered by Hoyle and Fowler (1964) and Tsuda
and Tsuji (1963). They will not be discussed here.

ACKNOWLEDGMENT

I acknowledge the help of J. Hauben, A. Liebman, and E. Milford, and Miss A. Weiswasser with
numerical calculations; of T. Psaropulous in drawing the figures; and of Mrs. Y. Lecavalier and M.
Yastishak in typing the manuscript.
I wish to thank Dr. R. Jastrow and the staff at the Institute for Space Studies for their hospitality.

APPENDIX A

Thermonuclear Reaction Rates in the Range $T_8 = 1$ to 50

For some $(\alpha, \gamma)(\alpha, n)(p, \gamma)$ reactions, e.g., $Mg^{24}(\alpha, \gamma)Si^{28}$, the level scheme of the
compound nucleus has been studied experimentally in a rather wide range of
energy and the widths Γ_α, Γ_p, Γ_γ and the statistical factor $\omega = (2J + 1)/(2I_t + 1) \times$
$(2I_i + 1)$ have been determined for all the resonances E_r in that range. In such a

case, a lower limit to the rate can be obtained by adding the effects of each resonance,

$$\frac{P}{\rho X_1} = \sum_{\substack{\text{all} \\ \text{resonances}}} \frac{Pi}{\rho X_1}$$

$$\log\left(\frac{P_i}{\rho X_1}\right) = 12.69 - \tfrac{3}{2}\log\mu - \log A_i + \log\left(\frac{\omega\Gamma_c\Gamma_n}{\Gamma}\right) - \tfrac{3}{2}\log T_8 - \frac{50.4E_r}{T_8} \qquad \text{(A1)}$$

Here, μ is the reduced mass and Γ_c, Γ_n the widths for the charged particle and neutral particle reactions, respectively. This rate is a lower limit because of the contribution of resonances lying outside the experimentally explored range.

The (α, γ) reactions have not, because of experimental difficulties, been brought below 1 MeV, and the resonances in that range are expected to dominate the rate at $T_8 < 3$ or so. In Figure 14, the positions of the lowest measured resonances in Si^{28} are plotted against the background of the Gamow peak to illustrate the situation.

The total capture rate can also be evaluated by combining the information on individual resonances with data obtained from optical models of the nuclei. Here, advantage is taken of the experimentally discovered regularity of the optical model parameters, a regularity best expressed in terms of the strength functions (ratio of averaged reduced width to level spacing). In opposition to the case of

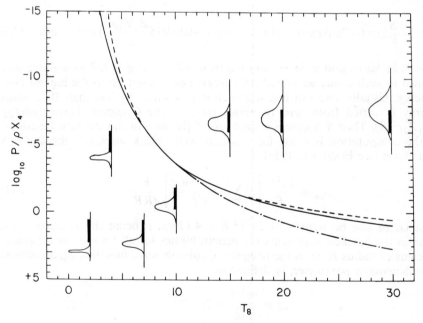

Figure 14. Rate of $Mg^{24}(\alpha, \gamma)Si^{28}$. The sum over individual resonances (– – –) is expected to be best in the range from $6 < T_8 < 10$. Below and above this range, unanalyzed resonances are expected to play a role and hence to increase the rate. The optical-model curve (— · —) is best in the range $T_8 = 3$ to about 10. Below this temperature, the randomness in the position of the levels makes it an upper limit. Above this temperature, the fact that Γ_α becomes larger than Γ_γ makes it again an upper limit (the optical model yields only the total capture rate). The intermediate curve in the higher-temperature range is based on the experimental gamma-strength function. It should be best there.

neutron reactions, alpha-reactions have so far failed to show structure in the individual partial wave (l) strength functions. The sum of these strength functions appears to be energy independent, at least in ranges wide enough to be of use in the evaluation of thermonuclear reaction rates.

Through the optical model, a computation of the capture cross-section can be made. For alpha-particle reactions, the nuclear potential is chosen to be of the form

$$V(r) = \frac{-(V_0 + iW_0)}{1 + \exp[(r - R_0)/a]} \tag{A2}$$

The best choice of parameters appears to be

$$V_0 = 50 \text{ MeV} \qquad\qquad W_0 = +10 \text{ MeV}$$
$$R_0 = (1.25 A_t^{\frac{1}{3}} + 1.6) \text{ f} \qquad a = 0.50 \text{ f}$$

where A_t is the mass number of the target nucleus (Vogt, private communication).

For proton reactions, the real potential has the same form, with $V_0 = +55$ MeV, $R_0 = 1.25 A_t^{\frac{1}{3}}, a = 0.5$ f. The imaginary potential has a Gaussian shape centered on the nuclear surface with $W_g = +4$ MeV.

Calculations have been made by Chalk River workers, and the results fitted by a formula of the form

$$\sigma = \frac{S}{E} \exp\left(-2\pi\eta\right) \exp\left(-gE\right) = \frac{S}{E} \exp\left(-0.98935 \frac{Z_1 Z_2 \mu^{\frac{1}{2}}}{E^{\frac{1}{2}}}\right) \exp\left(-gE\right) \tag{A3}$$

where σ is in barns and E in million electron volts, and g is left as a parameter determined in such a way as to make S energy-independent up to the highest possible energy. Usually, one can find a g such that S varies by less than 10% when the energy is varied from zero to about 60% of the Coulomb barrier energy ($B = Z_1 Z_2 e^2/R$). Then S becomes a measure of the sum of the strength functions.

If the computation is made for a square-well "black nucleus," the factor S takes the form (see Feshbach, 1953)

$$S_{\text{sw}} = \frac{2\pi^2 \lambdabar^2}{E} \left[\xi_0^2 \sum \frac{(2l + 1)G_0^2}{G_l^2} \right] \frac{1}{\pi K R} \tag{A4}$$

The first factor can be written as $2\pi^2 \lambdabar^2/E = 4.126/\mu$, μ being the reduced mass. The term in the brackets represents the penetrabilities for all l-waves for a square-well nucleus of radius R. G_l is the irregular Coulomb wave function. ξ_0^2, an almost energy-independent parameter, is defined as

$$\xi_0^2 \equiv \frac{y e^{2\pi\eta}}{\eta G_0^2} = 2y^{\frac{1}{2}} e^{4y^{\frac{1}{2}}}$$

where $y = 2R_b/R$, and $R_b = (\hbar^2/\mu Z_1 Z_2 e^2)$. The last term in equation (A4) is the value of the strength function for a black nucleus; $K \simeq 1.4$ f is the wave number inside the nucleus.

At energies much smaller than the Coulomb barrier and for $y^{\frac{1}{2}} > 4$ (as will be the case for our work on alpha-capture reactions and $C^{12} + C^{12}$), the factor $\Sigma(2l + 1)G_0^2/G_l^2$ becomes independent of energy. Its value is plotted in Figure 15.

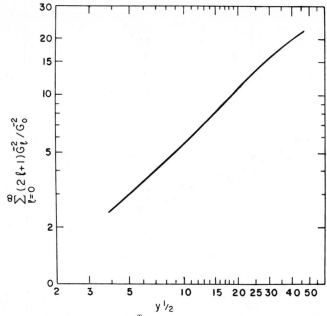

Figure 15. Graph of $\sum\limits_{\ell=0}^{\infty}(2l+1)G_l^{-2}/G_0^{-2}$ as a function of $y^{\frac{1}{4}}$ (see equation A3). The graph is valid for $y^{\frac{1}{4}} > 4$, and when the relative energy is less than one-half the Coulomb barrier energy ($B = Z_1 Z_2 e^2/R$).

Numerically, the value of S_{sw} is closely given by

$$\log S_{sw} = 0.1 + \tfrac{1}{2}\log\left(\frac{Z_1 Z_2}{\mu R}\right) + 0.458(Z_1 Z_2 \mu R)^{\frac{1}{2}} \tag{A5}$$

In the square-well formalism, the value of g is

$$g_{sw} = \frac{y^{\frac{3}{2}}}{6\eta^2} = 0.122\left(\frac{\mu R^3}{Z_1 Z_2}\right)^{\frac{1}{2}} \text{MeV}^{-1} \tag{A6}$$

with R in fermies, and μ in atomic mass units.

The results of optical-model calculations for alpha-capture by all the stable isotopes from C^{12} to S^{32} have been used to obtain the values of S_{om} (optical model). All the values are well represented by putting into the square-well formula equation (A5) the choice $R_s = (1.50A_1^{\frac{1}{3}} + 2.0)$f (Figure 16). The values of g obtained from the optical-model analysis have been plotted in Figure 16 and tabulated in Table V. It will be noted that they decrease rather rapidly with A_t, the mass of the target. A sort of effective radius can be obtained by inverting formula (A6)

$$R_g = 4.07\left(\frac{Z_1 Z_2}{\mu}\right)^{\frac{1}{3}} g^{\frac{2}{3}} \text{ f} \tag{A7}$$

The radius R_g needed to reproduce g is of the form $R_g = 0.7A_1^{\frac{1}{3}} + 4.0$. Vogt and Reeves (to be published) give an analysis of the meaning of these parameters in terms of the structure of the nuclear well.

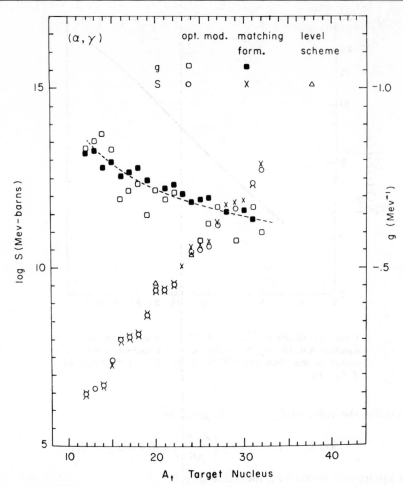

Figure 16. Values of the parameters S and g as a function of the target nucleus for (α, γ) reactions. The points refer to the results of optical-model calculation, to the matching formula equations (A5) and (A7), and for $Ne^{20}(\alpha, \gamma)$ and $Mg^{24}(\alpha, \gamma)$ to a determination of the strength function from the level scheme of the compound nucleus (calculations by A. Weiswasser).

Once S and g are found, the thermonuclear reaction rate becomes (S in MeV barns)

$$\log\left(\frac{P}{\rho\chi_1}\right) = 10.56 + \log\left(\frac{Z_1^{\frac{1}{3}}Z_2^{\frac{1}{3}}}{A_1\mu^{\frac{1}{3}}}\right) - \frac{2}{3}\log T_8 + \log S - 3.975\left(\frac{Z_1^2 Z_2^2 \mu}{T_8}\right)^{\frac{1}{3}}$$

$$\times (1 + 8.62 \times 10^{-3} g T_8)^{\frac{1}{3}} \tag{A8}$$

In Table V the values of S, g, and also the numerical constants necessary to compute the thermonuclear reaction rate are given (some of them have been slightly modified by the results of other techniques). To identify these constants, we rewrite

Table V. Parameters for the Computation of Thermonuclear Reaction Rates ($T_8 = 10^8\ °K$)*

	$\log S$ (MeV barns)	g (MeV^{-1})	M	N	P
$C^{12} + He^4$	6.45	0.82	16.6	30.11	0.07
$C^{13} + He^4$	6.55	0.83	16.7	30.25	0.07
$N^{14} + He^4$	7.22	0.78	17.4	33.72	0.08
$N^{15} + He^4$	7.29	0.79	17.5	33.88	0.08
$O^{16} + He^4$	7.95	0.75	18.1	37.20	0.08
$O^{17} + He^4$	8.04	0.77	18.2	37.35	0.08
$O^{18} + He^4$	8.14	0.78	18.4	37.47	0.08
$F^{19} + He^4$	8.73	0.74	18.9	40.67	0.09
$Ne^{20} + He^4$	9.44	0.72	19.7	43.75	0.09
$Ne^{21} + He^4$	9.39	0.72	19.6	43.88	0.09
$Ne^{22} + He^4$	9.47	0.73	19.7	43.99	0.09
$Na^{23} + He^4$	10.04	0.70	20.3	46.97	0.10
$Mg^{24} + He^4$	10.50	0.68	20.7	49.89	0.10
$Mg^{25} + He^4$	10.65	0.69	20.9	49.97	0.10
$Mg^{26} + He^4$	10.72	0.69	21.0	50.06	0.10
$Al^{27} + He^4$	11.25	0.67	21.5	52.91	0.10
$Si^{28} + He^4$	11.75	0.66	22.0	55.67	0.11
$Si^{29} + He^4$	11.82	0.66	22.1	55.75	0.11
$Si^{30} + He^4$	11.89	0.67	22.1	55.82	0.11
$P^{31} + He^4$	12.38	0.66	22.7	58.56	0.11
$S^{32} + He^4$	12.86	0.63	23.1	61.16	0.11

* Not to be used for C^{12} and O^{16}, see text.

equation (A8) in the form

$$\log\left(\frac{P}{\rho X_4}\right) = M - \tfrac{2}{3}\log T_8 - NT_8^{-\frac{1}{3}} - PT_8^{\frac{2}{3}} \tag{A9}$$

Although the optical model usually gives a fairly good representation of the data, individual nuclei may at times depart from their expected behavior. Then, both S and g could influence the rate to a great extent.

S can be evaluated directly from the experimental determination of the individual resonances. If the resonances do not overlap to the point where interference phenomena become important, the cross-section for individual resonances can be written in the Breit–Wigner fashion

$$\sigma = \pi \lambda^2 \frac{\omega \Gamma_c \Gamma_n}{(E - E_r)^2 + \Gamma^2/4} \qquad \Gamma = \Gamma_c + \Gamma_n + \dots \tag{A10}$$

Averaging over resonances in an energy range ΔE

$$\langle\sigma\rangle = \int \frac{\sigma\,dE}{\Delta E} = \sum 2\pi^2 \lambda^2 \frac{\Gamma_c \Gamma_n \omega}{\Delta E \Gamma} \tag{A11}$$

We define

$$\Gamma_{c,l} = \gamma_{cl}\xi_l^2\, e^{-gE}\, e^{-2\pi\eta} = \frac{y\, e^{2\pi\eta}}{\eta G_l^2} \tag{A12}$$

where $\gamma_{c,l}$ is the reduced width and ξ_l^2 an almost energy-independent factor representing angular momentum effects (l).

If $\Gamma_n > \Gamma_c$ all through the range ΔE, we have

$$\langle \sigma \rangle = \frac{S}{E} e^{-2\pi\eta - gE} \tag{A13}$$

with

$$S = (2\pi^2 \lambda^2 E)\frac{\Sigma \omega \xi_l^2 \gamma_c}{\Delta E} = 2\pi^2 \lambda^2 E \sum_l \frac{\langle \omega \xi_l^2 \gamma_{c,l} \rangle}{D_l} = [2\pi^2 \lambda^2 E]\xi_0^2(\gamma_c/D)$$
$$= \frac{4.126}{\mu}\xi_0^2(\gamma_c/D) \tag{A14}$$

Here, D_l is the average distance between resonances of same (l) with orbital angular momentum (l). The last term is a definition of $\xi_0^2(\gamma_c/D)$ based on the fact that the individual strength functions $\langle \gamma_c/D_l \rangle$ are known to be practically energy independent. Using the experimental values of $\Gamma_\alpha^{(\text{exp})}$, we calculate

$$\xi_0^2(\gamma_c/D) = \sum_{\substack{\text{all} \\ \text{res. in } \Delta E}} \frac{\Gamma_\alpha^{(\text{exp})} e^{2\pi\eta} e^{gE}}{\Delta E} \tag{A15}$$

On the other hand, when $\Gamma_c > \Gamma_n$

$$\langle \sigma \rangle = 2\pi^2 \lambda^2 \sum_{\Delta E} \frac{\omega \Gamma_n}{\Delta E} = \frac{4.126}{\mu} \left\langle \frac{\Gamma_n}{D} \right\rangle \frac{1}{E} = \frac{C}{E} \tag{A16}$$

with C in MeV barns. The total rate involves the integral of $\langle \sigma \rangle$ over all energies. However, the range of energies where $\Gamma_c < \Gamma_n$ brings a small contribution to the rate if the Gamow energy E_0 is above the energy (E^*) for which $\Gamma_n = \Gamma_c$. Then we write (N_0 is Avogadro's number)

$$\frac{P}{\rho X_1} = \frac{4N_0}{A_1} \frac{1}{(2\pi\mu kT)^{\frac{1}{2}}} C \, e^{-(E^*/kT)} \tag{A17}$$

or

$$\log\left(\frac{P}{\rho X_1}\right) = 10.00 - \log A_1 - \tfrac{1}{2}\log \mu - \tfrac{1}{2}\log T_8 + \log C - \frac{50.4E^*}{T_8} \tag{A18}$$

The value of E^* and C should really be obtained by trying to fit equation (A18) with the values obtained in equation (A1) by summing over all resonances. If the measurements of individual resonances are not available, then we may use the value of S obtained from the optical model calculations to get an average value of Γ_α/D, and by equating it to Γ_γ/D obtain E^*. That is,

$$S = 2\pi^2 \lambda^2 E \xi_0^2\left(\frac{\gamma}{D}\right)$$

$$\frac{\Gamma_\alpha}{D} = \xi_0^2\left(\frac{\gamma}{D}\right) e^{-2\pi\eta} e^{-gE} = \omega\left(\frac{\Gamma_\gamma}{D}\right) \qquad \text{for } E = E^* \tag{A19}$$

The use of optical models for thermonuclear reaction rates clearly implies that at least a few resonances (or at least one resonance with a more or less average reduced width) are to be found inside the Gamow peak for any temperature under consideration. Otherwise, the method may overestimate the rate by several orders of magnitude. Except for a few nuclei, (p, γ) or (α, γ) reaction thresholds are to be found in energy regions of the compound nucleus where the levels are fairly crowded. Typically, the spacing of levels will be of the order of 100 keV, but larger distances between two given levels are not uncommonly found. The position and width of the Gamow peak are given by

$$E_0 = 26.3(Z_1^2 Z_2^2 \mu)^{\frac{1}{3}} T_8'^{\frac{2}{3}} \text{ keV}$$

$$\Delta E_0 = 35(Z_1^2 Z_2^2 \mu)^{\frac{1}{6}} T_8'^{\frac{5}{6}} \text{ keV} \tag{A20}$$

$$T_8' = \frac{T_8}{1 + 8.6 \times 10^{-3} g T_8} \qquad g \text{ in MeV}^{-1}$$

Here, ΔE_0 is the full width at half maximum, g is defined in equation (A3). In Table VI, the values of E_0 and ΔE_0 (in keV) are given for typical protons and alpha-reactions.

Thus, in the low-temperature range, the widths of the Gamow peak become comparable to the average spacing between resonances, and, in the region $T_8 < 5$, important departures from the above assumption may be expected. If the corresponding level scheme is experimentally unknown, the determination of the rate becomes most uncertain. The rate obtained from optical-model calculations becomes essentially an upper limit. A lower limit, an expected value, and an estimation of the uncertainty can be obtained by including more nuclear statistics.

Experimentally, the probability that two neighboring nuclear levels will be found at a distance D larger than a preassigned value D_0 is well represented by the Wigner distribution formula

$$P(D > D_0) = \exp\left[\frac{-\Gamma}{4}\left(\frac{D_0}{\bar{D}}\right)^2\right] \tag{A21}$$

where $\bar{D}$ is the expectation value of D. The probability that two given resonances will be at $D > 2\bar{D}$ is about 0.04. Thus, the chances of finding one level at an energy E_r within $\Delta E = E_r - E_0 = 1.0\bar{D}$ of the Gamow peak E_0 is 96%. We define our upper limit to the rate as the value obtained if the level is at the center of the peak (the rate is then closely equal to our optical-model rate) and our lower limit as the value obtained if the level is at a distance $E_r = E_0 + 1.0\bar{D}$. The rate itself will be the geometric mean between these two values. The choice of 96% is arbitrary, to an extent. It is, however, in the spirit of the determination of uncertainties by

Table VI

	Z_1, Z_2					
	1, 6	1, 10	1, 16	2, 6	2, 10	2, 16
$\dfrac{\Delta E_0}{T_8^{\frac{5}{6}}}$	64	75	90	100	120	140
$\dfrac{E_0}{T_8^{\frac{2}{3}}}$	85	120	165	215	305	415

experimental physicists. Defining $u = \Delta E_0/E_0$, $v = (2.0\bar{D}/\Delta E_0)$, the ratio (f) of the lower limit to the upper limit is given by

$$f = \frac{P(E_r = E_0 + 1.0\bar{D})}{P(E_r = E_0)}$$

$$= \exp\left[(-\tfrac{4}{3}u^{-2})\left(1 + uv + \frac{2}{(1 + uv)^{\frac{1}{2}}} - 3\right)\right] \simeq e^{-v^2} \quad \text{(for } uv \ll 1)$$

(A22)

Consequently, $P_{\max} = P$ (optical model) as in equation (A9), $P_{\min} = P_{\max} f$, and we choose for the expected value and the uncertainty of the rate

$$\log P \pm \Delta \log P = [\log P (\text{opt. mod.}) + \tfrac{1}{2}\log f] \pm \tfrac{1}{2}\log f$$

The reduced widths themselves have a distribution given by

$$P(\gamma < \gamma_0) = \int_0^{x_0} e^{-x^2} \frac{2}{\sqrt{\pi}} dx \qquad x = \frac{\gamma}{\bar{\gamma}}$$

(A23)

(so that $\gamma = \bar{\gamma} \pm 0.9\bar{\gamma}$ with a probability of $\simeq 90\%$). The correct way of doing this problem would be to analyze the probability function of the ratio (γ/D). However, this is somewhat complicated for our purpose. With our previous choice of $D_0 = 3\bar{D}$, the chance of having two levels in the peak is itself 80%, so that the fluctuation in the strength function averaged over the Gamow peak should probably not reach a value of more than 3. A better estimate of $\log P$ should then be

$$\log P \pm \Delta \log P = [\log P (\text{opt. mod.}) + \tfrac{1}{2}\log f] \pm (0.5 + \tfrac{1}{2}\log f) \quad \text{(A24)}$$

In cases considered here, where the optical model has any interest [$O^{18}(\alpha, \gamma)Ne^{22}$, $Ne^{20}(\alpha, \gamma)Mg^{24}$, $Mg^{24}(\alpha, \gamma)Si^{28}$, etc.], the average level distance is about $100\,\text{keV}$. Then, $\tfrac{1}{2}\log f$ is about 0.8 at $T_8 = 1$, 0.3 at $T_8 = 2$, 0.15 at $T_8 = 3$, and 0.1 at $T_8 = 5$. Since we shall rarely need these reactions at temperature $T_8 < 3$, we have not considered this factor in the rates given in Table V.

We now discuss the individual reactions:

A-1. $C^{12}(\alpha, \gamma)O^{16}$

The rate is discussed in detail in Cartledge, Thibaudeau, and Reeves (1963) and also in Reeves (1964).

A-2. $C^{13}(\alpha, n)O^{16}$

An estimate using an experimental measurement of the low-energy nonresonant contribution has been made by Caughlan and Fowler (1964). They obtain a value $M = 17.1$, while the optical-model calculation yields 16.7.

A-3. $N^{14}(\alpha, \gamma)F^{18}$

The properties of the excited levels of F^{18} in the vicinity of 4.4 MeV can be inferred through an anlysis of the $Ne^{20}(d, \alpha)F^{18}$ and $O^{16}(He^3, p)F^{18}$ reactions (Enge and Wojtasek, 1959).

In particular, a level with $T = 1$ ($T =$ isotopic spin) should be identified by the fact that the $\alpha + F^{18}$ outcome would be forbidden. The relative intensities of alpha-groups from the $Ne^{20} + d$ reaction are shown in Table VII.

Table VII

Energy of the level MeV	Relative intensity
4.108	125
4.218	138
4.350	147
4.400	68
4.651	30
4.741	<7
4.844	70

Clearly, the 4.741 level cannot participate in the reaction. It is most likely at the $T = 1$ level. The 4.651 MeV level, however, is definitely active, although its contribution is about four times smaller than the contributions of the other levels. Since the outcoming α have more than 3 MeV, the Coulomb effect is not important. The reduction probably reflects an unusually small value of θ_α^2, or a high angular momentum of the excited state, or both.

The reaction rate can be written as

$$\log \left(\frac{P_{14,4}}{\rho X_4} \right) = 11.3 - \frac{12.45}{T_8} + \log \omega \Gamma_\alpha - \tfrac{3}{2} \log T_8$$

Following Brown (1962), we choose $R = 5.6 \, \mathrm{f}$ for the radius of interaction. We obtain for $\omega \Gamma_\alpha$ of this level

$$\log \omega \Gamma_\alpha = -13.4 + \log (\omega \theta_\alpha^2) \qquad (A25)$$

The average value of $(\omega \theta_\alpha^2)$ for levels of slightly higher energy is $\simeq 0.06$. Here, because of the low experimental yield, we choose $\omega \theta_\alpha^2 = 0.02$ and obtain

$$\log \left(\frac{P_{14,4}}{\rho X_4} \right) = -3.8 - \frac{12.45}{T_8} - \tfrac{3}{2} \log T_8 \qquad (A26)$$

This contribution to the total rate is dominant in the range $0.4 < T_8 < 1.7$. At $T_8 > 1.7$, the contribution from the 4.844 MeV level becomes dominant. Using the same experimental data, we obtain $\omega \theta_\alpha^2 = 0.05$ and

$$\log \left(\frac{P_{14,4}}{\rho X_4} \right) = 2 - \frac{22.18}{T_8} - \tfrac{3}{2} \log T_8 \qquad (A27)$$

At temperatures $T_8 > 3$, the rate obtained from Table V should become valid.

A-4. $O^{16}(\alpha, \gamma)Ne^{20}$

Almqvist and Kuehner and also Evans $et\ al.$ (to be published, 1964) report the following results from measurements of some properties of excited levels of Ne^{20}: For the 5.63(3−) level,

$$\frac{\Gamma \gamma}{\Gamma} = 0.077 \pm 0.008$$

$$\Gamma = 1.9 \times 10^{-3} \, \mathrm{eV}$$

$$(2l + 1) \frac{\Gamma_\alpha \Gamma_\gamma}{\Gamma} = (1.0 \pm 0.1) \times 10^{-3} \, \mathrm{eV}$$

For the 5.80 $(1-)$ level

$$\frac{\Gamma\gamma}{\Gamma} < 3 \times 10^{-4}$$

The capture rate becomes

$$\frac{P_{16,4}}{\rho X_4} = \frac{(2.1 \mp 0.2) \times 10^2}{T_8^{\frac{3}{2}}} \times 10^{-45.9/T_8} \qquad (A28)$$

in the range $2 < T_8 < 8$. The photodisintegration rate is

$$\frac{P_{16,4}^{\text{dis.}}}{\rho X_4} = 10^{12.15} \times 10^{-28.4/T_9} \qquad (A29)$$

This last rate is good to within 10% at $T_8 < 8$ and is most likely valid to $T_9 < 4$ (within 50%).

A-5. $O^{18}(\alpha, \gamma)Ne^{22}$; $O^{18}(\alpha, n)Ne^{21}$

The rate can be computed from Table V. The neutron branching ratio can be obtained by multiplying the rate of $O^{18}(\alpha, \gamma)Ne^{22}$ by the ratio of neutron to total emission rates as discussed in Appendix B.

A-6. $Ne^{20}(\alpha, \gamma)Mg^{24}$

The experimental data cover the range $E_{\text{CM}} = 2.073$ to 3.269 MeV, corresponding to $E^* = 11.387$ to 12.583 MeV in the Mg^{24} compound nucleus in which 11 resonances are reported. The value of $\xi_0^2(\gamma/D)$ was found to be $10^{9.47}$ and $S = 10^{9.56}$, as compared to $S = 10^{9.31}$ when calculated with the optical model.

A-7. $Mg^{24}(\alpha, \gamma)Si^{28}$

Here, we have information in the range $E_{\text{CM}} = 1.31$ MeV to 3.73, corresponding to $E^* = 11.296$ to 13.714 in the Si^{28} compound nucleus. We have $\xi_0^2(\gamma/D) = 10^{10.33}$ and $S = 10^{10.41}$. The optical model S is $10^{10.56}$. At temperatures above $T_8 = 16$ (where $\Gamma_\gamma < \Gamma_\alpha$), we obtain from the experimental data $(\Gamma_\gamma/D) = C = 1.4 \times 10^{-5}$. The best choice of E^*, both from the experimental data and from the calculation over the sum of the resonances, is $E^* = 1.63$ MeV. This yields

$$\log\left(\frac{P}{\rho X_4}\right) = 4.15 - \frac{82.1}{T_8} - \tfrac{1}{2}\log T_8 \qquad (A30)$$

In Figure 14, the rate computed by summing over the resonances is shown, together with the low-temperature approximation (from Table V), and the high-temperature approximation. As discussed before, the rate obtained by summing over the resonances is expected to underestimate the rate at both ends. In the graph, the energy range where we have information on individual resonances is shown with respect to the Gamow peak (calculations by Jay Hauben).

A-8. $C^{12} + C^{12}$

The Chalk River group (Vogt, 1964) has recently studied very thoroughly (experimentally and theoretically) the nuclear reactions between C^{12} and a few light nuclei (C^{12}, N^{14}, O^{16}). The elastic scattering and the capture cross-section data have been improved to about 20% accuracy. The best fit to the data is obtained with the following set of parameters: $V_0 = 50$ MeV, $W_0 = 10$ MeV, $R_0 = 5.77$ F, $a = 0.40$ F. In Figure 17, the capture cross-section thus obtained is compared with the experiments. Good agreement is obtained if the resonances are averaged over.

The lower range of the optical-model calculation has been fitted by a formula of the form (σ in barns, E in MeV)

$$\log \sigma = 17.30 - \log E - \frac{37.87}{E^{\frac{1}{2}}} - 0.35 E \qquad (A31)$$

The resulting value of σ is plotted in Figure 17. The difference between the optical-model values and equation (A31) is less than 3% in the range 3.5 to 5.0 MeV. In the same range the cross section varies by a factor of more than four hundreds. The thermonuclear reaction rate can be written as

$$\log \left(\frac{P}{\rho X_c} \right) = 26.37 - \frac{36.55 (1 + 0.070 \, T_9)^{\frac{1}{3}}}{T_9^{\frac{1}{3}}} - \frac{2}{3} \log T_9 \qquad (A32)$$

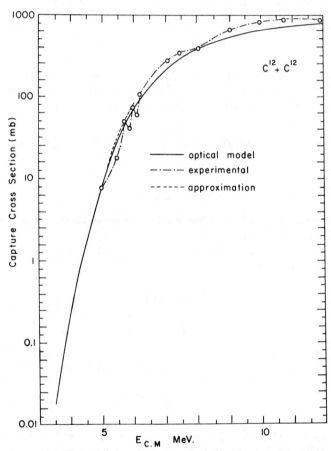

Fig. 17. The $C^{12} + C^{12}$ total reaction rate. The $(—\cdot—)$ curve represents the experimental results. The solid curve is the result of the optical model. The dashed curve (merging in the solid curve at low energy) is obtained from the fitting equation (A31).

In the range of stellar interest ($T_9 \simeq 0.7$), equation (A32) gives a result which is about a factor of 2 smaller than the result given by Reeves (1962) or by Hoyle and Fowler (1964).

The uncertainty about the result has been greatly reduced. In view of the fact that the experimental results are good to 20% accuracy, that these results are well reproduced by the optical model, and that equation (A31) matches to within a few per cent the behavior of the optical-model curve up to 0.60 of the Coulomb-barrier energy, it is difficult to see how the rate equation (A32) could be wrong by more than a factor of 3.

A-9. Na$^{23}(p, \alpha)$Ne^{20}

We have detailed information for incident proton energy (center of mass) varying from 100 keV to 778 keV. The thermonuclear reaction rate has been computed by summing over the resonances and also by using the average reduced value of $\omega\Gamma_\alpha\Gamma_p/\Gamma$ as calculated from the first fourteen resonances. Using the method described in Appendix A, we get

$$\sum_{res} \frac{\omega\Gamma_\alpha\Gamma_p}{\Gamma} e^{2\pi\xi} e^{gE} = 10^{2.45} \tag{A33}$$

The Coulomb factor used in equation (A33) is the proton factor. We neglect the alpha Coulomb factor since the Q value for Na$^{23}(p, \alpha)$Ne^{20} is already 2.3 MeV. This approximation should be reasonably good. The rate is then given by the formula

$$\log\left(\frac{P_{23.1}}{\rho X_1}\right) = 14.00 - 19.43\, T_8^{-\frac{1}{3}} - \tfrac{2}{3}\log T_8 - 0.02\, T_8^{\frac{2}{3}} \tag{A34}$$

APPENDIX B

Production of Neutrons Through Endothermic Branching of Thermonuclear Reactions Between Charged Particles

A. The fractional number of reactions F taking place with relative energies higher than a threshold energy E_t is

$$F = \frac{\int_{E_t}^{\infty} \sigma(E)n(E)v\, dE}{\int_0^{\infty} n(E)\sigma(E)v\, dE}$$

$$= \frac{\int_{E_t}^{\infty} \exp\left(-E/kT' - a/E^{\frac{1}{2}}\right) dE}{e^{-\tau}\Delta E_0\sqrt{\pi/2}} \tag{B1}$$

where $a/E^{\frac{1}{2}} = 2\pi\eta$, $\tau = 3E_0/kT'$, $E_0^{\frac{3}{2}} = akT'/2$, $\Delta E_0 = 4E_0/\sqrt{\tau}$. T' has been defined in Appendix A. Here, the integral in the denominator has been evaluated with the usual method of replacing the integrand by a properly adjusted Gaussian. The treatment of the numerator requires a little more attention. Define $E/E_0 = x$, $E_t/E_0 = x_t = 1 + u_t$ (we consider only $u_t > 0$), and $y = x_t - x$. Develop the integrand in the numerator of equation (B1) in a power series in y (including terms up to y^2)

$$\frac{E}{kT'} + \frac{a}{E^{\frac{1}{2}}} = x_t + \frac{2}{x_t^{\frac{1}{2}}} + y\left(1 - \frac{1}{x_t^{\frac{3}{2}}}\right) + y^2 \frac{3}{4} \frac{1}{x_t^{\frac{5}{2}}} + \dots \tag{B2}$$

Inserting this development in equation (B1) and integrating, we get

$$F = \frac{x_t^{\frac{3}{4}}}{2}[1 - \varphi(z_t)]\exp\left[\frac{\tau}{9}\left(x_t^{\frac{3}{2}} - 5x_t - \frac{5}{x_t^{\frac{1}{2}}} + 9\right)\right] \tag{B3}$$

where

$$z_t = \frac{\tau^{\frac{1}{2}}}{3}(x_t^{\frac{3}{4}} - x_t^{-\frac{1}{4}})$$

and

$$\varphi(z_t) = \int_0^{z_t} \frac{2}{\pi^{\frac{1}{2}}}\exp(-z^2)\,dz$$

as tabulated, e.g., in Jamke–Emde, page 24.

Some useful approximations are:

for $u_t \ll 1$

$$x_t^{\frac{3}{2}} - 5x_t - \frac{5}{x_t^{\frac{1}{2}}} + 9 \simeq \tfrac{15}{8}u_t^3 - \tfrac{45}{32}u_t^4$$

$$F \simeq 1 - \varphi(z_t)$$

for $u_t \ge 0.5$

$$F \simeq \frac{kT}{\Delta E_0}\frac{\exp\left[-\tau/3(x_t + 2/x_t^{\frac{1}{2}} - 3)\right]}{1 - x_t^{-\frac{3}{2}}} \tag{B4}$$

B. The average number of neutrons g emitted per collision is

$$g = \frac{\displaystyle\int_{E_t}^{\infty} \sigma(E)n(E)v\,dE(\Gamma_n/\Gamma)}{\displaystyle\int_0^{\infty} \sigma(E)n(E)v\,dE} \tag{B5}$$

where Γ_n/Γ is the ratio of the neutron width to the total width of the levels of compound nucleus, and E_t, here, becomes the threshold energy for neutron emission. The ratio (Γ_n/Γ) is small only for $(E - E_t)$ less than a few keV. In the rest of the energy range, its value remains roughly constant. Depending upon the reaction under consideration, it takes values ranging from $(\Gamma_n/\Gamma) = 1$ to $\simeq 0.1$. Neglecting the lowest part of the range, the average number of neutrons emitted per collision becomes

$$g \simeq (\overline{\Gamma_n/\Gamma})F$$

B-1. $O^{18}(\alpha, n)Ne^{21}$ $(Q = -700\text{ keV})$

In the range of interest $(\Gamma_n/\Gamma) \simeq 1$, since Γ_α is still fairly small because of Coulomb effects and no other channel is open $[Q(\alpha, p) \simeq -5.7\text{ MeV}]$. As a function of temperature, we obtain

T_8	x_t	z_t	F	g
2.0	1.78	3.27	$\simeq 10^{-3}$	$\simeq 10^{-3}$
3.0	1.36	1.38	0.05	0.05
3.5	1.22	0.85	0.17	0.17
4.0	1.12	0.44	0.31	0.31

Here, g is also the number of neutrons produced per each O^{18} nucleus in the gas. We notice that at higher temperature this number will be doubled by the $Ne^{21}(\alpha, n)Mg^{24}$ reaction.

B-2. $Ne^{22}(\alpha, n)Mg^{25}$ ($Q = -482$ keV)

For the same reasons as in the previous case, we expect $\overline{(\Gamma_n/\Gamma)} \simeq 1$, since we have $Q(\alpha, p) = -3.5$ MeV.

T_8	x_t	z_t	F	g
1.5	1.26	1.23	0.07	0.07
2.0	1.04	0.20	0.38	0.38

Again, g is the number of neutrons per Ne^{22} nuclei. Higher temperatures may double the crop of neutrons [by $Mg^{25}(\alpha, n)Si^{28}$] if enough He^4 atoms are still present by then.

B-3. $C^{12}(C^{12}, n)Mg^{23}$ ($Q = -2.603$ MeV)

The experimental evidence (Bromley, 1960) gives a branching ratio of 0.05 (fairly constant) from 10 MeV down to 5 MeV in the center-of-mass system. In the lower energy range, two channels are still widely open [(C^{12}, α), $Q = 4.619$ MeV; (C^{12}, p), $Q = 2.230$ MeV]. It is probably reasonable to assume that the value $(\Gamma_n/\Gamma) = 0.05$ is valid at lower energies, although this may be uncertain by a factor of 2 or 3. To evaluate F, we use $E_0 = 2.418(T'_9)^{\frac{2}{3}}$, where $T'_9 = T_9/(1 + 0.070\, T_9)$, as discussed in Appendix A (E_0 in MeV).

T_9	x_t	z_t	F_{app}	$g/2$	F_{exact}
0.6	1.56	2.82	5.9×10^{-4}	2×10^{-5}	5.5×10^{-4}
0.7	1.41	2.02	8.8×10^{-3}	1.5×10^{-4}	7.1×10^{-3}
0.8	1.30	1.41	0.045	1×10^{-3}	0.036
0.85	1.25	1.18	0.079	2×10^{-3}	0.065
0.9	1.20	0.97	0.12	3×10^{-3}	0.105
1.00	1.13	0.58	0.24	0.005	0.215
1.10	1.06	0.28	0.37	0.01	0.373
1.20				0.012	0.487

As each collision takes two C^{12}, we have listed in column $g/2$ the number of neutrons per C^{12} in the gas. In the last column, the exact value of F (computer-integrated) is given (calculations by E. Milford).

APPENDIX C
Table of Q Values for Various Reactions

Target	Q values, MeV				
	(p, γ)	(p, α)	(p, n)	(α, γ)	(α, n)
C^{12}	1.943	−7.557	−18.390	7.161	−8.507
C^{13}	7.549	−4.064	−3.004	6.357	2.214
N^{14}	7.291	−2.922	−5.931	4.404	−4.737
N^{15}	12.125	4.964	−3.543	4.012	−6.430
O^{16}	0.598	−5.218	—	4.730	−12.145
O^{17}	5.597	1.193	−3.544	7.347	0.588
O^{18}	7.992	3.980	−2.450	9.667	−0.700
F^{19}	12.844	8.114	−4.031	10.465	−1.949
Ne^{20}	2.454	−4.132	−16.100	9.314	−7.221
Ne^{21}	6.743	−1.750	−4.305	9.885	2.555
Ne^{22}	8.790	−1.675	−3.624	10.616	−0.482
Na^{23}	11.693	2.379	−4.842	10.098	−2.971
Mg^{24}	2.287	−6.859	−14.800	9.986	−7.193
Mg^{25}	6.301	−3.142	−5.044	11.133	2.655
Mg^{26}	8.272	−1.826	−4.797	10.650	0.036
Al^{26}	7.471	−1.871	—	10.417	−0.910
Al^{27}	11.581	1.594	−5.598	9.664	−2.652
Si^{28}	2.735	−7.700	−14.580	6.946	−8.135
Si^{29}	5.585	−4.832	−5.743	7.111	−1.532
Si^{30}	7.286	−2.378	−5.030	7.917	−3.504
P^{31}	8.863	1.917	−6.218	6.998	−5.572
S^{32}	2.285	−4.211	−13.780	6.641	−8.628
S^{33}	5.218	−1.526	−6.358	6.792	−2.002
S^{34}	6.367	−0.631	−6.203	7.213	−4.629
Cl^{35}	8.506	1.865	−6.763	7.213	−5.866
S^{36}	8.402	0.544	−1.920	6.809	−3.066
A^{36}	1.890	−4.355	—	7.044	−8.685

REFERENCES

E. Almqvist and J. A. Kuehner (1964) (To be published).

S. Biswas and C. E. Fichtel (1964), *Astrophys. J.* **139**: 941.

D. A. Bromley, J. A. Kuehner, and E. Almqvist (1960), *Phys. Rev. Letters* **4**: 365.

R. E. Brown (1962), *Phys. Rev.* **125**: 347.

E. M. Burbidge, G. R. Burbidge, W. A. Fowler, and F. Hoyle (1957), *Rev. Mod. Phys.* **29**: 547.

A. G. W. Cameron (1959), *Astrophys. J.* **130**: 916.

A. G. W. Cameron (1959), *Astrophys. J.* **130**: 452.

A. G. W. Cameron, *Nuclear Astrophysics* (To be published).

W. Cartledge, M. Thibaudeau, and H. Reeves (1963), Publications of the Institute for Space Studies, New York.

G. R. Caughlan and W. A. Fowler (1963), *Astrophys. J.* **136**: 453.

G. R. Caughlan and W. A. Fowler (1964), *Astrophys. J.* **139**: 1180.

H.-Y. Chiu (1961), *Phys. Rev.* **123**: 1040.

D. D. Clayton, W. A. Fowler, T. E. Hull and B. A. Zimmerman (1961), *Ann. Phys.* **12**: 331.

W. Deinzer and E. E. Salpeter (1964) (To be published).

H. Enge and J. H. Wojtasek (November 1959), *M.I.T. L.N.S. Progr. Rep.*

E. Evans *et al.* (1964) (To be published).

H. Feshbach, M. M. Shapiro, and V. F. Weisskopf (1953), *Tables of Penetrabilities of Charged Particle Reactions*, N.Y.O.-3077.

L. Goldberg, E. A. Müller, and L. H. Aller (1960), *Astrophys. J. Suppl.* **5**: 1–138.

J. L. Greenstein and G. Wallerstein (1964), *Astrophys. J.* **139**: 1163; see also this volume, p. 425.

C. Hayashi, R. Hoshi, and D. Sugimoto (1962), *Progr. Theoret. Phys. Suppl.* 22.

E. Hofmeister, R. Kippenhahn, and A. Weigert (1964), *Ztschr. Astrophys.* **59**: 215; see also this volume, p. 263.

F. Hoyle and W. A. Fowler (1964), *Astrophys. J. Suppl.* 91.

C. L. Inman and M. A. Ruderman (1964), Publications of the Institute for Space Studies.

P. D. Parker, J. N. Bahcall, and W. A. Fowler (1964), *Astrophys. J.* **139**: 602.

P. D. Parker and R. W. Kavanagh (1963), *Phys. Rev.* **131**: 2578.

H. Reeves (1964), *Stars and Stellar Systems, Vol. 8* (To be published).

H. Reeves and E. E. Salpeter (1959), *Phys. Rev.* **116**: 1505.

S. Sakashita and M. Nishida (1964), *Progr. Theoret. Phys.* **31**: 227.

E. E. Salpeter (1959), *Astrophys. J.* **129**: 608.

E. E. Salpeter (1954), *Austral. J. Phys.* **7**: 373.

M. Schmidt (1959), *Astrophys. J.* **129**: 243.

M. Schwarzschild (1958), *Structure and Evolution of Stars*, Princeton University Press, Princeton, New Jersey.

M. Schwarzschild and H. Selberg (1962), *Astrophys. J.* **136**: 150.

P. A. Seeger, W. A. Fowler, and D. D. Clayton (To be published).

H. Tsuda and H. Tsuji (1963), *Progr. Theoret. Phys.* **30**: 34.

H. Tsuda (1963), *Progr. Theoret. Phys.* **29**: 29.

E. W. Vogt, D. McPherson, J. Kuehner, and E. Almqvist (1964) (To be published).

G. Wallerstein, J. L. Greenstein, P. Parker, H. L. Helfer, and L. H. Aller (1963), *Astrophys. J.* **137**: 280.

RADIATIVE ABSORPTION AND OPACITY CALCULATIONS*

Arthur N. Cox

This report on photon absorption is presented from the viewpoint of the user who wants to apply absorption coefficients and opacities to stellar atmosphere and interior structure calculations. Even though a brief theoretical review is presented, the emphasis is on the methods of calculations and a study of the results.

The flow of radiative energy through matter is given by the equation of transfer

$$\frac{\partial I_v(s)}{\rho \partial s} = -(\varkappa_v' + \sigma_v)I_v(s) + \sigma_v \int I_v(s')p(\cos \Theta)\frac{d\omega}{4\pi} + \varkappa_v' B_v(T) \tag{1}$$

with $I_v(s)$ the specific radiation intensity in direction s at photon frequency v, ρ the matter density, $\varkappa_v'$ the absorption of the material for photons of energy hv, corrected for stimulated emission by

$$\varkappa_v' = \varkappa_v[1 - \exp(-hv/kT)] \tag{2}$$

and σ_v is the scattering coefficient. The attenuation of the radiation beam is the first term on the right-hand side, whereas the source terms consist of the scattering into the beam from all other directions s' at angles from s of Θ with the phase function $p(\cos \Theta)$ and the thermal emission given by $\varkappa_v' B_v(T)$. The scattering process may have an angular dependence, but in our equation the photon energy change by scattering is assumed to be negligible. The thermal emission used is actually only the spontaneous emission, because the stimulated emission of photons of energy hv is allowed for by reducing the true absorption coefficient $\varkappa_v$ to $\varkappa_v'$. The other factor of the spontaneous emission is the Planck function

$$B_v(T) = \frac{2hv^3}{c^2} \frac{1}{\exp(hv/kT) - 1} \tag{3}$$

The cgs units for the equation of transfer are normally used, in which case the units of $\varkappa_v'$ and σ_v are cm^2/g.

The radiative energy flux in direction x and angle θ from s is defined as

$$F_{v,x} = \int I_v(s) \cos \theta \, d\omega \tag{4}$$

For the deep layers of a star, the specific intensity I_v is almost isotropic and almost equal to its value in thermodynamic equilibrium. Therefore, the diffusion approximation that

$$I_v(s) = B_v(T) - \frac{1}{\rho(\varkappa_v' + \sigma_v)} \frac{\partial B_v(T)}{\partial s} \tag{5}$$

* Work performed under the auspices of the United States Atomic Energy Commission.

is often used. For these cases, the flux can be evaluated with the result

$$F_x = - \frac{ac}{3\varkappa\rho} \frac{\partial T^4}{\partial x} \tag{6}$$

if a suitable Rosseland mean of the absorption and scattering coefficients is made. This mean is

$$\frac{1}{\varkappa} = \frac{\int_0^\infty [1/(\varkappa_v' + \sigma_v)][\partial B_v(T)/\partial T] \, dv}{\int_0^\infty [\partial B_v(T)/\partial T] \, dv} \tag{7}$$

or, in reality, a transparency mean. Actually, the diffusion equation is an approximation to the exact equation of transfer only if the scattering is negligible or has certain forms, such as being isotropic or having the Rayleigh phase function.

There are a number of atomic processes that can either absorb or scatter photons. Absorption coefficients for these processes are correctly given only by quantum mechanical calculations. However, the highest accuracy for these coefficients is compromised for computing speed, and the resulting data are thought to be at least of useful accuracy.

Usually, the most important process of absorption is the photoelectric ejection of an orbital electron from a bound atomic level. This process requires bound electrons and photons of enough energy to move the electron to a free state. Any energy in excess of the ionization energy is given as kinetic energy to the free electron. The semiclassical Kramers formula for the absorption and the quantum mechanical correction, the Gaunt factor, are given in the opacity chapter in volume eight of the compendium *Stars and Stellar Systems*. Bound–free absorption is caused simultaneously by electrons in many bound levels of different ions or elements.

The second most important process in stars is free–free absorption, which is important at higher temperatures after most atoms are ionized. A free electron absorbs a photon and changes the electron orbit to one more energetic in the field of a nuclear charge. For this process, a semiclassical formula and a Gaunt factor are used. No sudden absorption jumps or edges occur with the free–free process, because all orbit changes are allowed.

The most important photon scattering process is free-electron scattering. At high temperatures, this Thomson or Compton scattering dominates radiation transfer because all electrons are ionized and free–free processes are important only at the higher densities.

A process which has only recently been considered in astrophysics is the bound–bound absorption. An incoming photon is absorbed, and a bound electron is moved to a more energetic bound state. This quite discontinuous process requires bound electrons and depends on the spectral line position, shape, and width. At high temperatures, many ionic configurations can exist simultaneously, and each configuration has its own spectrum. This configuration-splitting can take a simple line or its normal fine-structure splitting and produce a broad band of lines which appreciably affects the Rosseland (transparency) mean opacity.

Together with the line absorption, it is possible to scatter coherent photons with the bound electrons. In this process, the bound electron moves, before or after the photon arrives, to a higher bound level and then returns to its original level.

All resonance lines are considered to scatter photons with the Rayleigh phase function and are therefore treated differently from absorption lines in the equation of transfer. Note that the high photon energy scattering goes to Thomson scattering and the low photon energy scattering is Rayleigh scattering.

In astrophysics, the negative ion of hydrogen has a large bound–free and free–free absorption at stellar surface temperatures. The extra electron, bound by 0.75 eV energy, absorbs energy according to rather complicated quantum mechanical rules. Therefore, unlike the previously discussed processes, tables of the monochromatic absorption by the bound–free and free–free processes are used for opacity calculations.

Three molecular absorption calculations can be considered. These are

$$H + H^+ + h\nu \rightarrow H_2^+$$
$$H + H + h\nu \rightarrow H_2$$
$$H_2 + e + h\nu \rightarrow H_2 + e^*$$

The last process is really free–free absorption by H_2^-, even though a bound state apparently does not exist. Results to date do not consider the absorption due to the last process nor do they even consider that H_2 can form, depleting the abundance of H and H^- and causing H_2 Rayleigh scattering.

Electron conduction is an efficient means of energy transport in degenerate matter. The conduction effect is added to the radiation transport effects (i.e., the opacity) to give a "total" opacity in these calculations. The Rosseland flux formula can then be used to include both radiation diffusion and electron conduction. Line-opacity data only are presented in this paper.

METHOD OF CALCULATION

In order to calculate the absorption and emission of photons by matter, it is necessary to know the material properties, or the equation of state, because, for example, bound–free and bound–bound absorptions and bound–bound scattering require a knowledge of the number of electrons in each bound electron level (occupation numbers).

The Saha ionization formula is used to determine the relative number of atoms in adjacent ionization stages ($x_{i,j+1}/x_{i,j}$). The form used is

$$\frac{x_{i,j+1}}{x_{i,j}} = \frac{B_{i,j+1}}{B_{i,j}} \exp\left[-\frac{\chi_{ij} - E_0}{kT} - \eta + \frac{(j+1)e^2}{R_D kT} \right] \qquad (8)$$

with $B_{i,j}$ the partition function for element i with j electrons missing, χ_{ij} the ionization energy for the conversion from j to $j + 1$ electrons ionized,

$$E_0 = \frac{3Rhc}{5} \sum_i \frac{x_i y_i^2}{y r_i} \qquad (9)$$

kT the temperature in energy units with k being Boltzmann's constant, η the electron degeneracy parameter, e the electronic charge, and R_D the Debye radius. The E_0 is a correction to η, which here is calculated using Fermi–Dirac integrals, $F_{1/2}(\eta)$, involving only the electron kinetic energy.

The partition functions, or number of ways the atoms in an ionization stage can exist, are given by the general formula

$$B_{i,j} = \sum_k g_{ijk} \exp\left[-\frac{\varepsilon_{ijk}}{kT} - \frac{y_i}{r_i^3 kT}(\langle r_0^2 \rangle - \langle r_k^2 \rangle)Rhc \right] \qquad (10)$$

with g_{ijk} the degeneracy of state k at energy ε_{ijk} above the ground state, corrected by free-charge perturbations computed from an ion sphere model with uniform charge. The energy correction involves the average number of free electrons from element i, y_i; the radius of its ion sphere containing y_i electrons, r_i; the average of the square of the orbital radius $\langle r_k^2 \rangle$; the average of the square of the ground-state electron orbital radius $\langle r_0^2 \rangle$; and the Rydberg energy $Rhc = 13.60$ eV. The energy-level depression due to the free charges, where the partition function sum is terminated, is given by the term $(j + 1)e^2/R_D$ in the Saha equation, and the relative shifting of the level positions appears in the partition-function formula. E_0 is the negative of the average potential energy of the free-electron cloud with N_e electrons of mass m per unit volume, and

$$N_e = \frac{4\pi}{h^3}(2mkT)^{3/2}F_{1/2}(\eta) \qquad (11)$$

Solution of these ionization equations involves formation of two polynomials

$$P_i = \frac{\text{no. nuclei of element } i}{x_{i0}}$$

$$= 1 + \frac{x_{i1}}{x_{i0}} + \frac{x_{i2}}{x_{i1}}\frac{x_{i1}}{x_{i0}} + \cdots + \frac{x_{iZ}}{x_{iZ-1}}\cdots\frac{x_{i1}}{x_{i0}} \qquad (12)$$

$$Q_i = \frac{\text{no. electrons from element } i}{x_{i0}}$$

$$= \frac{x_{i1}}{x_{i0}} + 2\frac{x_{i2}}{x_{i1}}\frac{x_{i1}}{x_{i0}} + \cdots + Z\frac{x_{iZ}}{x_{iZ-1}}\cdots\frac{x_{i1}}{x_{i0}} \qquad (13)$$

One notes that

$$y_i = \frac{Q^i}{P_i} \qquad (14)$$

and with

$$\sum_i x_i y_i = y \qquad (15)$$

one can get

$$N_e = y\frac{\rho N_0}{\mu_c} \qquad (16)$$

with N_0 Avogadro's number and

$$\mu_c = \sum_i x_i \mu_{ci} \qquad (17)$$

the cold (neutral) atomic weight of the average mixture atom. The individual abundances in each ionization stage, including the H^- stage, can be found from the known ratios between stages and

$$x_{i0} = \frac{1}{P_i} \tag{18}$$

Occupation numbers are given by

$$x_{ijk} = \frac{v_k g_{ijk}}{B_{ij}} \exp(\varepsilon'_{ijk}/kT) \tag{19}$$

with v_k the number of electrons in the level denoted by state k, and ε'_{ijk} the energy of that state.

The normal equation of state calculation starts with a given temperature and density. A guess is made for the number of free electrons, and then η, E_0, R_D, and energy level corrections are computed. Computing the partition functions and their solution gives y_i and y. The new N_e is then used to make a second solution for y_i and y. Convergence occurs unless the density corrections are excessively large, in which case the theory is not valid anyway. After convergence, solution for occupation numbers and other equation of state data needed for absorption calculations can be made.

For higher temperatures ($T > 10^6$ °K), it is possible to use simpler equations where only the atomic levels of an average atom, not the levels for each ionization stage, are considered. From statistical mechanics, the number-fraction of element i in state k, averaged over all ionization stages, is

$$x_{ik} = \frac{g_{ik}}{\exp[(E_{ik}/kT) - \eta] + 1} \tag{20}$$

with g_{ik} the statistical weight taken as $2n^2$, with n the principal quantum number. The energy is

$$\frac{E_{ik}}{Rhc} = -\frac{Z^{*2}}{n_{ik}^2} + \frac{y_i}{r_i}\left[3 - \frac{\langle r_k^2 \rangle}{r_i^2}\right] + E_0 \tag{21}$$

as given by the Bohr hydrogen-like atom with the ion-sphere energy corrections. The E_0 correction is included here to correct η. The effective charge is

$$Z_{ik}^* = Z_i - \sum_{k' \neq k} x_{ik}\sigma_{kk'} - x_{ik}\left(1 - \frac{1}{g_{ik}}\right)\sigma_{kk} \tag{22}$$

using the Slater-type screening constants. Average screening due to electrons in the same shell is given here by the second term.

The equation of state solution in this case is made by assuming that at first all electrons are free, i.e., $y_i = Z_i$. The energies are put in the occupation-number formula with an η computed assuming complete ionization, and x_{ik} values result. The number of bound electrons is

$$n_b = \sum_i x_i \sum_k x_{ik} \tag{23}$$

and

$$y = \sum_i x_i(Z_i - \sum_k x_{ik}) \tag{24}$$

Better energy levels can then be computed with the occupation numbers known, and another solution made. Iterative solution will then give this simpler equation of state for absorption calculations.

For the bound–bound (line) absorption, some special procedures are followed to obtain occupation numbers for special atomic configurations in which lines occur. The strength of each component of the greatly split line depends on the configuration j' abundance, $x_{ij'k}$.

A partition function for the entire element i is

$$B_i = \sum_{j'} g_{ij'} \exp\left[-E_{ij'}/kT + \eta \sum_k (v_k - x_{ik})\right] \tag{25}$$

with $g_{ij'}$ the number of ways one can get v_k electrons into the number of spaces in level k, and v_k corrected, for convenience only, by the average occupation in the level. Now

$$\frac{E_{ij'}}{Rhc} = -\frac{9}{5}\frac{y_i^2}{r_i} + \sum_k v_k\left(-\frac{Z_{ik}^{**2}}{n_k^2} - y_i\frac{\langle r_k^2 \rangle}{r_i^3} + E_0\right) \tag{26}$$

which differs from the electron level energy by including the first term for the free-electron-cloud potential energy, and by using

$$Z_{ik}^{**} = Z_i - \sum_{k' < k} \sigma_{kk'}v_{k'} - \frac{1}{2}(v_k - 1)\sigma_{kk} \tag{27}$$

The effective charge for the entire atom is used to obtain this energy of the entire atom. The σ_{kk} are twice as large as the Slater screening constants, and the $\frac{1}{2}$ compensates for them.

Occupation numbers are now computed according to the formula

$$x_{ij'k} = \frac{v_k g_{ij'}}{B_i} \exp\left[-E_{ij'}/kT + \eta \sum_k (v_k - x_{ik})\right] \tag{28}$$

These configuration occupation numbers are a factor in the line strength for all the lines that arise from configuration j'.

In opacity calculations, it is convenient to use a dimensionless absorption coefficient

$$D(u) = \frac{\varkappa \rho u^3}{A} \tag{29}$$

with

$$u = \frac{h\nu}{kT} \tag{30}$$

$$A = \frac{2^4}{3\sqrt{3}}\frac{he^2}{mc}\frac{N}{kT} \tag{31}$$

where there are N average atoms in the mixture per unit volume. The opacity integral is then

$$I = \int_0^\infty \frac{W(u)\,du}{D(u)} \tag{32}$$

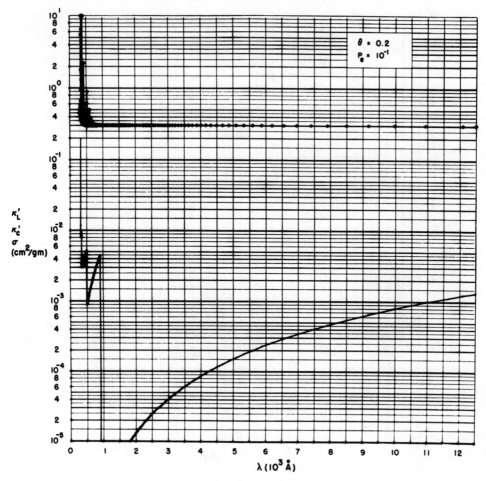

Figure 1. Monochromatic absorption and scattering coefficients as a function of wavelength for $\theta = 5040/T(°K)$ and electron pressure P_e.

with

$$W(u) = \frac{15}{4\pi^4} \frac{u^7 e^{2u}}{(e^u - 1)^3} \tag{33}$$

and

$$D(u) = \sum D_{bf} + D_{ff} + D_s + D_l + D_{cs} + D_{H^-} + D_m \tag{34}$$

is the sum of all D from the various atomic and molecular processes described before. The line D is

$$D_l(u) = \sum_{i,j,k} x_i x_{ij} x_{ijk} f_{kk'} P(k') \pi b(u) \frac{3\sqrt{3}}{2^4} u^3 \tag{35}$$

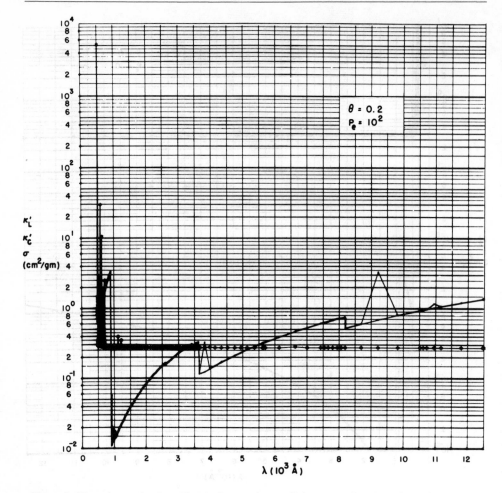

Figure 2. Monochromatic absorption and scattering coefficients as a function of wavelength for $\theta = 5040/T(°K)$ and electron pressure P_e.

where $f_{kk'}$ is the f-value or quantum mechanical correction for the absorption, $P(k')$ is the probability that there is a space in level k' for the electron to go into, and $b(u)$ is the line shape expressing how the absorption varies with u or v.

The availability of a space in level k' is given by

$$P(k') = \frac{2(2l_{k'} + 1) - x_{ijk'}}{2(2l_{k'} + 1)} \tag{36}$$

The line shape is taken as either a Lorentz profile

$$\pi b(u) = \frac{\delta_u/4\pi}{(u - u_c)^2 + (\delta/4\pi)^2} \tag{37}$$

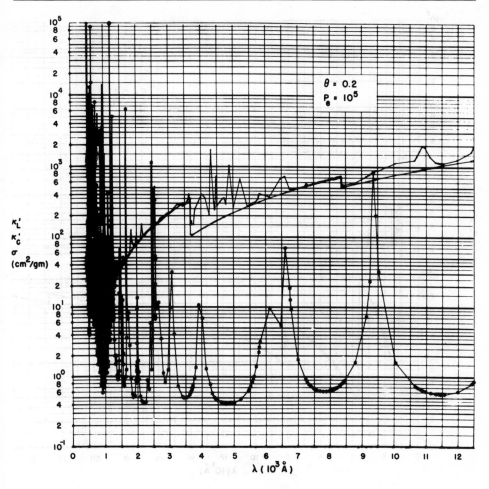

Figure 3. Monochromatic absorption and scattering coefficients as a function of wavelength for $\theta = 5040/T(°K)$ and electron pressure P_e.

or the resonance shape

$$\pi b(u) = \frac{8\pi e^2 kT}{3mc^3 h} \frac{u^4[1 + (x_{ijk}/x_{ijk})\{P(k)/P(k')\}]}{[(u^2 - u_c^2)^2 + u^2\{(4\pi e^2 kT/3mc^2 h)u_c^2\}^2](1 - e^{-u})} \tag{38}$$

For resonance lines where an excited electron has no other lower level to go to except the ground state, the resonance shape is used, except in the core where the Lorentz shape gives a larger D_l. Note the scattering does not have any correction for stimulated emission as for the absorption, and the correction factor in $W(u)$ is cancelled by a factor in D_l. The Lorentz line width is

$$\frac{\delta_u}{4\pi} = 0.0134\rho \frac{y}{\sum_i x_i \mu_{ci}} \frac{n_{ik}^4 + n_{ik'}^4}{(kT)^{\frac{3}{2}} Z_{ik}^{*2}} \tag{39}$$

with kT in kilovolts.

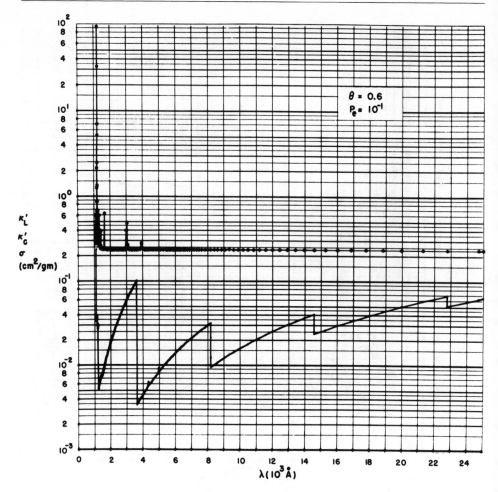

Figure 4. Monochromatic absorption and scattering coefficients as a function of wavelength for $\theta = 5040/T(°K)$ and electron pressure P_e.

For hydrogen and once-ionized helium, the approximate Stark line-shape theory of Griem is used. Tests show that even though the Griem theory gives about the same shape that the Lorentz formula gives in the wings, accurate opacity calculations require these special shapes for such strong lines.

ABSORPTION COEFFICIENTS AND OPACITIES

Typical absorption coefficients for the Aller mixture ($X = 0.596$, $Y = 0.384$, $Z = 0.020$) are given in Figures 1 to 8 as a function of wavelength. Each figure, for different $\theta = 5040/T(1/\deg K)$ and $\log P_e$ (dyn/cm^2) values, plots the line absorption $\varkappa_L'$ and the continuous absorption $\varkappa_C'$, both corrected for stimulated emission, and the scattering coefficient σ. Prominent features are the constant Thomson limit at high temperature, absorption edges such as the Lyman jump, and the numerous

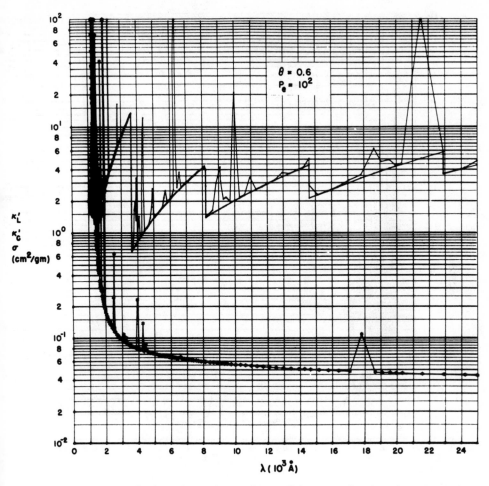

Figure 5. Monochromatic absorption and scattering coefficients as a function of wavelength for $\theta = 5040/T(^\circ K)$ and electron pressure P_e.

absorption and resonance scattering lines. As the electron pressure P_e increases, more lines and edges appear. At the lowest temperature, the H⁻ bound–free and free–free absorption appears. While some lines are missing and others have fairly crude shapes, these plots should be useful for construction of stellar atmospheres.

Opacities of the Aller mixture are plotted as a function of temperature for various densities in Figures 9 and 10. A characteristic peak appears in the opacity near the hydrogen ionization temperature. A rapid decrease occurs at lower temperatures, and a more gradual decrease to the Thomson limit appears at higher temperatures. The irregularity in the opacity is due to the appearance and disappearance of some effects such as depletion of electrons in shells at certain temperatures. The opacity increases with density almost everywhere. Since simple opacity laws do not fit the data, it is recommended that an entire opacity table be put in the machine for stellar structure calculations and linear interpolation for $\log \varkappa_L(\log \rho, \log T)$ used.

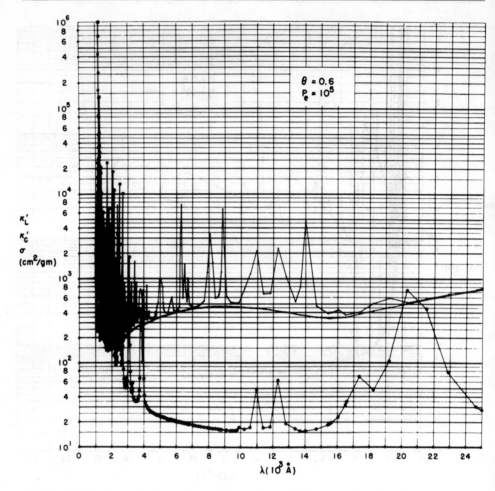

Figure 6. Monochromatic absorption and scattering coefficients as a function of wavelength for $\theta = 5040/T(°K)$ and electron pressure P_e.

Because the lines increase the opacity so much at temperatures below 10^6 °K, it is important to know if the opacities depend critically on the uncertain line shapes. Figure 11 is a plot of opacities versus density for various temperatures where the lines are most important. The opacities have been computed assuming that the lines have the widths described earlier, widths three times larger for all lines except those of hydrogen and helium, and widths given by the Griem–Stark theory for all lines. Continuous opacities have also been plotted to show how lines do greatly increase opacities. The conclusions that can be drawn are that these radiative opacities are probably accurate to 30%, and that line width uncertainties are not overwhelming.

   A graphic demonstration of the line effects is given in Figure 12, where contours of the ratio of the radiative opacity to the continuous opacity are plotted on the temperature–density plane. Typical solar interior conditions are also given. The maximum line effect occurs at 10^5 °K and 10^{-4} g/cm^3, in the middle of the hydrogen

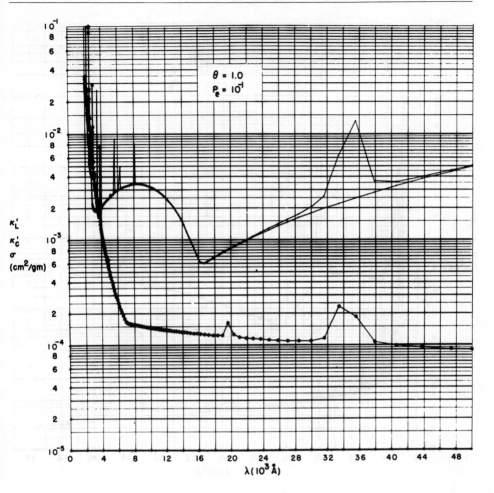

Figure 7. Monochromatic absorption and scattering coefficients as a function of wavelength for $\theta = 5040/T(°K)$ and electron pressure P_e.

convection region in a solar-type star. However, the effect of lines at higher temperatures, where most of the mass resides and where thermonuclear reactions occur, is not severe.

Another interesting analysis of opacity results is the effect on opacities of conversion of hydrogen to helium. Figure 13 shows that linear interpolation in log $\varkappa_L(X)$ gives very accurate opacities over wide ranges of X, at least in regions where transmutation of hydrogen to helium can occur. Use of opacities in stellar-evolution calculations requires only two tables—one for the primordial hydrogen content and one where the hydrogen has all been depleted.

Frequently, it is interesting to know which absorption or scattering process is dominant at a given temperature and density. Four processes are investigated in Tables I to IV. The fraction of the opacity due to one process is difficult to define because of the nature of the Rosseland mean. The absorptions and scatterings are

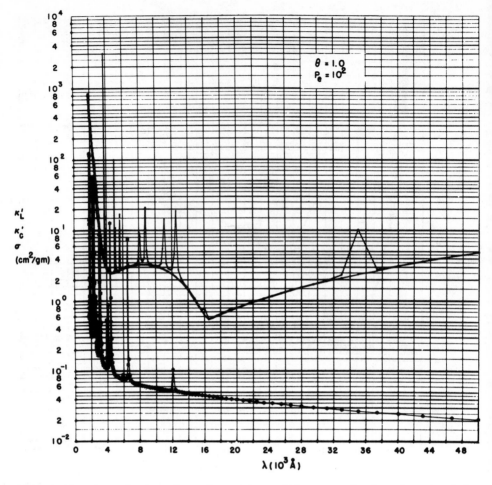

Figure 8. Monochromatic absorption and scattering coefficients as a function of wavelength for $\theta = 5040/T(°K)$ and electron pressure P_e.

additive only monochromatically. Two expressions are given for the fraction of the line opacity due to a given process or group of processes:

$$1 - \frac{\varkappa_{L \text{ all but the processes}}}{\varkappa_L}$$

and

$$\frac{\varkappa_{L \text{ processes}}}{\varkappa_L}$$

Only if the line opacities were additive would these expressions be equal. Note that the line opacity with all effects is greater than the sum of its parts. The tables give only indications of which process are important; quantitative values of the fractions are impossible to obtain.

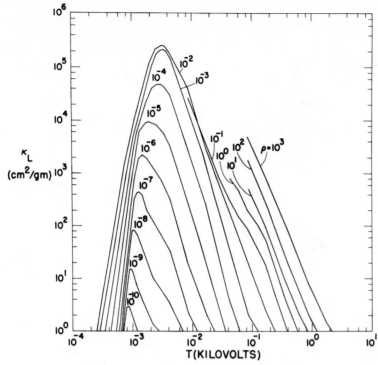

Figure 9. Line opacities versus temperature for various densities (higher opacities).

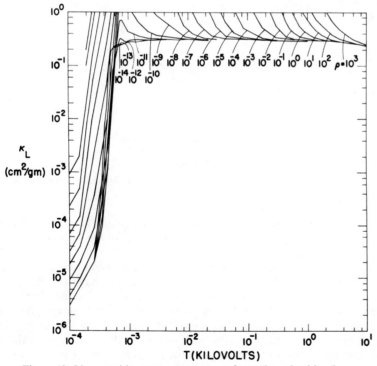

Figure 10. Line opacities versus temperature for various densities (lower opacities).

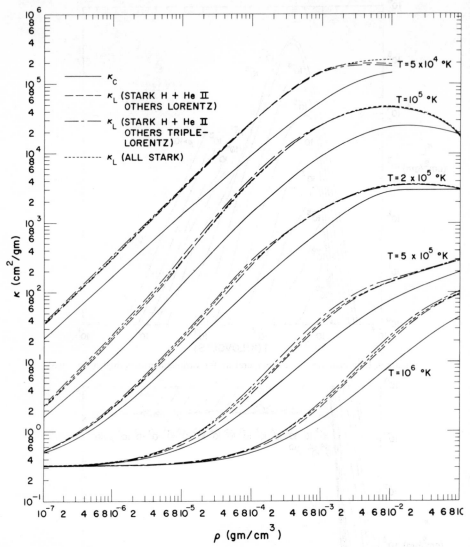

Figure 11. Continuous and line radiative opacities versus density for various temperatures.

The effects of free-electron scattering on the opacity are given as a function of temperature and density in Table I. Scattering dominates the opacity at the low densities above 5×10^6 °K.

Table II shows that where the free-electron scattering is not important, hydrogen and helium free–free absorption dominates. The density dependence of this nucleus–electron–photon interaction makes the effect small at low density. The bound–free absorption of the iron group at 10^6 to 10^7 °K is not as important as one might expect.

The more abundant heavier elements, carbon, nitrogen, oxygen, and neon, give bound–free and free–free absorptions as shown in Table III. Except at the few-million-degree and low-density region, these elements do not appear to be important

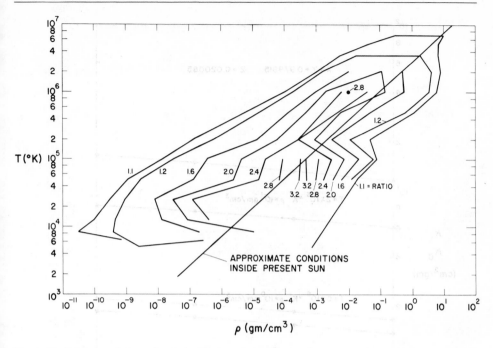

Figure 12. Contours of the ratio of the line opacity to the continuous opacity on the temperature–density plane.

Table I. Fraction of the Line Opacity due to Electron Scattering

$$\left(1 - \frac{\varkappa_{\text{all but scattering}}}{\varkappa}\right)\Big/\left(\frac{\varkappa_{\text{scattering only}}}{\varkappa}\right)$$

T(deg K)	ρ(g/cm³)					
	10^{-1}	10^0	10^1	10^2	10^3	10^4
5×10^7	1.0000	0.9998	0.999	0.99	0.92	0.67
	0.9997	0.9977	0.984	0.91	0.70	0.36
2×10^7	0.998	0.98	0.93	0.80	0.48	0.160
	0.993	0.94	0.79	0.50	0.20	0.065
1×10^7	0.96	0.77	0.55	0.36	0.118	0.027
	0.91	0.64	0.32	0.14	0.036	0.011
5×10^6	0.53	0.23	0.144	0.072	0.0155	0.0040
	0.44	0.14	0.059	0.020	0.0054	0.0025
2×10^6	0.028	0.0123	0.0109	0.0043	0.00124	0.00084
	0.019	0.0063	0.0037	0.0013	0.00070	0.00077

Note: $(\varkappa_{\text{all}} - \varkappa_{\text{all but scattering}}) > \varkappa_{\text{scattering only}}$, or $\varkappa_{\text{all}} > \varkappa_{\text{all but scattering}} + \varkappa_{\text{scattering only}}$.

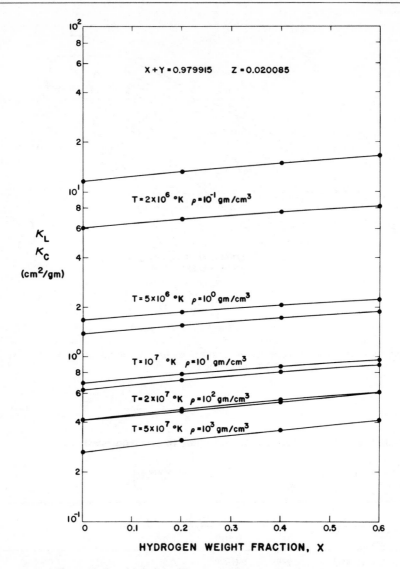

Figure 13. Opacities of the Aller metal mixture versus hydrogen weight fraction. Upper line is the line opacity and lower line is the continuous opacity for each temperature–density point.

contributors to the opacity. These tables show that at the center of the sun, half of the opacity is due to electron scattering and about half is due to free–free absorption in the field of hydrogen and helium atoms.

Between 10^4 °K and 10^5 °K, Table IV shows that bound–free and bound–bound absorption by hydrogen atoms is quite important. This is a surprising result, because one would expect complete ionization of hydrogen. Apparently, the great abundance of hydrogen allows its lines and absorption edges to influence radiation transport even when it is mostly stripped of electrons.

Table II. Fraction of the Line Opacity due to Hydrogen and Helium Free–Free Absorption

$$\left(1 - \frac{\varkappa_{\text{all but H, He free–free}}}{\varkappa}\right)\Bigg/\left(\frac{\varkappa_{\text{H, He free–free only}}}{\varkappa}\right)$$

T(deg K)	$\rho(\text{g/cm}^3)$					
	10^{-1}	10^0	10^1	10^2	10^3	10^4
5×10^7	0.000210	0.001747	0.01218	0.0623	0.227	0.49
	0.000010	0.000096	0.00094	0.0087	0.065	0.28
2×10^7	0.00417	0.0250	0.101	0.29	0.60	0.75
	0.00024	0.0023	0.019	0.12	0.43	0.71
1×10^7	0.0283	0.095	0.208	0.51	0.78	0.87
	0.0027	0.019	0.097	0.39	0.74	0.86
5×10^6	0.091	0.119	0.28	0.70	0.85	0.91
	0.017	0.053	0.22	0.65	0.85	0.90
2×10^6	0.249	0.221	0.50	0.85	0.90	0.92
	0.022	0.067	0.38	0.84	0.89	0.92

Note: $(\varkappa_{\text{all}} - \varkappa_{\text{all but H, He free–free}}) > \varkappa_{\text{H, He free–free only}}$, or $\varkappa_{\text{all}} > \varkappa_{\text{all but H, He free–free}} + \varkappa_{\text{H, He free–free only}}$.

Table III. Fraction of the Line Opacity due to Carbon, Nitrogen, Oxygen, and Neon

$$\left(1 - \frac{\varkappa_{\text{all but C, N, O, Ne}}}{\varkappa}\right)\Bigg/\left(\frac{\varkappa_{\text{C, N, O, Ne only}}}{\varkappa}\right)$$

T(deg K)	$\rho(\text{g/cm}^3)$					
	10^{-1}	10^0	10^1	10^2	10^3	10^4
5×10^7	0.0000320	0.000241	0.00153	0.0067	0.0214	0.051
	0.0000021	0.000020	0.00018	0.0014	0.0087	0.037
2×10^7	0.00099	0.00608	0.0210	0.042	0.069	0.11
	0.00011	0.00095	0.0061	0.024	0.059	0.11
1×10^7	0.0174	0.064	0.100	0.105	0.10	0.097
	0.0036	0.020	0.062	0.093	0.10	0.094
5×10^6	0.28	0.40	0.36	0.16	0.11	0.077
	0.12	0.25	0.32	0.16	0.11	0.072
2×10^6	0.81	0.84	0.54	0.14	0.090	0.067
	0.46	0.65	0.45	0.13	0.088	0.066

Note: $(\varkappa_{\text{all}} - \varkappa_{\text{all but C, N, O, Ne}}) > \varkappa_{\text{C, N, O, Ne only}}$, or $\varkappa_{\text{all}} > \varkappa_{\text{all but C, N, O, Ne}} + \varkappa_{\text{C, N, O, Ne only}}$.

Table IV. Fraction of the Line Opacity due to Hydrogen Bound–Free and Bound–Bound Absorption

$$\left(1 - \frac{\varkappa_{\text{all but }H_{bb,bf}}}{\varkappa}\right)\Big/\left(\frac{\varkappa_{H_{bb,bf}\text{ only}}}{\varkappa}\right)$$

T(deg K)	$\rho(\text{g/cm}^3)$			
	10^{-6}	10^{-5}	10^{-4}	10^{-3}
1×10^5	0.514	0.490	0.419	0.305
	0.090	0.052	0.035	0.034
5×10^4	0.74	0.80	0.79	0.59
	0.28	0.25	0.20	0.15
2×10^4	0.78	0.80	0.81	0.80
	0.62	0.72	0.76	0.70
1×10^4	0.63	0.42	0.27	0.20
	0.45	0.26	0.15	0.11

Note: $(\varkappa_{\text{all}} - \varkappa_{\text{all but }H_{bb,bf}}) > \varkappa_{H_{bb,bf}\text{ only}}$, or $\varkappa_{\text{all}} > \varkappa_{\text{all but }H_{bb,bf}} + \varkappa_{H_{bb,bf}\text{ only}}$.

For accurate calculations of stellar atmosphere and interior structure, modern absorption data are necessary. The data presented here show the current state of knowledge in this field. More absorption, scattering, and opacity calculations are needed for other mixtures of astrophysical interest, and such calculations are in progress at Los Alamos and elsewhere.

ENERGY TRANSPORT BY
TURBULENT CONVECTION*

Edward A. Spiegel

INTRODUCTION

In studies of the physics of stellar structure, one of the basic quantities is the heat flux through each layer of the star. Generally, this flux is predominantly radiative, although there are some rare exceptions, such as neutrino losses directly from extremely hot, dense stellar interiors and possibly mechanical flux in the outer layers of some stars. Under certain conditions, however, energy transfer by convection dominates radiative transfer, and some understanding of this phenomenon is therefore necessary for the construction of stellar models of many spectral classes.

To discuss convective heat flux we must naturally begin by asking, under what conditions can we expect convection to occur at all in a star? This question was already answered to sufficient accuracy for astrophysical purposes in the last century, and it was first taken up in the literature of astronomy by K. Schwarzschild. The approach was to consider a static stellar model and to subject a small, arbitrary region in it to a vertical displacement. This hypothetical displacement was assumed to maintain pressure equilibrium and to occur instantaneously, hence adiabatically. The test for instability then consists of comparing the density of the displaced gas with that of its new surroundings. If, after an upward displacement, the displaced gas is lighter than its surroundings, it will rise even further, while if it is denser it will sink back toward its original position. The former case is clearly unstable, and it occurs whenever

$$\left| \frac{d\bar{T}}{dr} \right| > \frac{g}{c_p} \tag{1}$$

where $d\bar{T}/dr$ is the mean vertical temperature gradient, g is the acceleration of gravity (assumed locally constant), and c_p is the specific heat at constant pressure (e.g., see Landau and Lifshitz, 1959). The quantity $-g/c_p$ defines the critical temperature gradient for which the displaced gas is in neutral equilibrium and is called the adiabatic gradient

$$\left(\frac{dT}{dr} \right)_{AD} = -\frac{g}{c_p} \tag{2}$$

Now, condition (1) is fulfilled in some layers of almost every kind of star. One important reason for this was pointed out by Unsöld (1930), who was the first in modern times to recognize the possible importance of convection for the outer layers of stars. The effect discussed by Unsöld is the decrease in the value of

* Prepared with the financial support of the Air Force Office of Scientific Research under Grant No. AF-AFOSR 62-386.

$(dT/dr)_{AD}$ which occurs in regions where an abundant element is partially ionized. In such regions, the ionization energy can be quite large, and so, therefore, is the heat capacity.

Condition (1) can also be fulfilled when the gradient, in the absence of convection, becomes large. In general, this can occur in two situations. First, there is the possibility that in the outer layers of cool stars the opacity can become very high (generally because of absorption by H^-), so that a large temperature gradient is required to carry the stellar luminosity by radiative transfer. Then, in stellar interiors, when the nuclear processes are quite sensitive to temperature (as in the carbon cycle), very large temperature gradients arise. Hence, there is a variety of circumstances for which we must consider the possibility that heat transfer by convection may be important.

This then leads us to a second question, namely, given that convection can occur in a region of a star, what is the measure of its importance relative to radiative heat transfer? To derive a usable criterion we must make estimates of both fluxes. The radiative flux, of course, is the better understood of the two. For, except in the outermost layers of a stellar atmosphere, the radiative flux is given quite adequately by the Eddington approximation,

$$\mathcal{F} = -K\frac{d\overline{T}}{dr} \tag{3}$$

where K is a radiative conductivity based on an appropriate mean opacity. The convective flux, on the other hand, is more complicated and can be shown to be

$$\Phi = \overline{\rho c_p w \theta} \tag{4}$$

where ρ is the local density, θ is a local temperature fluctuation, w is the vertical velocity, and the bar signifies an average over a sphere of radius r. (Since $\rho c_p \theta$ is an energy density and w a velocity, it is clear that Φ is a flux.) In calculating stellar models, one crudely simplifies equation (4) to

$$\Phi \cong \bar{\rho}\bar{c}_p \theta_{rms} w_{rms} \tag{4a}$$

Next, the average temperature fluctuation is estimated by

$$\theta_{rms} \cong -A\left(\frac{d\overline{T}}{dr} + \frac{g}{c_p}\right)l \tag{5}$$

where l is some unknown length scale and A is a number of order unity.

The convective heat transport now takes the form

$$\Phi = -K_{eddy}\left(\frac{d\overline{T}}{dr} + \frac{g}{c_p}\right) \tag{6}$$

where the eddy conductivity is

$$K_{eddy} = A\bar{\rho}\bar{c}_p w_{rms} l \tag{7}$$

(It should be stressed that Φ is not proportional to the temperature gradient, but rather to its superadiabatic excess, since convective instability depends on this excess.)

The model of convection underlying equations (6) and (7) is that there are moving blobs or eddies, analogous to the particles of kinetic theory. In this picture, originated by Taylor and Prandtl, l is analogous to a mean free path and is interpreted as the distance a blob of fluid travels before it breaks up and loses whatever identity it may have. It is usual to equate this length to the size of the blob, because of the qualitative observation in laboratory turbulence that such blobs, or eddies, travel about one diameter before breaking up. The length l is called the mixing length. [This concept was first put to astrophysical use by Biermann (1945).]

We can now write a criterion for the convective heat transfer to be important. It is that

$$\frac{\Phi}{\mathscr{F}} = \frac{wl}{\kappa}\left[\frac{(d\bar{T}/dr) + (g/c_p)}{d\bar{T}/dr}\right] \gtrsim I \tag{8}$$

where κ, defined by

$$\kappa = \frac{K}{\rho c_p} \tag{8a}$$

is a radiative diffusivity. The coefficient in this criterion

$$P_e = \frac{wl}{\kappa} \tag{9}$$

is known as the Péclet number in fluid-dynamics literature and as Unsöld's Γ in the astrophysics literature. It is the thermal analogue of the more familiar Reynold's number

$$R_e = \frac{wl}{v} \tag{10}$$

where v is the kinematic viscosity.

Now, since $(d\bar{T}/dr)$ is negative,

$$\frac{(d\bar{T}/dr) + (g/c_p)}{d\bar{T}/dr} \leqslant 1 \tag{11}$$

hence, a necessary condition for equation (8) to be true is

$$P_e = \frac{wl}{\kappa} > 1 \tag{12}$$

Although fulfillment of requirement (12) is not enough to ensure that convective transfer is important, we can certainly say that when requirement (12) is not satisfied, i.e., when

$$P_e \ll 1 \tag{12a}$$

we can neglect convective transfer. Of course, to use even this condition we must know something about w and l, and this requires a knowledge of the dynamics of convective turbulence. However, a variety of crude estimates—mainly based on mixing-length arguments—exists, and these show that convective transfer can be important in the outer layers of main sequence stars of types later than F in the cores of early-main-sequence stars, in the envelopes of red giants, and throughout stars in pre-main-sequence contraction. Thus, as anticipated above, the problem of computing convective heat transfer must be confronted in many phases of the theory of stellar evolution, and this explains the reason for the present discussion in this volume.

At present, calculations of stellar models employ equations (6) and (7) for the evaluation of convective heat flux. To these equations must be added an expression for w, which may be derived by estimating the work done by the buoyancy forces over the distance l. This kind of estimate is quite rough, and the uncertainty is supposed to be taken up in the value of l, which also appears in the expression for w. Moreover, nonadiabatic effects may also be included (Öpik, 1950, and Vitense, 1953) and an expression for Φ in terms of l be derived. Of course, this procedure is uncertain, and attempts to improve it in details have been made (e.g., Spiegel, 1963), but these do not help to decide on an optimum choice of the function l. A variety of estimates exist, but none of these appear to have a firm foundation. Moreover, experimental determination is impossible since the conditions of laboratory convection, while useful for testing many theoretical ideas, cannot provide an appropriate value of l. This is true because, in the laboratory, convection is limited by the apparatus size, while in stellar convection other factors seem to be important. The upshot is that, for want of a suitable alternative, stellar models are now being calculated with the mixing-length theory, and these calculations would be much more trustworthy if some reliable way of prescribing l were found. But we must stress that this is just a stopgap procedure, and what is ultimately required is a sounder theory of convection. In any case, whether our aim is the immediate one of an improved estimate for l, or the long-range one of seeking a reliable theory, it is clear that we must examine the fundamental theory of convection more carefully. Therefore, I shall here attempt to present a discussion of more fundamental approaches to the theory of turbulent convection. In doing this, I hope to make clear that, although we are far from having anything like a complete theory of stellar convection, we can now treat certain limiting cases with sufficient accuracy to hope for improved estimates for stellar convection in the foreseeable future.

PRELIMINARIES ON THERMAL CONVECTION

Boussinesq Equations

All theories of turbulent convection which have been attempted so far are based on an approximation which is usually credited to Boussinesq (1903). In this approximation, the fluid density is taken to be constant except when density fluctuations give rise to the differential buoyancy forces that drive the motion. Physically, this means that two basic effects are neglected in current theories. The first is the variation of the background density with position; the second is the presence of rapid fluctuations of density associated with sound waves. In turn, this means that two parameters of the general problem are taken to be small: One is the ratio of the thickness of the convection zone to the density scale height, and the other is the ratio of characteristic frequencies in turbulent convection to those of random acoustic motions.

The Boussinesq approximation does not hold in stars, though it is quite a good representation of convection in the laboratory. However, the approximation has the theoretical advantage that it lays bare the basic difficulties of the convection problem, namely, the nonlinear, stochastic character of the equations. It therefore seems sensible to treat the Boussinesq case first, and to try to understand laboratory convection before attempting the more general problem of astrophysical convection. This, at any rate, has been the universally adopted point of view, but the work on turbulent convection by astrophysicists has naturally always been slanted toward the possibility of generalizing to non-Boussinesq convection.

A formal derivation of the Boussinesq equations is too lengthy to be given here (see Spiegel and Veronis, 1960, and Mihaljan, 1962). However, the equations can be found in many places (e.g., Chandrasekhar, 1961), and are

$$\frac{\partial u_i}{\partial t} + u_j \frac{\partial u_i}{\partial x_j} = -\frac{1}{\rho_0} \frac{\partial p}{\partial x_i} + \left(\frac{\rho - \rho_0}{\rho_0}\right) g_i + \nu \nabla^2 u_i \tag{13}$$

$$\frac{\partial T}{\partial t} + u_i \frac{\partial T}{\partial x_i} = \kappa \nabla^2 T \tag{14}$$

$$\frac{\partial u_i}{\partial x_i} = 0 \tag{15}$$

$$\rho = \rho_0[1 - \alpha(T - T_0)] \tag{16}$$

Here, u_i is the velocity vector [sometimes written (u, v, w)], and $g_i = g\delta_{i3}$, where g is the acceleration of gravity and the summation convention has been used. Equation (16) is the Boussinesq equation of state in which α is the coefficient of thermal expansion, T_0 is a reference temperature which may be taken as the average temperature of the fluid, and ρ_0 is the corresponding density, which is taken as the overall density of the fluid except in the buoyancy term in equation (13). These equations describe convection in a plane-parallel layer of fluid of infinite horizontal extent. We can then outline an idealized experiment whose results we shall attempt to describe on the basis of these equations.

Let the layer of fluid be bounded by two rigid plates with separation d. We shall require these plates to be constructed of material with extremely high thermal conductivity and shall assume that they are maintained at constant temperatures. These conditions can usually be realized in the laboratory. We shall also assume the rather artificial condition that the plates are completely slippery. This latter assumption (sometimes known as the free-boundary condition) simplifies the calculations and hopefully does not introduce any gross qualitative errors. These conditions can be expressed as follows:

Rigidity of boundaries:

$$w = 0 \quad \text{at} \quad z = 0, d \tag{17}$$

High conductivity of boundaries:

$$T = \begin{cases} T_L = \text{constant} \quad \text{at} \quad z = 0 \\ \\ T_U = \text{constant} \quad \text{at} \quad z = d \end{cases} \tag{18}$$

Slipperiness of boundaries:

$$\frac{\partial u}{dz} = \frac{\partial v}{\partial z} = 0 \quad \text{at} \quad z = 0, d \tag{19a}$$

Condition (19a) is the condition that no tangential stress be exerted by the boundaries. Equation (15) then implies that

$$\frac{\partial^2 w}{\partial z^2} = 0 \quad \text{at} \quad z = 0, d \tag{19}$$

Equations for the Mean Quantities

In speaking of mean properties of a turbulent fluid, we should properly refer to averages over an ensemble of systems identical to the system described in the preceding section. However, since the system described is statistically homogeneous in horizontal directions, it seems natural to assume that averages over the horizontal coordinates are equivalent to averages over an ensemble of identical systems. Such horizontal averages are convenient to work with and are used in most discussions of convective turbulence. For example, the temperature can be expressed as a sum of an averaged plus a fluctuating term as follows:

$$T(\mathbf{x}, t) = \bar{T}(z) + \theta(\mathbf{x}, t) \tag{20}$$

where z is the vertical coordinate and the overbar denotes horizontal average. In equation (20), $\bar{T}$ has no time dependence, since we are treating a statistically steady system.

With this averaging convention, we can conveniently look at averages of the equations of motion to see how the mean properties of the convecting system behave. The simplest result of this kind, obtained from the mass conservation equation, is

$$\bar{w} = \text{constant} \tag{21a}$$

where w is the vertical component of velocity. Since we have assumed that the fluid is bounded by rigid, horizontal plates, the constant in equation (21a) must be zero. Therefore, mass conservation implies that

$$\bar{w} = 0 \tag{21}$$

It is also clear from the symmetry of the problem that any mean horizontal velocity can be transformed away, and hence we can write

$$\bar{\mathbf{u}} = 0 \tag{22}$$

where $\mathbf{u}$ is the total velocity field. Thus, the velocity is a purely fluctuating quantity in thermal convection.

The horizontal average of the momentum equation is the equation of hydrostatic equilibrium

$$\frac{d\bar{p}}{dz} = -g\rho_0 \tag{23}$$

In general, the pressure should include a contribution from the turbulence, $\overline{\rho \mathbf{u}^2}$, but this can be shown to be negligible in the Boussinesq approximation.

From the average of the heat equation, we obtain, after an integration,

$$-K\frac{d\bar{T}}{dz} + \rho_0 c_p \overline{w\theta} = \text{constant} \tag{24a}$$

where c_p is the specific heat and K is the conductivity of the fluid. This equation simply tells us that the sum of the conductive and convective fluxes must be constant in a statistically steady system. It is usually rewritten in the form

$$\kappa\beta + \overline{w\theta} = \mathcal{H} \tag{24}$$

where $\beta = -(d\bar{T}/dz)$, $\kappa = K/\rho c_p$, and $\mathcal{H}$ is the total heat flux divided by ρc_p.

Now, in the stellar problem, we would like to find $\overline{T}(z)$ by integrating β inward from the surface of the star. But, as equation (24) shows, we cannot fully determine the mean structure of the system, i.e., β, without finding $\overline{w\theta}$, and this means that we must study the dynamics of turbulent motion to find w and θ. Thus, we are forced to go far beyond the realm of these simple mean-structure equations in order to evaluate the single mean quantity $\overline{w\theta}$. To do this, we must consider the equations for w and θ themselves.

Equations for the Fluctuating Quantities

The Boussinesq equations of motion, with the mean equations subtracted out, are

$$\frac{\partial \mathbf{u}}{\partial t} - \nu\nabla^2\mathbf{u} - g\alpha\theta + \nabla\varpi = -(\mathbf{u}\cdot\nabla\mathbf{u} - \overline{\mathbf{u}\cdot\nabla\mathbf{u}}) \tag{25}$$

$$\frac{\partial\theta}{\partial t} - \kappa\nabla^2\theta - \left(\beta - \frac{g}{c_p}\right)w = -(\mathbf{u}\cdot\nabla\theta - \overline{\mathbf{u}\cdot\nabla\theta}) \tag{26}$$

and

$$\nabla\cdot\mathbf{u} = 0 \tag{27}$$

where

$$\varpi = \frac{p - p_0}{\rho_0} \tag{27a}$$

An essential feature of these equations is that there are two kinds of nonlinear terms. First, there are the terms which I have written on the right-hand sides of the equations. These are quadratic in fluctuating quantities and give rise to the highly stochastic nature of the problem. Second, there is the term which is nonlinear because of the distortions of the mean temperature field resulting from convective heat transfer. The term βw is the prototype of the kind of nonlinear term which the astrophysicist must grapple with, because, in non-Boussinesq systems, the effects on density due to the convection feed back on the system, and many terms of this kind, which are bilinear in averaged and fluctuating quantities, appear in the governing equations. The different role of the two kinds of nonlinear terms will be discussed in more detail later on.

It should be mentioned here, though, that the physical distinction between the two kinds of nonlinearity is already made clear in the mixing-length studies. There, velocities are calculated from buoyancy forces arising in the fully convective model, so that the distorted temperature and density fields are taken into account. The fluctuation interaction is then added secondarily by the mixing-length assumption itself.

Nondimensional Parameters

The number of parameters characterizing the experimental system described above is rather large, but if we nondimensionalize the equations, we can combine them into two independent governing parameters. The nature of these two parameters depends on the units we choose for the nondimensionalization, and I shall indicate what the conventional choices are.

One way to discover appropriate nondimensional parameters is to indicate, for each process of interest, a characteristic time. For the driving force, gravity, we must consider the inverse free-fall time across the layer. But a fluid element freely falling through the fluid itself feels a reduced acceleration. The reduction factor is approximately $\Delta\rho/\rho_0$, where $\Delta\rho$ is the density change across the system. In a Boussinesq system $\Delta\rho = -\alpha\rho_0\Delta T$, where ΔT is the temperature difference between the lower and upper boundaries $(= T_l - T_u)$. Hence, the free-fall time is

$$\tau_g = \left(\frac{g\Delta\rho}{\rho_0 d}\right)^{-\frac{1}{2}} = \left(\frac{g\alpha\Delta T}{d}\right)^{-\frac{1}{2}} \tag{28}$$

Similarly, the thermal and viscous dissipative processes have characteristic times

$$\tau_\kappa = \frac{d^2}{\kappa} \tag{29}$$

and

$$\tau_\nu = \frac{d^2}{\nu} \tag{30}$$

These three times characterize the efficiency of all processes depending on external parameters. From these, we can construct two nondimensional numbers. The two which are customarily employed are the Rayleigh number

$$R = \frac{\tau_\nu\tau_\kappa}{\tau_g^2} = \frac{g\alpha\Delta T d^3}{\kappa\nu} \tag{31}$$

and the Prandtl number

$$\sigma = \frac{\tau_\kappa}{\tau_\nu} = \frac{\nu}{\kappa} \tag{32}$$

For a given experimental setup, R and σ are fixed, and, in principle, all average properties of the flow may then be predicted.

The Rayleigh number is frequently written in the form

$$R = \frac{g\alpha\beta_0 d^4}{\kappa\nu} \tag{32a}$$

where β_0 is the mean temperature gradient $(= \Delta T/d)$. Here, I have given the form for a liquid. For a gas, we must include the fact that the buoyant acceleration is not directly proportional to the temperature gradient, but to its excess over the adiabatic gradient (as mentioned in the introduction). For a gas, then, we must replace β_0 by $\beta_0 - g/c_p$ in the definition of the Rayleigh number (Jeffreys, 1930).

To give an idea of the values of the parameters in the stellar problem, let me just mention that at the edge of the solar convection zone, $R \approx 10^{12}$ and $\sigma \approx 10^{-9}$. In this estimate, the parameters have been evaluated locally, d has been taken as the local scale height rather than the thickness of the convection zone, and a radiative diffusivity has been used for κ.

From the foregoing estimates, it is clear that the viscosity plays a secondary role in astrophysical convection, and because of this it is frequently possible, for many purposes, to describe the system in terms of a single parameter $\Lambda = \sigma R$. An alternative limit which is of theoretical interest is that of very large σ, in which case the parameter $G = R/\sigma$ is sometimes used.

Once the Rayleigh number and Prandtl number are specified, we can attempt to predict several features of the convection. The one which is most directly measurable is the total heat transfer $\mathscr{H}$. This is frequently put in a nondimensional form called the Nusselt number

$$N = \frac{\mathscr{H}}{\kappa(\Delta T/d)} = \frac{\mathscr{H}}{\kappa\beta_0} \tag{33}$$

which is the ratio of total heat transfer in the turbulent state to that in the absence of turbulence.

It should be noted that the Reynolds and Péclet numbers mentioned earlier are not taken as basic parameters in convection theory, although they are used in other fluid-dynamics studies. The reason is that they are based on velocities, and there are no prescribed velocities in the convection problem as outlined here. Hence, although it is frequently useful to talk in terms of R_e and P_e, they can only be estimated but never fixed *a priori*, and therefore they do not play as fundamental a role in the theory as R and σ do.

Stability Theory

In the introduction, I outlined the derivation of a general criterion for convective instability. It amounted to asking whether the buoyancy force on a displaced parcel of fluid is restorative or not. Actually, the instability criterion derived in this way is not a sufficient condition. For not only must the buoyancy force be in the right direction to promote instability, it must be sufficiently large to overcome the viscous force which opposes the motion. Thus, a more complete instability condition is that the ratio of the buoyancy force to the viscous force should be sufficiently large (e.g., Spiegel, 1960). This ratio is closely related to the Rayleigh number, and the condition for instability turns out to be

$$R \geq R_c \tag{34}$$

where R_c is some constant.

The first theoretical determination of R_c was given by Rayleigh (1916) for just the experimental situation described above. His procedure was to consider disturbances of the fluid having infinitesimal amplitude and to see how they evolve in time. In an unstable situation, such a calculation applies only in the first instant of the motion, since the amplitude of a disturbance grows exponentially in time. But the mathematics of stability theory is very useful for the general problem, and I shall outline it here.

If we treat fluctuating quantities as small and neglect their products, we find first of all that the convective heat transfer is negligible, so that, according to equation (24), β is a constant, namely, $\beta = \beta_0 = \Delta T/d$. Hence, the mean state in stability theory is just the undisturbed static state that exists in the absence of convection. The equations for the fluctuating quantities then are

$$\frac{\partial \mathbf{u}}{\partial t} - \nu\nabla^2\mathbf{u} - \mathbf{g}\alpha\theta + \nabla\varpi = 0 \tag{35}$$

$$\frac{\partial \theta}{\partial t} - \kappa\nabla^2\theta - \beta w = 0 \tag{36}$$

and

$$\nabla \cdot \mathbf{u} = 0 \tag{37}$$

These stability equations are simple to treat, but I shall discuss them here in a general terminology which will be useful in the discussion of the turbulent case (see Ledoux, Schwarzschild, and Spiegel, 1961).

With the aid of equation (37), we can solve for the pressure by taking the divergence of equation (35). This gives a Poisson equation:

$$\nabla^2 \varpi = g\alpha \frac{\partial \theta}{\partial z} \tag{38}$$

Then, we can eliminate ϖ from the stability equations to obtain an integro-differential equation for the quantity

$$U = \begin{pmatrix} u \\ v \\ w \\ \theta \end{pmatrix} \tag{39}$$

This equation has the form

$$\frac{\partial U}{\partial t} = HU \tag{40}$$

where H is a time-independent linear operator. Equation (40) admits separable solutions of the form

$$U = e^{\eta t} f(x, y)\Omega(z) \tag{41}$$

The quantity η is a separation constant known as the "growth rate," and it turns out that f satisfies the membrane equation

$$\nabla_1^2 f = -k^2 f \tag{42}$$

where

$$\nabla_1^2 = \frac{\partial^2}{\partial x^2} + \frac{\partial^2}{\partial y^2} \tag{43}$$

and k^2 is another separation constant.

With solutions of the form (41), we obtain an eigenvalue equation of the form

$$H\Omega = \eta\Omega \tag{44}$$

where the boundary conditions (17) to (19) must be added. H now depends on k, and for each value of k, a sequence of eigenvalues $\eta(k, n)$ is obtained, where $n = 1, 2, 3, \ldots$. Similarly, a sequence of eigenfunctions $\Omega^{(k,n)}$ is found in general, but in the present simple example, the Ω are independent of k. The index n characterizes the complexity of the vertical structure of Ω and generally is the number of zeros (less one) of Ω.

In most treatments, the form of solution of f which is adopted is

$$f = e^{ilx + imy} \qquad (45)$$

where

$$l^2 + m^2 = k^2 \qquad (45a)$$

Thus, k is a horizontal wavenumber and k^{-1} characterizes the horizontal scale of a particular disturbance.

An interesting feature of the solutions is that for each value of k and n there are three eigenvalues which we shall label η_+, η_0, and η_-. For $R > 0$, η_0 and η_- are negative definite, while η_+ can have positive or negative value, depending on the values of k and n. This distinction among the eigenvalues can be understood by examining the correlation coefficient

$$j = \frac{\langle U_3^{(k,n)} U_4^{(k,n)} \rangle}{\langle [U_3^{(k,n)}]^2 \rangle^{\frac{1}{2}} \langle [U_4^{(k,n)}]^2 \rangle^{\frac{1}{2}}} \qquad (46)$$

where the angular brackets denote space averages. For the η_+, η_0, and η_- modes the values of j are $+1$, 0, and -1, respectively. Now, j measures the correlation between vertical velocity and temperature fluctuation and thus indicates the effect of buoyancy forces. For an η_- mode, an upward velocity is associated with a negative temperature fluctuation since $j = -1$. But a negative temperature fluctuation implies a positive density fluctuation and thus a downward buoyancy force. Therefore, buoyancy forces work against the motion of η_- modes, and these cannot be unstable, hence $\eta_- < 0$. The η_0 modes have $j = 0$ and are not coupled to the gravitational acceleration. Examination of the solution shows that the modes have purely horizontal motion which resembles a network of vortices with vertical axes. If η_0 modes are once excited, they will be viscously damped, and hence $\eta_0 < 0$.

The most important modes are the η_+ modes. They are always driven by the buoyancy forces (for $R > 0$), and the sign of η_+ depends on whether these forces can overcome the stabilizing effects of dissipation and boundaries. It has been shown by Pellew and Southwell (1940) that the principle of the exchange of stabilities holds, that is, that neutral stability is characterized by the condition $\eta_+ = 0$, and no growing oscillation can occur. Rayleigh's initial study of convective stability thus was aimed at finding the lowest value R_c of R for which $\eta_+ = 0$. Indeed, the results of stability investigations are usually put in these terms.

In Figure 1 are shown the loci of the curves $\eta_+ = 0$ in the R-k plane for several values of n. In this diagram, a nondimensional form of k is used, namely, $a = kd$. We see that for each n the curve has a single minimum occurring at values $R = R_c^{(n)}$ and $a = a_c^{(n)}$. The value R_c ($= R_c^{(1)}$) is called the critical Rayleigh number, and a_c ($= a_c^{(1)}$) is called the wave number of maximum instability. The values of R_c and a_c are not dependent on the Prandtl number.

It is useful to imagine a third axis in Figure 1 which is η_+. For each value of n the η_+ surfaces are ridges above the R-a plane for values of R and a above the marginal stability curves. To aid in this visualization, I have plotted curves of η_+ versus a schematically in Figure 2. These curves represent a cut in the R-k-η_+ space on a plane $R = $ constant, and the curves are drawn for a fixed value of σ. One point to notice is that for any given R ($> R_c$) there exists, for each n, a band of a with $\eta_+ > 0$. This band represents the range of horizontal scales which is unstable.

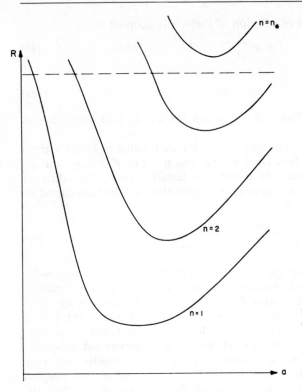

Figure 1. Loci of marginal stability curves in the R–a plane ($a = kd$). These are the curves $\eta_+ = 0$. Each curve is for a given vertical mode described by n. For given R (indicated by dashed line), there is a lowest value of n ($= n_*$) such that marginal stability cannot occur.

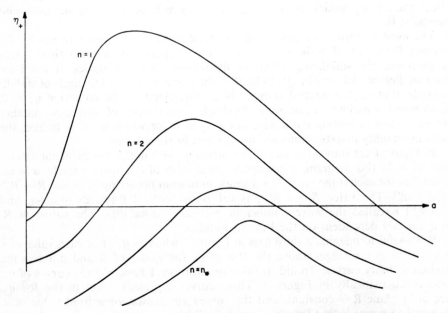

Figure 2. η_+ as a function of a ($= kd$) for different n. n_* is the smallest value of n for which η_+ is always negative. The curves are drawn for a given value of R and σ.

It should also be noted that if one is interested in a particular value of the Rayleigh number, as indicated by the dashed horizontal line in Figure 1, then there is some particular value of n ($= n_*$), such that no modes with $n \geq n_*$ can be unstable.

This completes the account of linear theory which is needed for our further discussions. However, I would like to point out that a great number of physical effects have been added to the problem since Rayleigh's first investigation (e.g., Chandrasekhar, 1961), and a rather large amount of literature now exists for the linear theory.

THE DYNAMICS OF TURBULENT CONVECTION

Some Experimental Results and Dimensional Arguments

Before proceeding to a detailed discussion of the physics of convection, I should like to mention some elementary aspects of the problem in this subsection. Perhaps the most interesting of these are the general features which are known from experiments.

As I pointed out earlier, the stability of a fluid layer heated from below is characterized by the Rayleigh number defined by equation (31). Experimentally and theoretically, it is known that convection can occur once the Rayleigh number exceeds a critical value $R_c \approx 10^3$. The experiments reveal that when the Rayleigh number is increased, so is the total heat transport. For Rayleigh numbers somewhat above 10^3, but below 10^5, the heat transport follows the law

$$\mathscr{H} \propto R^{\frac{1}{4}} \tag{47a}$$

When the Rayleigh number is raised sufficiently in excess of 10^5, the heat transport law becomes

$$\mathscr{H} \propto R^{\frac{1}{3}} \tag{47b}$$

In both (47a) and (47b), the exponents are reliable to a few percent, at least for typical laboratory values of σ ($\gtrsim 1$). [For a fuller discussion see Jakob (1949), Malkus (1954a), and Globe and Dropkin (1959).] In the range of the Prandtl number σ covered by the experiments, the heat transport also depends weakly on the Prandtl number, in the sense that $\mathscr{H}$ decreases with σ for fixed R (Globe and Dropkin, 1959).

More detailed measurements, especially of $\overline{T}$, have also been made, but the amount of exact information of this kind is disappointingly small. The main qualitative result is that for increasing R, the temperature profile becomes increasingly nonlinear, and that for very large R, a distinct thermal boundary layer develops. This is illustrated by the two temperature profiles in Figure 3. Profile (a) is the linear profile which is found in the absence of convective motions; profile (b) schematically illustrates the profile in a highly turbulent situation. An account of the determination of such profiles is given by Townsend (1959).

It is now possible to understand some of these results qualitatively on the basis of simple arguments based on equation (24), which says that the sum of the convective and conductive fluxes $\overline{w\theta}$ and βw is constant. At large Rayleigh numbers, we can expect violent motion and hence that the convective heat transfer term should be large over the main body of the fluid. However, at the boundaries, where w and θ must vanish, the situation is different. Thus, $\overline{w\theta}$ can be expected to be large in the midregions and small near the boundaries and, from equation (24), we would

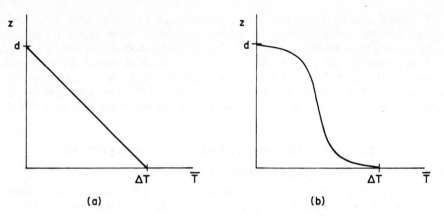

Figure 3. (a) The mean temperature profile in the absence of motion. (b) The mean temperature profile in a fluid with turbulent heat transport.

then conclude that the temperature gradient β is large near the boundaries and relatively small away from the boundaries. This would then lead us to a profile qualitatively like that illustrated in Figure 3b. The important feature is that the value of the temperature gradient at the boundary then fixes $\mathscr{H}$, since $\overline{w\theta} = 0$ on the boundaries. Thus, if β_B is the boundary value of the gradient,

$$\mathscr{H} = \kappa\beta_B \tag{48}$$

With this qualitative picture, we can rationalize the experimental result (47b) for large R. From dimensional considerations, we should suspect that the Nusselt number N is a function only of R and σ and therefore that

$$\mathscr{H} = \kappa\frac{\Delta T}{d}N(R, \sigma) \tag{49}$$

In any suitably restricted range of R and σ, we might approximate equation (49) by

$$\mathscr{H} = \kappa\frac{\Delta T}{d}N_0 R^b \sigma^c \tag{50}$$

where N_0, b, and c are constants. Moreover, as we have seen, for very large R the form of β suggests the presence of distinct thermal boundary layers whose structure fixes β_B, hence $\mathscr{H}$. Once these boundary layers are sharply delineated, they should be insensitive to the precise flow pattern in the midregions of the fluid. Hence, we can expect that $\mathscr{H}$ should be insensitive to the exact separation of the boundaries. To make $\mathscr{H}$, as given by equation (50) independent of d, we must choose $b = \frac{1}{3}$, and this reproduces the experimental result (47b). Thus, we see that the development of distinct thermal boundary layers implies the $R^{\frac{1}{3}}$ law for large R, as has been pointed out by Priestley (1959).

Now consider what happens when we reduce the viscosity for a fixed value of R. It is difficult to imagine how the heat transport could depend on the viscosity for very small viscosities. Indeed, it has always been assumed by astrophysicists that $\mathscr{H}$ is independent of viscosity in stars, where $\sigma \ll 1$, and this assumption seems physically reasonable. It suggests that when σ is very small we should set $b = c$, and this partly rationalizes the observed decrease of $\mathscr{H}$ with σ that I mentioned previously (b itself is always non-negative). In particular, for very efficient convection with $\sigma \ll 1$, we would expect

$$\mathscr{H} = \kappa \frac{\Delta T}{d} N_0 \Lambda^{\frac{1}{3}} \tag{51}$$

Of course, it is clear that equation (51) must break down as $\Lambda \to 0$, since it implies $\mathscr{H} \to 0$ in that limit. But it is not hard to show that $\mathscr{H}$ must be at least as large as $\kappa \Delta T/d$. This implies that the $R^{\frac{1}{3}}$ dependence does not hold in the limit $\Lambda \to 0$, and we shall see later that this is borne out by other considerations. This means that there are two regimes of low Prandtl number: low Prandtl number with $\sigma \ll 1$; and very low Prandtl number with $\Lambda = \sigma R \ll 1$.

The case of large Prandtl number cannot be dealt with as easily, and I should like to defer it until later. The case of small R we shall not discuss at all, although much theoretical work has been concentrated here. For R just above R_c, we have (among others) the nonlinear perturbation theory of Malkus and Veronis (1958), while the $R^{\frac{1}{3}}$ law has been reproduced by the use of laminar boundary layer theory (Pillow).

The Fluctuation Interaction

The equations for the fluctuating quantities, (25) to (27), are especially difficult to treat because of the terms quadratic in fluctuating quantities, which we shall call the "fluctuation interactions." These interactions have been the object of intense scrutiny over the past few decades, but their role in actual physical situations is still not completely understood. A great deal has been written on the subject, and it would be impossible to attempt a complete summary here. I shall just make a few remarks about the problem and mention some promising recent work.

To simplify the discussion, let us write equation (25), the equation for the velocity in the form

$$\frac{\partial \mathbf{u}}{\partial t} - \nu \nabla^2 \mathbf{u} = -\nabla \varpi - \mathbf{u} \cdot \nabla \mathbf{u} + \mathbf{f} \tag{52}$$

where we have written $\mathbf{f}$ for the buoyancy force. Now, in practice, $\mathbf{f}$ is described by equation (26), but let us assume for the sake of discussion that we can prescribe $\mathbf{f}$ as some fixed mechanical stirring and that there is no feedback on $\mathbf{f}$ from $\mathbf{u}$. Let us also assume that the flow is incompressible, so that equation (27) holds,

$$\nabla \cdot \mathbf{u} = 0 \tag{27}$$

Since $\mathbf{u}$ is a stochastic variable in highly turbulent flows, we are not interested so much in the detailed structure of the solution as in average properties, such as the mean kinetic energy $\langle \mathbf{u}^2 \rangle$. (In this section the averages we speak of will be ensemble averages denoted by angular brackets.) Ideally, we should like to know a probability distribution for the velocity or, equivalently, the set of stochastic moments $\langle u_i u_j u_k \ldots \rangle$. The trouble is that if we form an equation for $\langle \mathbf{u}^2 \rangle$ from equation (52) this contains an unknown third-order moment. We can also form an equation for this third-order moment from equation (52), but it contains a fourth-order moment, and so on. Clearly, the set of equations for the stochastic moments form an infinite hierarchy, as is quite familiar in statistical mechanics, and some approximation or additional assumption is needed to truncate the system.

I should now like to mention some attempts to close off this infinite system, but before doing this, let us introduce Fourier representation, which has a number of advantages for this discussion. We can write

$$\mathbf{u} = \sum_{\mathbf{k}} \mathbf{v}(\mathbf{k}, t) \exp(i\mathbf{k} \cdot \mathbf{x}) \tag{53}$$

In equation (53), the sum is over all $\mathbf{k}$ satisfying the boundary conditions, and, for a finite system, we can imagine that these boundary conditions are periodic, although more refined ways of handling the normalization problem exist (Batchelor, 1953).

From equations (27) and (53), we obtain

$$\mathbf{k} \cdot \mathbf{v}(\mathbf{k}, t) = 0 \tag{54}$$

Thus, the Fourier amplitude of velocity is orthogonal to the wave vector, and the concept of a turbulent eddy arises partly from this result, with the eddy size $\pi/|\mathbf{k}|$.

The Fourier transform of equation (52) is (see Batchelor, 1953)

$$\frac{\partial v_i(\mathbf{k})}{\partial t} + \nu k^2 v_i(\mathbf{k}) = i k_l P_{ij}(\mathbf{k}) \sum_{\mathbf{k}'} v_j(\mathbf{k} - \mathbf{k}') v_l(\mathbf{k}') + P_{ij} \hat{\mathbf{f}}_j \tag{55}$$

where the projection operator

$$P_{ij}(\mathbf{k}) = \delta_{ij} - \frac{k_i k_j}{k^2} \tag{56}$$

ensures that equation (54) is satisfied, and $\hat{\mathbf{f}}$ is the transform of $\mathbf{f}$. A striking feature of equation (55) is that the fluctuation interaction in wave-number space is such that modes interact in triads where the interacting wave vectors always form a triangle.

Just as the stochastic properties of $\mathbf{u}(\mathbf{x}, t)$ may be described by the set of averaged moments, the stochastic properties of $\mathbf{v}(\mathbf{k}, t)$ may be described by its moments. The most important of these is the second moment, whose contracted form is the power spectrum

$$E(\mathbf{k}) = \langle v_i(\mathbf{k}) v_i(\mathbf{k}) \rangle \tag{57}$$

An equation for E can be derived from equation (55); this has the form

$$\frac{\partial E}{\partial t} + 2\nu k^2 E = T_r + \langle |P_{ij} v_i \hat{f}_j| \rangle \tag{58}$$

where T_r is a trilinear, averaged quantity. We see that the kinetic energy in each mode is altered by viscous damping, net transfer to or from other modes (T_r is a sum over all other pairs of modes which can interact with $\mathbf{k}$), and input from the driving force. In the convection problem, as in most real physical problems, the input term is complicated, and, for many years, students of turbulence theory have avoided the difficulties of input by restricting their considerations to large $|\mathbf{k}|$. The argument has been that for sufficiently small scales (large $|\mathbf{k}|$), the input from external forces is negligible and the level of excitation is fixed by transfer from large scales and possibly viscous damping. For these small scales, the geometry of the experiment is thought not to matter, and it has been assumed that the small-scale properties of the flow are isotropic. This leads to considerable mathematical simplification, but even in this approximation no complete solution of the equations has been found.

The most important early work in the problem of turbulent convection were the dimensional arguments of Kolmogoroff (see Batchelor, 1953) and approximations for T_r based on physical reasoning (e.g., Heisenberg, 1948). Unfortunately, these theories are of little use in the convection problem, where the main concern is with the largest scales of motion. Indeed, Heisenberg's transfer function gives physically unmeaningful results for small $|\mathbf{k}|$ in the convection problem (Spiegel, 1962a), although it must be pointed out that the work of Kolmogoroff and Heisenberg has received some vivid experimental confirmation for very small scales (Grant, Stewart, and Moilliet, 1962).

The reason for the failure of the Heisenberg transfer function is that it implies that any wave number $|\mathbf{k}|$ receives energy from lower wave numbers only. But when $|k|$ becomes of the order of d^{-1} where d is a dimension of the system, this cannot be correct. It is possible to construct phenomenological transfer functions which do not present this difficulty (e.g., Kraichnan and Spiegel, 1962), but any phenomenological theory introduces unknown parameters and, in the long run, such an approach is quite limited.

Much more promising are the statistical theories which have recently been developed. To describe the ideas in these, let us write equation (55) schematically as

$$\dot{v}_i + \lambda_i v_i = \mathscr{A}_i(t) \qquad \text{(no summation)} \tag{59}$$

where v_i now represents the Fourier amplitude for the wave vector $\mathbf{k}_i$, the dot means time differentiation, $\lambda_i = \nu k_i^2$, $\mathscr{A}_i$ represents the fluctuation interaction, and where, for simplicity, we have omitted the driving force. Equation (59) is just the Langevin equation with the difference that the driving term $\mathscr{A}$ for each mode is a functional of the amplitudes of all the other modes; that is, $\mathscr{A}$ is of the form

$$\mathscr{A}_i = \sum_{l,m} A_{ilm} v_l v_m \tag{60}$$

where the sum is over all (l, m) such that

$$\mathbf{k}_i - \mathbf{k}_l + \mathbf{k}_m = 0 \tag{61}$$

and the A_{ilm} are some appropriate coefficients. Thus, we cannot treat $\mathscr{A}$ as a random function of time, as in the Brownian motion problem, but there is a strong temptation to make a variety of stochastic assumptions about $\mathscr{A}$.

The first of the stochastic approaches which has been explored is the so-called cumulant-discard approximation. In this approximation, although it is recognized that $\mathscr{A}$ and v are not stochastically independent, it is suggested that their higher stochastic moments can be treated as if they were stochastically independent. In the lowest approximation, this is equivalent to saying that the fourth moment of the actual velocity field is related to the second moment as in a joint normal distribution. In this form, the approximation was first used by Millionshchikov (1941), and later applications have been made by Heisenberg (1948), Chandrasekhar (1951), and Proudman and Reid (1954). Higher-order approximations are possible, but I know of no attempts at this.

The cumulant-discard approach is open to criticism, as Kraichnan (1957) has pointed out. Briefly, his argument is as follows. The correlations among any finite number of Fourier amplitudes must certainly vanish as the size of the system tends to infinity and the number of allowed modes correspondingly increases. (This property which Kraichnan calls the "weak dependence principle" has been known for some time, and it applies only to homogeneous turbulence.) In assuming a quasi-random property of $\mathscr{A}$, however, one throws away a very large number of such small correlations, and the result is the neglect of a finite quantity in the equations of motion. Just how large an error this introduces is not clear in general, but in isotropic turbulence theory the cumulant discard often leads to unphysical results.

Kraichnan (1959) has gone beyond his criticism of the cumulant-discard approximation and has proposed a more satisfactory approximation [although not all authors hold this view—see Proudman (1962)]. In particular, Kraichnan suggests that the weak statistical interdependence that can occur among modes is chiefly due to the direct interactions among the triads whose wave vectors form triangles. Suppose, then, that we consider a triad of wave vectors $(\mathbf{k}_1, \mathbf{k}_2, \mathbf{k}_3)$ satisfying equation (61). In the terminology we are using here, we can write the three Langevin equations for the members of this triad as a set of three equations to be considered together. We must then rewrite the forcing terms $\mathscr{A}$ by removing the terms in each which involve the direct couplings among members of the triad. Thus, we obtain a set of three nonlinear equations in three unknowns, each of which contains an as yet unknown forcing function. These three simultaneous equations have the form

$$\dot{v}_\alpha + \lambda_\alpha v_\alpha - A_{\alpha\beta\gamma} v_\beta v_\gamma = \mathscr{A}'_\alpha \qquad \text{(no summation)} \tag{62}$$

where the Greek indices range over the three members of the triad, and the suffixes α, β, and γ are all different. Here

$$\mathscr{A}'_\alpha = \mathscr{A}_\alpha - A_{\alpha\beta\gamma} v_\beta v_\gamma \qquad \text{(no summation)} \tag{63}$$

Now equation (62) represents three equations for the three members of the triad $(\mathbf{k}_1, \mathbf{k}_2, \mathbf{k}_3)$, and it describes their direct interaction exactly. Kraichnan's idea is now to make stochastic assumptions about $\mathscr{A}'$, but to treat the rest of equation (62) exactly. Thus, in his theory, the direct interactions among triads are precisely taken into account, but this interaction is allowed to take place in the randomizing background of all other open-ended chains of interactions. It is then possible to obtain an integrodifferential equation for E, whose solution is by no means simple, but the problem at least has been reduced to a determinate one, which has now been solved for the case of isotropic turbulence with satisfying results. Since, as seems likely, the case of isotropic turbulence is the most sensitive to the approximation used for the fluctuation interaction, there are grounds for optimism here. The solution of the equation for fully developed turbulent convection now seems possible.

One final remark may be in order here. It may be asked whether the entire problem should not be tackled numerically, starting with the raw equations themselves. However, existing machines do not seem adequate for the task of directly solving the equations of motion for a truly turbulent flow (Corrsin, 1959), and it may be some time before they are. In the approximate theories, however, the quantities calculated are much smoother in space and time than the raw quantities, and much coarser grids can be used in solving the approximate equations. It is therefore necessary to continue to resort to such approximations, apart from their obvious interest in clarifying the physical processes.

Finally, in closing this section, I would like to re-emphasize that this discussion is as incomplete as it is schematic, and does not include, for example, the interesting new work by Edwards (1964), which makes approximations in analogy with the Fokker–Planck approach.

Formulation of the General Problem

As I have pointed out, there are two kinds of nonlinear interaction terms in the convection equations, and the fluctuation interactions are by far the more difficult to treat. This aspect of the theory makes the problem rather different from the theory of isotropic turbulence. There, only the fluctuation interactions occur, and therefore the theory of isotropic turbulence is very sensitive to the kind of approximations that are made for the interactions. In the convection problem, on the other hand, the fluctuation interactions represent only one form of coupling among modes, and other important ways of distorting the motions occur. (An exception is the case of very small σ, which will come up later.) It may well be, therefore, that the results obtained in convection theory are not as sensitive to the nature of the approximation one uses for the fluctuation interaction as they are in isotropic turbulence theory.

The situation may be likened to that of investigating equilibrium statistical mechanics by use of the Boltzmann equation. There, one has the feeling that so long as the approximation used for the collision term is physically reasonable in some sense, the details of the approximation are not always crucial. For example, even with a hard-sphere approximation, one may expect to achieve the Boltzmann distribution. Indeed, the complete neglect of collisions in the Vlasov equations still seems to leave a good deal of useful information in the equations in certain problems. If we can reasonably expect parallel situations in convection theory, then we can be hopeful that a theory of convection can be based on relatively crude approximations to the fluctuation interactions. With this expectation, Ledoux, Schwarzschild, and I (1961) undertook the approach I shall now outline. Although our explicit results have so far been quite meager, we have at least been able to prescribe a specific program which has the advantage of bringing out the nature of the problem of obtaining reasonable solutions. In this section, I should like to outline the proposed program.

It is convenient here to use a normal-mode representation which consists of the set of eigenfunctions of the following equations:

$$\frac{\partial u}{\partial t} - \nu\nabla^2 u - g\alpha\theta + \nabla\varpi = 0 \tag{64}$$

$$\frac{\partial \theta}{\partial t} - \kappa\nabla^2\theta - \left(\beta + \frac{g}{c_p}\right)w = 0 \tag{65}$$

$$\nabla \cdot \mathbf{u} = 0 \tag{66}$$

Equations (64) and (65) are just equations (25) and (26) with the fluctuation interactions omitted. Moreover, if β were constant, equations (64), (65), and (66) would be just the equations of stability theory. However, β is supposed to be the actual temperature gradient in the fully turbulent state, which, according to equation (5), is

$$\beta = \frac{\mathscr{H} - \overline{w\theta}}{\kappa} \tag{67}$$

For the moment, let us proceed as if we knew β and consider equations (64) to (66), with the analogy to stability theory in mind. In this case, it is even possible to prove that the principle of the exchange of stabilities holds for this more general system with arbitrary β and the boundary conditions we have been using (Spiegel,

1962b). (However, the principle has not been established for the more realistic conditions which call for the vanishing of horizontal motion on the boundaries, and this feature of the problem makes the free-boundary condition attractive.)

As before, the equations are separable, and an eigenvalue equation of the form of equation (44) results. Our procedure will be to assume that the eigenfunctions of this equation form a complete set and to write these functions, as in the discussion of stability theory, as

$$U_i^{(l,m,n,j)}(x, y, z) = e^{ilx + imy}\Omega_i^{(l,m,n,j)}(z) \qquad (i = 1, 2, 3, 4) \tag{68}$$

(The four components of U_i refer to the three velocity components and the temperature fluctuation.) The general velocity–temperature field can then be expressed by an expansion of the form

$$U_i(\mathbf{x}, t) = \sum_{j=-1}^{+1} \sum_{n=1}^{\infty} \int_{-\infty}^{+\infty} dl \int_{-\infty}^{+\infty} dm A_i^{(l,m,n,j)}(t) U_i^{(l,m,n,j)}(\mathbf{x}) \tag{69}$$

Substitution of this expansion into the full equations yields an equation of the form

$$\left[\frac{\partial}{\partial t} + \eta_j(l, m, n)\right] A_i^{(l,m,n,j)} = \sum_{\substack{l',m',n',j' \\ l'',m'',n'',j''}} P_{ipq} A_p^{(l',m',n',j')} A_q^{(l'',m'',n'',j'')} \tag{70}$$

The nonlinear term on the right-hand side is indicated only schematically here. It is, as before, a bilinear term in which the two factors contribute amplitudes from all wave numbers and phases.

Although equation (70) is of the same general form as equation (55), it is much more complicated in detail. First, we no longer have the selection rule that only modes whose wave vectors form a triangle can interact; here the couplings are more general. Moreover, P and η are both functionals of β, which is not really known at this stage. It may then be asked, why should one prefer a representation of this kind in favor of the Fourier representation? The answer is that if we had used the Fourier representation, the left-hand side of equation (70) would be much more complicated than it is here. In particular, the term βw in the temperature equation would give rise to a convolution. The point is that to gain simplicity in the terms linear in fluctuating quantities we have selected a representation which makes the fluctuation interactions complicated. The real advantage of this appears when we confront the non-Boussinesq case and the general equations all contain variable coefficients. In that case, if we use the present kind of representation, the equation for the amplitude A will take the same form as equation (66). Hence, the formulation given here seems to be generalizable to non-Boussinesq convection, and that is a prime advantage for astrophysical work. It should also be pointed out that since the fluctuation interactions are to be approximated anyway, it seems advantageous to cast the other terms in a form in which they, at least, can be treated simply.

Suppose we now formulate an equation for the power spectrum

$$\mathscr{E}(l, m, n, j) = \langle A_i^{(lmnj)} \tilde{A}_i^{(lmnj)} \rangle \tag{71}$$

where $\tilde{A}$ is the adjoint solution of equation (44). Since the motion is statistically steady, equation (70) leads to an equation of the form

$$2\eta_j(lmn)\mathscr{E}(l, m, n, j) = \mathscr{T}(l, m, n, j) \tag{72}$$

where $\mathscr{T}$ is trilinear in the A and represents the net transfer from (l, m, n, j) to all other modes. In order to solve this equation we should now introduce some approximation for $\mathscr{T}$, as is done in the theory of isotropic turbulence, and this is the point of the program. But even if such an approximation were simple to handle, the problem would still be difficult, because we do not know β, and hence η, a priori. The equations therefore must be solved iteratively; this is a formidable task, since the number of modes is so large. Progress in this direction has been accordingly slow.

Although equation (72) is difficult to solve, its meaning is quite clear, and it suggests a convenient picture of the dynamics of turbulent convection. It states simply that the net energy fed into a given mode by buoyancy and dissipation must be balanced, on the average, by its energy exchange with all other modes.

Let us consider first the η_+ modes. These are of two kinds, those with $\eta_+ > 0$, which we shall call "unstable," as in stability theory, and those with $\eta_+ < 0$, which we shall call "stable." The unstable η_+ modes gain energy from the potential energy resident in the instability through the action of the buoyancy forces. To maintain the energy balance, the energy must be drained to other modes. The energy drain occurs through the nonlinear transfer, which can be thought of as two separate processes. First, there is a degradation of scale, which is just the usual cascade of isotropic turbulence. In this process, some of the energy from the unstable η_+ modes goes to the stable η_+ modes, which, by virtue of their small scale, lose the energy through dissipative processes. The fluctuation interactions also have the important effect of shifting the phase of the modes and thus generating η_- modes. (The phase mixing also generates η_0 modes, but we shall omit these to simplify this account.) The stable η_+ modes and the η_- modes all have negative η and are damped by the linear terms. They owe their existence solely to excitations by the nonlinear interactions with unstable η_+ modes. In Figure 4, I have drawn a flow diagram for the energy, indicating fluctuation interactions by curved arrows.

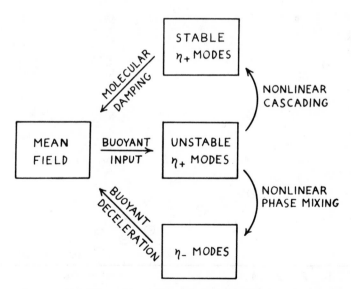

Figure 4. Energetics of modes in convective turbulence. Net energy transfers are indicated by the arrows. The curved arrows represent the fluctuation interactions.

On the basis of this qualitative discussion of the energy balance, we can sketch the power spectrum of the turbulent velocity–temperature field. In doing this, we may refer to Figure 2, showing the dependence of η_+ on n. We notice that there is a value n_* such that $\eta \lesssim 0$ for all $n \geq n_*$. Hence, only η_+ modes with $n < n_*$ are directly excited by buoyancy forces, and these will have largest amplitude. However, η_- modes will also have nonzero amplitude for $n < n_*$ because of fluctuation interactions. For $n \geq n_*$, η_+ and η_- modes will have rapidly diminishing amplitudes for increasing n, but there will still be finite excitation. Since temperature and velocity fluctuations at high wave number are excited by separate interactions (i.e., by $\mathbf{u} \cdot \nabla \mathbf{u}$ and $\mathbf{u} \cdot \nabla \theta$), they can be excited with the same or the opposite sign with almost equal frequency. In other words, little phase information will be transferred from mode to mode. This qualitative remark is strengthened by the localness of the nonlinear transfer; phase information is not likely to be transferred over any great distance in (l, m, n)-space. It seems likely, therefore, that η_+ and η_- modes will have comparable amplitudes for $n > n_*$. In Figure 5, we illustrate these remarks with a plot of the power spectra as a function of n. (Figure 5 is illustrative and does not represent the results of actual computation.)

Similar remarks may be made about the heat-transfer spectrum of steady-state thermal turbulence. There is, however, the difference that because of their phase, the η_- modes convect heat in the downward direction; their presence therefore causes diminution in the convective heat transfer. In this way, the fluctuation interactions diminish convective heat transfer by exciting η_- modes.

Figure 6 shows a schematic plot of the convective transfer in each mode as a function of n. The curve of net heat transfer shown is the linear combination of the transfers in the two kinds of modes. Since, for $n \geq n_*$, the amplitudes of η_+ and η_- modes tend to be nearly equal, the net heat transfer drops toward zero for $n \geq n_*$.

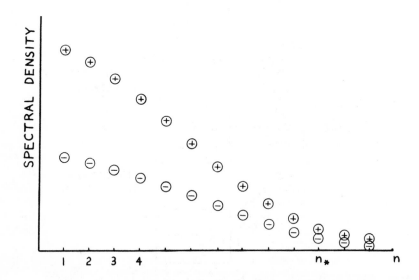

Figure 5. The spectral density $\mathscr{E}$ of modes of varying n for fixed k, R, and σ in turbulent convection. $\oplus$ represents η_+ modes; $\ominus$ represents η_- modes.

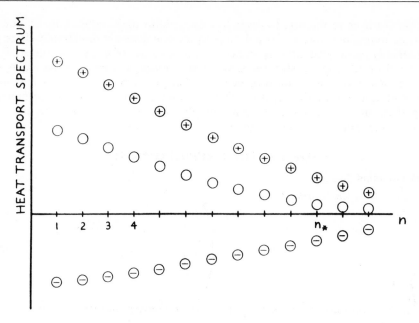

Figure 6. The heat transport for different vertical modes for fixed k, R, and σ. $\oplus$ represents η_+ modes; $\ominus$ represents η_- modes; $\bigcirc$ represents the sum of transports of η_+ and η_- modes.

With this general picture in mind, we can restate the program for treating the problem of convective turbulence. We begin by assuming a form for β and computing the η and the corresponding eigenfunctions. Then, with a suitable approximation for the fluctuation interactions, we can attempt to compute relative amplitudes of the modes. With these we can then compute a new β from equation (63), which puts us in a position to repeat the process until it converges and completes the prescription of the procedure.

Having stated the program, we must point out some possible reservations. The most obvious is the question of convergence of the procedure. It seems clear that unless the first guess for β is reasonably accurate, a large amount of computation may be required; indeed, the iteration may not even converge. However, as we shall see in the next section, certain cases can be already handled with reasonable accuracy, and by moving slowly away from these, we may hope to avoid convergence difficulties. But even more worrisome is the question of ultimate accuracy. In this respect, as I mentioned, success of Kraichnan's approximation in the isotropic case is cause for optimism; indeed, it seems hopeful that even cruder approximations may suffice for some purposes. Perhaps the most sensitive question regarding accuracy is that of the effect of treating only a limited number of modes, which is certainly necessary in any calculation. Our choice of representation was made with this limitation directly in mind, and we are hopeful that in this representation a relatively small number of modes will suffice. Indeed, if this is true, it may even be possible to handle the fluctuation interactions numerically without approximation, even though this did not seem possible in the case of isotropic turbulence.

Of course, this discussion provides nothing more than a prospectus, but it is one which is becoming more and more possible. Indeed, the program as formulated here carries over to the non-Boussinesq case without major alteration. Of course, in the compressible case, we would have to include acoustic modes as well as the ones already mentioned, and these have complex values of η. Even in the study of convection in a pulsating star, the same kind of discussion can also be made, but now the pulsational modes must be included. The problem in that case is analogous to that of studying nonlinear, nonradial pulsations, and may also be formulated along these lines, with the additional complication that pulsational modes have time-periodic amplitudes and complex η.

SOME FURTHER APPROXIMATIONS

Weak-Coupling Theory

In terms of the four-vector

$$
U = \begin{pmatrix} u \\ v \\ w \\ \theta \end{pmatrix}
\tag{40}
$$

the general equations for fluctuating quantities in convection theory are of the form

$$
\frac{\partial U_i}{\partial t} + H_{i\mu}U_\mu = -\lambda[B_i\{U_i\} - \overline{B_i\{U_i\}}]
\tag{73}
$$

where B_i is a bilinear functional of U_i representing the fluctuation interactions and H is the integrodifferential operator discussed earlier. We should recall that H contains buoyancy, viscous, and conduction terms, but these are expressed for the fully turbulent state in which the temperature gradient is

$$
\beta = \frac{\mathcal{H} - U_3 U_4}{\kappa}
\tag{74}
$$

Thus, as before, the nonlinearity due to distortions of the mean state by turbulence is contained in β and the explicit fluctuation interactions are represented by B_i.

The coefficient λ in equation (73) is actually unity, and is introduced there as a fictitious coupling constant. If we were to nondimensionalize equation (73), λ would then actually appear as some real nondimensional parameter, such as a Reynolds number of the fluctuations. In the present statement, λ is introduced as some ordering parameter in preparation for a perturbation expansion for small fluctuation interaction. But why should such an expansion seem reasonable? There are actually several reasons, and I should like to list a few here.

1. In isotropic turbulence theory, a similar expansion in Reynold's number has been attempted, which leads to a formal-series solution of the problem. The formal series probably is not convergent, or at best converges extremely slowly. Nevertheless, by regrouping terms in this series, one can recover the results of more exact treatments such as Kraichnan's (Kraichnan, 1961, and Wyld, 1961).

2. The fluctuation interactions are the ones which make the problem indeterminate, and so must be approximated. An expansion in coupling constant seems to be the simplest approximation possible; this is also suggested by other branches of physics, such as many-body theory. (I have already suggested the analogue to the

Vlasov equations.) Indeed, the general formulation in the previous section is based on the notion that the detailed nature of the fluctuation interactions are, in some sense, of secondary importance, and, as already pointed out, this reasoning underlies the current formulation of mixing-length theory.

3. Malkus (1954b) has put forward the hypothesis that in convection the heat transport is maximum, subject to the constraints implied by the equations of motion. Although his calculations based on this hypothesis are somewhat inconsistent (Spiegel, 1962b), he has obtained results which bear a striking resemblance to those obtained in certain of the existing experiments. Moreover, Howard (1963) has shown that it is actually possible to find upper bounds on the heat transport which are in fair agreement with laboratory results for $\sigma \gtrsim 1$. But the discussion in the past section indicated that the fluctuating interactions tend to diminish heat transport. Hence, an expansion about zero fluctuating interactions would be expected to at least duplicate the successes of the maximum heat transport theories.

I shall not go into any of the details of the λ-expansion here, but I do want to indicate some features of the zero-th order solution. If we write

$$\left.\begin{array}{l} U_i = U_i^{(0)}\lambda^0 + U_i^{(1)}\lambda^1 + \cdots \\[2mm] \beta = \beta^{(0)}\lambda^0 + \beta^{(1)}\lambda^1 + \cdots \end{array}\right\} \quad (75)$$

etc.

and substitute these expansions into equations (73) and (74) and the continuity equation, we obtain for the lowest-order terms

$$\frac{\partial U_i^{(0)}}{\partial t} + H_{ij}^{(0)} U_j^{(0)} = 0 \tag{76}$$

$$\frac{\partial U_\alpha^{(0)}}{\partial x_\alpha} = 0 \tag{77}$$

and

$$\beta^{(0)} = \frac{\mathscr{H}^{(0)} - \overline{U_3^{(0)} U_4^{(0)}}}{\kappa} \tag{78}$$

These are just equations (64) to (67), which formed the basis of our discussion of turbulent dynamics. However, they are to be treated differently here. In the previous section, the program suggested was to: (1) use equations (76) and (77) to find $U^{(0)}$ for an assumed $\beta^{(0)}$, (2) obtain relative amplitudes using an approximation to the fluctuation interactions, and then (3) compute a new $\beta^{(0)}$ with equation (78). But here, step (2) is to be omitted, and the joint solution $U_i^{(0)}$ and $\beta^{(0)}$ can be found without further approximation (in principle). Once these zero-th order equations are solved, they can be used to solve the first-order equations, which are only slightly more difficult, and so on. But the solution to equations (76) to (78) already poses a difficult problem.

From our previous discussion, we know that $U_i^{(0)}$ will have an infinite number of solutions of the form

$$U_i^{(l,m,n,j)} = |lmnj\rangle = e^{\eta t + ilx + imy}\Omega_i^{(l,m,n,j)}(z) \tag{79}$$

and some appropriate linear combination of these must be selected. Moreover, equation (78) must be satisfied and, since we are interested only in statistically steady flows, β must be independent of time. This last condition permits a considerable simplification. For, if we take an arbitrary linear combination of solutions of the form (79), the one with the largest positive value of η will dominate after sufficient time elapses. (As we saw before, only real values of η need be considered.) Therefore, no solutions with positive η can be included. Moreover, solutions with negative η will die away, since there is nothing to keep them excited. Thus, only solutions with $\eta = 0$ need be considered, and the only ones which satisfy this condition are marginally stable η_+ modes.

From Figures 1 and 2, we can get some idea of which modes must be included in this restricted solution, based on marginally stable modes. We see that, in general, for a given value of R, there are at most two values of a ($= kd$) which can be admitted for each n up to n_*, where $n_* - 1$ is the highest value of n which can be unstable. Thus, we can expect to have a general solution which is a linear combination of up to $2(n_* - 1)$ wave numbers. Moreover, the majority of solutions will be unstable, since R, in general, will be greater than R_c. (It must be kept in mind that the R_c determined here is not the value obtained in stability theory, since R_c is a function of β in this nonlinear case.) Thus, only solutions with $R = R_c\{\beta\}$ are stable to infinitesimal disturbances, and it is likely that only one such solution exists. Such a solution has only one term of the form of equation (79). There is also the interesting feature that steady-state solutions of equations (76) to (78) do not involve the Prandtl number, hence the λ^0-solutions are Prandtl-number independent.

Recently a very exciting contribution to the problem has been made by Herring (1963, 1964), who has managed to solve equations (76) to (78) numerically by integrating the time-dependent equations forward in time. He has determined the steady solution by calculating the evolution of a system which initially has more than one mode and finds that, for the boundary conditions we have been using, there is always a single mode which emerges victorious and, whenever this one is included initially, the system evolves to a steady state which includes just this mode. By repeating the calculation for all Rayleigh numbers, the general solution can thus be constructed numerically.

The temperature profiles which Herring has found from his solutions are similar to those which have so far been obtained by experiment, but they show some bumps representing negative temperature gradients which seem to be unrealistic. The total heat transfer obtained is proportional to $R^{\frac{1}{3}}$ for large R, which also agrees with experiments. Unfortunately, as I have already mentioned, no aspect of the steady solution depends on the Prandtl number, whereas our preliminary discussions have already shown that the correct heat transport does depend on Prandtl number. It is interesting, however, that Herring's heat transports show reasonable agreement with those determined experimentally at high Prandtl number.

Herring's solutions are really quite interesting as an example of an approximate solution which seems to have some bearing on experimental results. They show very clearly the value of separating out the fluctuation interactions and perturbing about the fully developed state. But a number of questions remain unanswered. In particular, the physical reason for the Prandtl-number restrictions of the solutions is not clear, and it would be of interest to compute the next order terms in the λ-expansion to see how these modify the picture.

Very Low Prandtl Number

The Prandtl number in the sun is about 10^{-9}. The reason for this small value is that the kinematic viscosity in the sun is primarily due to atomic processes (radiative viscosity being negligible), while the conductivity is overwhelmingly radiative for large-scale motions. Both viscosity and conductivity are, dimensionally, a length times a speed. So for stars in general, the Prandtl number is

$$\sigma = \frac{c_S l_A}{c_L l_p} \tag{80}$$

where c_S and c_L are the speeds of sound and light, respectively, and l_A and l_p are atomic and photon mean free paths. Hence, the small value of σ in the sun is typical for all stars. This suggests that for the purposes of astrophysics, it is of interest to investigate the case of low Prandtl number (e.g., Ledoux, Schwarzschild, and Spiegel, 1961).

One way to approach the low Prandtl-number limit is to put the equations of motion in nondimensional form, so that the Prandtl number and Rayleigh number appear explicitly. Then, all quantities in the equations can be expanded in Taylor series in σ. The leading terms then give the equations for low Prandtl number (Spiegel, 1962a). In this limit, the expression for the total heat flux becomes

$$N = \frac{\mathscr{H}}{\kappa \beta_0} = \frac{\beta}{\beta_0} + 0(\sigma^2) \tag{81}$$

where $\beta = \Delta T / d$. Thus, in this limit, the flux is purely conductive, which means that β must be constant over the fluid. The normal modes we discussed earlier are then appropriate combinations of trigonometric functions.

The heat equation (26) reduces to

$$\kappa \nabla^2 \theta = \beta w \tag{82}$$

where the quantities are here dimensionful. The fluctuation interaction $\mathbf{u} \cdot \nabla \theta - \overline{\mathbf{u} \cdot \nabla \theta}$ has therefore disappeared, which means that there is no eddy conductivity and, as we saw, no convective heat flux. The momentum and continuity equations (25) and (27) are unaltered, however. Thus, the only nonlinear effect left in the low Prandtl-number limit is the fluctuation interaction, $\mathbf{u} \cdot \nabla \mathbf{u}$ (the eddy viscosity). The problem is therefore reduced to that of homogeneous, anisotropic turbulence.

We can see at once why the weak-coupling theory of the last section cannot apply here. In this low Prandtl-number limit, the fluctuation interaction $\mathbf{u} \cdot \nabla \mathbf{u}$ is the only nonlinear term, while in the weak-coupling expansion, it should be secondary to the distortions of β. It also follows that, in this low σ case, any results obtained are quite sensitive to the approximation used for the fluctuation interaction, again, since it is the only nonlinear term. On the other hand, since $\mathbf{u} \cdot \nabla \theta - \overline{\mathbf{u} \cdot \nabla \theta}$ is not important, there is no phase mixing and η_- modes may be omitted. This simplifies the calculations, and it was possible to obtain some preliminary results using the approximations from Heisenberg (1948) and Kovasznay (1948) in the two papers I referred to. With these, the next-order corrections to the heat transport can be estimated, and we obtain

$$N = 1 + \gamma \left(\frac{\Lambda}{R_c}\right)^2 + \cdots = 1 + \left(\frac{\gamma R^2}{R_c^2}\right)\sigma^2 + \cdots \tag{83}$$

where γ is a pure number whose value depends on the spectrum of the convective turbulence and therefore on the approximation introduced for the eddy viscosity.

The low Prandtl limit is valid only if the second term on the right in equation (83) is negligible, that is, if

$$\Lambda = \sigma R \ll \frac{R_c}{\gamma^{\frac{1}{2}}} \equiv R_* \qquad (84)$$

Earlier, we have seen that if $\sigma \ll 1$, then N should depend only on Λ, and that we must distinguish two cases: low Prandtl number ($\sigma \ll 1$) and very low Prandtl number ($\Lambda \ll R_*$). The latter case is the one we have treated here and, in this, convective heat transfer is negligible.

For the edge of the convection zone in the sun, with the local scale height as a characteristic length $\Lambda \approx 10^3 \approx R_c$, so convective transport is only marginally important, since $\gamma^{\frac{1}{2}} \approx 1$. This is just the condition which we would expect to define the edge of an important convective region. Of course, inside the solar convective zone, $\Lambda \gg R_c$, and our very low Prandtl-number theory is just not valid in the solar convection zone. In a B0 star, on the other hand, convection occurs but never disturbs the radiative equilibrium (Traving, 1955), since the B0 star satisfies the very low Prandtl-number limit. In that case, we can expect convective instability and turbulence, but convective transfer is completely negligible. Another way to look at it is to compute P_e from our solution. We find that $\Lambda \ll R_*$ implies $P_e \ll 1$, where P_e is the Peclet number mentioned earlier. As we saw then, this implies negligible convective flux.

In the case $\Lambda > R_*$ but $\sigma \ll 1$, we can, as we saw, expect that $\mathcal{H}$ depends only on Λ. Since convective transfer can now be important, we expect boundary layers to form, and from equation (51) we conclude that

$$N = \left(\frac{\Lambda}{\Lambda_0}\right)^{\frac{1}{3}} \qquad (85)$$

where Λ_0 is a constant. But as Λ decreases below R_*, N must approach the value given in equation (83). This permits us to estimate Λ_0 by insisting that equations (85) and (83) give the same value of N at $\Lambda = R_*$. We obtain, for $\gamma \approx 1$,

$$\Lambda_0 = \tfrac{1}{8} R_* = \frac{R_c}{8\gamma^{\frac{1}{2}}} \approx \begin{cases} 80, & \text{free boundaries} \\ 210, & \text{rigid boundaries} \end{cases} \qquad (86)$$

as a very crude estimate of the transport coefficient for low Prandtl number, which is nevertheless close to the measured value for Mercury (Globe and Dropkin, 1959; see also the mixing-length estimates of Kraichnan, 1962).

High Prandtl Number

Having seen that the theory of convection can be simplified by looking at the low Prandtl-number limit, we are naturally led to look into the opposite limit, high Prandtl number. But, the interest in this case is only that it may be illuminating as regards the general theory, whereas the low Prandtl-number case was prompted by direct astrophysical interest.

The equations may be derived in a way analogous to that described for the low Prandtl-number case. The difference here is that the simplification now occurs in equation (25), the momentum equation. This becomes

$$-\nu\nabla^2\mathbf{u} - \mathbf{g}\alpha\theta + \nabla\varpi = 0 \tag{87}$$

while all the other equations remain unaltered. We can combine this simplified equation with equation (27) to obtain

$$\nu\nabla^4 w = -g\alpha\nabla_1^2\theta \tag{88}$$

The point here is that turbulent viscosity $\mathbf{u}\cdot\nabla\mathbf{u}$ is not important and, as for low σ, there is only one fluctuation interaction $\mathbf{u}\cdot\nabla\theta - \overline{\mathbf{u}\cdot\nabla\theta}$. But the situation differs from that of low σ in that convective heat transport can be important and so the temperature profile can be highly nonlinear. Hence, the normal modes are quite complicated, and the full nonlinear problem is more difficult than the low σ case.

We can attempt to get some idea of the kind of behavior to expect in this limit with some simple approximations. Suppose we replace the fluctuation interaction term in equation (26) by an eddy conductivity term $\kappa_e\nabla^2\theta$. Then equation (26) becomes

$$\frac{\partial\theta}{\partial t} = \kappa(1 + q)\nabla^2\theta - \beta w = 0 \tag{89}$$

where $q = \kappa_e/\kappa$, and where, for simplicity, I am considering only liquids and thus have omitted $(dT/dz)_{\text{AD}}$. (We might as well do this, since there are no gases with high σ.)

Now q is, at the moment, an unknown parameter, and this begins to sound like the usual drawback of mixing-length theory. However, we have an additional possibility, since we expect that if the eddy conductivity idea is to make sense, then the convective heat transfer should be given by

$$\overline{w\theta} = \kappa_e\beta \tag{90}$$

Thus, with equation (24), we find

$$1 + q = \frac{\mathscr{H}}{\kappa\beta} \tag{91}$$

and equation (89) becomes

$$\frac{\partial\theta}{\partial t} - \frac{\mathscr{H}}{\beta}\nabla^2\theta - \beta w = 0 \tag{92}$$

Equations (88) and (92), together with equation (24),

$$\mathscr{H} = \kappa\beta + \overline{w\theta} \tag{24}$$

now define a determinate set of equations for a mean velocity and temperature field. They resemble the weak-coupling equations in that the only statistically steady solutions are marginally stable on the mean temperature profile. But in the steady state, the Prandtl number drops out of the problem, and this is in accord with experimental results showing $\mathscr{H}$ to be very insensitive to σ for large σ and with mixing-length estimates (Kraichnan, 1962).

The approximation used here for the eddy conductivity is crude, but is useful in bringing out the result that as $\sigma \to \infty$, $\mathscr{H}$ approaches some asymptotic value independent of σ. The experimental asymptotic value is not far below the upper bound on the allowed heat transport established by Howard (1963), and is also close to the value Herring obtained from his solution of the weak-coupling approximation. Thus, it appears that we are in a position to calculate many of the gross features of high Prandtl-number convection with good accuracy. On the other hand, details of the turbulent flow can not yet be obtained in this way, since the weak-coupling equations, given in this section, and Howard's equations, though all different, give only steady nonturbulent solutions. In spite of this, it is quite tempting to try to take advantage of the success of high Prandtl-number theory for the stellar case. One possibility is the following (see also Unno, 1961).

Suppose we consider the full, non-Boussinesq equations for convection in a thick convection zone and formally carry out the weak coupling expansion for the procedure of this section. We then obtain a set of equations which can be solved with a reasonable amount of effort. From what we have seen above, it seems safe to assume that these solutions are a good representation of high Prandtl-number convection in a compressible atmosphere. We can then also use the mixing-length equations to treat this high Prandtl-number case and choose the mixing length to make the two calculations agree. Thus, it is possible to tabulate the mixing length for a variety of effective temperatures and surface gravities.

It is generally supposed that the optimum choice for the mixing length is of the order of the largest scale consistent with the geometry of the system. (In a stellar convection zone, this limit is generally taken to be a scale height.) If this is true, then the mixing length should not be sensitive to such microscopic parameters as the Prandtl number. Hence, we should be able to use the mixing lengths found in the manner just outlined, for low σ as well.

There are, of course, flaws in this reasoning. The chief one is that the nature of the flow at high and low σ may be quite different. But this objection may not be serious if the characteristic lengths are the same in both cases, and we have no indication that they would not be. At any rate, this provides one method of taking advantage of the recent developments in convection theory. But it is only a crude sample, and I believe that one can now hope to do much better within a very few years.

ACKNOWLEDGMENT

I should like to thank Dr. D. W. Moore for his comments on the manuscript of this talk.

REFERENCES

G. K. Batchelor (1953), *The Theory of Homogeneous Turbulence,* Cambridge University Press, England.
L. Biermann (1942), *Ztschr. Astrophys.* **21**: 339.
H. Boussinesq (1903), *Théorie analytique de la chaleur, Vol. 2,* Gauthier-Villars, Paris, p. 172.
S. Chandrasekhar (1951), *Proc. Roy. Soc.* A **210**: 18.
S. Chandrasekhar (1961), *Hydrodynamic and Hydromagnetic Stability,* Clarendon Press, Oxford, England.
S. F. Edwards (1964), *J. Fluid Mech.* **18**: 239.
S. Globe and D. Dropkin (1959), *J. Heat Transfer* **81**: 24.
H. L. Grant, R. W. Stewart, and A. Moilliet (1962), *J. Fluid Mech.* **12**: 241.
W. Heisenberg (1948), *Ztschr. Phys.* **124**: 628.
J. R. Herring (1963), *J. Atmos. Sci.* **20**: 325.
J. R. Herring (1964), *J. Atmos. Sci.* **21**: 277.
L. N. Howard (1963), *J. Fluid Mech.* **17**: 405.
M. Jakob (1949), *Heat Transfer,* John Wiley and Sons, New York.
H. Jeffreys (1930), *Proc. Cambridge Phil. Soc.* **26**: 170.

L. Kovasznay (1948), *J. Aeronaut. Soc.* **15**: 745.

R. H. Kraichnan (1957), *Phys. Rev.* **107**: 1485.

R. H. Kraichnan (1959), *J. Fluid Mech.* **5**: 497.

R. H. Kraichnan (1961), New York University, Institute of Mathematical Sciences, Division of Electromagnetic Research, Report No. HSN-3.

R. H. Kraichnan (1962), *Phys. Fluids* **5**: 1374.

R. H. Kraichnan and E. A. Spiegel (1962), *Phys. Fluids* **5**: 583.

Landau and Lifshitz (1959), *Fluid Mechanics, Vol. 8*, Pergamon Press, London.

P. Ledoux, M. Schwarzschild, and E. A. Spiegel (1961), *Astrophys. J.* **133**: 184.

W. V. R. Malkus (1954a), *Proc. Roy. Soc. Astron.* **225**: 185.

W. V. R. Malkus (1954b), *Proc. Roy. Soc. Astron.* **225**: 196.

W. V. R. Malkus and G. Veronis (1958), *J. Fluid Mech.* **4**: 225.

J. M. Mihaljan (1962), *Astrophys. J.* **136**: 1126.

M. Millionshchikov (1941), *Dok. Akad. Nauk. Uz. SSR* **32**: 615.

E. J. Öpik (1950), *Monthly Notices Roy. Astron. Soc.* **110**: 559.

Pellew and Southwell (1940), *Proc. Roy. Soc.* A **176**: 312.

A. F. Pillow, "The Free Convection Cell in Two Dimensions," NACA Document No. 16464.

C. H. B. Priestley (1959), *Turbulent Transfer in the Lower Atmosphere*, University of Chicago Press, Chicago, Illinois.

I. Proudman (1962), "Mechanique de la turbulence," *Centre Nat. Rech. Sci.*, Paris, p. 107.

I. Proudman and W. H. Reid (1954), *Trans. Roy. Soc. London* **247**: 163.

Lord Rayleigh (1916), *Phil. Mag.* **32**: 529; *Collected Papers* **6**: 432.

E. A. Spiegel (1960), *Astrophys. J.* **132**: 716.

E. A. Spiegel (1962a), *J. Geophys. Res.* **67**: 3063.

E. A. Spiegel (1962b), "Mechanique de la turbulence," *Centre Nat. Rech. Sci.*, Paris, p. 181.

E. A. Spiegel (1963), *Astrophys. J.* **138**: 216.

E. A. Spiegel and G. Veronis (1960), *Astrophys. J.* **131**: 442.

A. A. Townsend (1959), *J. Fluid Mech.* **5**: 209.

G. Traving (1955), *Ztschr. Astrophys.* **36**: 1.

W. Unno (1961), *Proc. Astron. Soc. Japan* **13**: 276.

A. Unsöld (1930), *Ztschr. Astrophys.* **1**: 138.

E. Vitense (1953) *Ztschr. Astrophys.* **32**: 135.

H. W. Wyld, Jr. (1961), *Ann. Phys.* **14**: 143.

NEUTRINOS IN ASTROPHYSICS

Hong-Yee Chiu

INTRODUCTION

Traditionally, astrophysics has dealt with the interaction of photons and electrons, and only in recent years has nuclear physics been taken seriously. As physics progresses, new particles and new interactions must be included in astrophysical considerations. Recently, it has been proven that the weak interaction has important consequences in stellar processes and possibly also in cosmic processes.

Feynman and Gell-Mann (1958) suggested a theory of weak interactions in which the interaction is caused by the self-interaction of a current J with itself:

$$H = J^*J \tag{1}$$

where the current J is given by

$$J = G[\psi_e(1 + i\gamma_5)\gamma_\nu\psi_{\nu_e} + \psi_p(1 + i\gamma_5)\gamma_\nu\psi_n + \psi_\mu(1 + i\gamma_5)\gamma_\nu\psi_{\nu_\mu}] \tag{2}$$

and J^* is the Hermitian conjugate of J, G is the weak-interaction coupling constant, and

$$\frac{GM_p^2}{\hbar c} = 10^{-5}(\pm 1\%) \tag{3}$$

ψ_x is the wave function for particle x, γ_ν is the Dirac matrices, and $\gamma_5 = i\gamma_1\gamma_2\gamma_3\gamma_4$.

Symbolically, we can write

$$H = J^*J = (e_{\nu_e})(e_{\nu_e}) + (pn)(e_{\nu_e}) + (\mu\nu_\mu)(e_{\nu_e}) + \cdots \tag{4}$$

where each term represents a combination that will appear in a possible reaction, provided that all conservation laws (such as the lepton conservation law, etc.) are satisfied. For example, the term $(pn)(e_{\nu_e})$ will give rise to

$$p + e^- \rightarrow n + \nu_e$$
$$n \rightarrow p + e^- + \bar{\nu}_e \tag{5}$$

Both reactions have been observed (the first one is the inverse beta-decay process, the second one is the ordinary beta-decay process of a neutron into a proton, an electron, and a neutrino). The term $(\mu\nu_\mu)(e\nu_e)$ represents μ-decay and $\mu\nu_e$-scattering $(\mu + \nu_e \rightarrow e + \nu_\mu)$ (not observed yet). The term that is of the greatest astrophysical importance is the $(e\nu_e)(e\nu_e)$ term.

Although there are many versions of the weak-interaction theory, none of them has provided a simpler description of the experimental phenomena than the Feynman and Gell-Mann theory, which we will use throughout this paper. The results of some of the processes we describe would not be the same if another theory of weak interaction was used. Our choice of the Feynman and Gell-Mann theory is not based on the simplicity of the theory (since it has more than once been proven

175

that nature is not simple), but because of its close agreement with experimental facts and its sound physical basis. Many of the more subtle features of the Feynman and Gell-Mann theory have been demonstrated to be correct; among them are the prediction of the rate of the decay of a π-meson into an electron and a neutrino (Anderson and Lattes, 1957), and the prediction that the current J possesses certain conservation properties analogous to those of the electromagnetic current (Lee, Mo, and Wu, 1963). From a theoretical point of view, the current–current inter-action hypothesis can be demonstrated to be a consequence of another hypothesis —the postulated existence of an intermediate boson (denoted by W) (Lee and Yang, 1960). Experimental searches for W have not been successful; this is probably because the mass of the W-particle is very large ($\gtrsim 2$ BeV).

At the time of writing, no experimental evidence exists indicating either the presence or the absence of the term $(ev_e)(ev_e)$ in nature, although it is currently believed that this term must exist. The reason that the $(ev_e)(ev_e)$ process is important in astrophysics is obvious. Inside stars, the heavier nucleons provide the source of gravitational field for binding, while the lighter electrons interact more freely with matter and radiation; through the medium of electrons, a photon can be converted into a neutrino. There are three neutrino processes that are most important in ordinary stars:

$$\gamma + \gamma \rightleftarrows e^- + e^+ \to v_e + \bar{v}_e \quad \text{(pair-annihilation process)}$$

$$\gamma \text{ (plasmons)} \to v_e + \bar{v}_e \quad \text{(plasma process)}$$

$$\gamma + e^- \to e^- + v_e + \bar{v}_e \quad \text{(photoneutrino process)} \tag{6}$$

In addition, during the collapse phase and inside neutron stars, other neutrino processes (especially the URCA process) may also be important.

In general, neutrino processes are important only in the later stages of stellar evolution, when the internal temperature is above 10^9 °K. Neutrino processes, therefore, need not be included in the discussion of main-sequence stars whose temperature is of the order of 1.5×10^7 °K; however, in the core of red giants, it has been demonstrated that neutrino processes already contribute about 20% of the energy loss (Chiu, 1963). The temperature–density regime where most stars are found can roughly be divided into three domains—in each of which one of the previously mentioned neutrino processes is dominant—as shown in Figure 1. We shall give expressions for the neutrino processes, and also a short table of the total neutrino rates, which will be useful for estimates and for stellar-evolution calcula-tions.

NEUTRINO PROCESSES

Pair-Annihilation Process (Chiu and Morrison, 1960)

When the temperature of a star is close to T_0 ($= 5.95 \times 10^9$ °K) given by

$$mc^2 = kT_0 \tag{7}$$

(all notations have their usual meanings, the cgs system will be used throughout this paper, and m is the electron mass), photons will have sufficient average energy so that in collision processes, electron pairs will be created:

$$\gamma + \text{other particles} \rightleftarrows e^- + e^+ \tag{8}$$

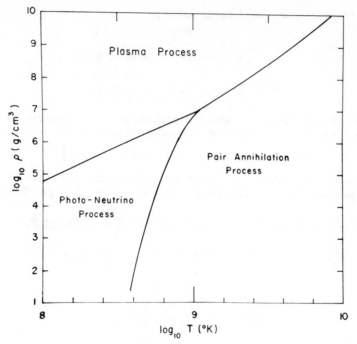

Figure 1. Domains of dominance of the three most important neutrino
processes.

Because the radiation and electron pairs are confined, thermodynamic equilibrium
is established for the above process in a time less than 10^{-12} sec. The Feynman and
Gell-Mann theory predicts that electron pairs can annihilate into neutrino pairs:

$$e^- + e^+ \rightarrow \nu_e + \bar{\nu}_e \tag{9}$$

The cross-section for process (9) in the center-of-mass system is given by

$$\sigma = \frac{G^2 m^2}{3\pi} \frac{[(E_T/mc^2)^2 - 1)]}{(v/c)} = \sigma_0 \frac{[(E_T/mc^2)^2 - 1)]}{(v/c)} \tag{10}$$

where E_T is the total energy of the e^- and e^+, including their rest-mass energy.
The general expression for the cross-section in an arbitrary system was (as far as
we know) first given by Chiu and Stabler (1961). Inside ordinary stars, neutrinos
have a mean free path far exceeding their dimension, hence one does not need to
consider the inverse process to process (9). The case when the mean free path of
neutrinos is small compared with stellar dimensions occurs only during stellar
collapse (Bahcall, 1964a, Bahcall and Frautschi, 1964, and Chiu, 1964).

The total energy-loss rate (in ergs/g-sec) is given by

$$-\frac{dU}{dt} = \frac{1}{\rho} \int n_-(p_-) n_+(p_+) \sigma(E_- + E_+) \, d^3\mathbf{p}_- \, d^3\mathbf{p}_+ \tag{11}$$

where $n_-(p_-)$ and $n_+(p_+)$ are the Fermi distribution functions for electron and
positron, respectively, and E_- and E_+ are the energy of electrons and positrons,

respectively. From statistical mechanics, we have

$$n_-(p_-) = \frac{1}{1 + \exp\left[(E_- - \mu)/(kT)\right]} \frac{2}{h^3} \tag{12}$$

$$n_+(p_+) = \frac{1}{1 + \exp\left[(E_+ + \mu)/(kT)\right]} \frac{2}{h^3} \tag{13}$$

where μ is the chemical potential, which is determined by the temperature T and the density ρ of the medium through the following implicit integral representation:

$$\rho = (\langle A \rangle / \langle Z \rangle) m_p \left[\int n_-(p_-)\, dp_- - \int n_+(p_+)\, dp_+ \right] \tag{14}$$

By transforming n_- and n_+ to a different Lorentz frame, we can use the simple expression equation (10) for σ. The resulting energy-loss rate is (Chiu, 1961a)

$$-\frac{dU}{dt} = \frac{1}{\rho} U_0 \{ 8\langle E_+ \rangle \langle E_-^2 \rangle + 7N_+ \langle E_- \rangle - 2\langle E_-^2 \rangle \langle 1/E_+ \rangle + 5N_- \langle 1/E_+ \rangle$$

$$+ 8\langle E_- \rangle \langle E_+^2 \rangle + 7N_- \langle E_+ \rangle - 2\langle E_+^2 \rangle \langle 1/E_- \rangle + 5N_+ \langle 1/E_- \rangle \} \tag{15}$$

where

$$\langle E_\pm^n \rangle = \frac{h^3}{2} \int_0^\infty n_\pm \frac{E_\pm^n}{mc^2} x_\pm^2 \, dx_\pm \tag{15a}$$

$$N_\pm = \frac{h^3}{2} \int_0^\infty n_\pm x_\pm^2 \, dx_\pm \tag{15b}$$

$$x = \frac{p_\pm}{mc} \tag{15c}$$

$$U_0 = \frac{32\pi^2}{3} (\sigma_0 c)(mc^2) \left(\frac{mc}{h}\right)^6 = 1.88 \times 10^{19} \text{ ergs/cm}^3\text{-sec.} \tag{15d}$$

$(-dU/dt)$ has been tabulated to five decimal places (Chiu, 1961a). The asymptotic formulas for (dU/dt) are

$$\frac{dU}{dt} = \frac{4.3 \times 10^{15}}{\rho} T_9^9 \qquad\qquad T_9 \gg 6 \qquad \text{nondegenerate} \tag{16}$$

$$\frac{dU}{dt} = \frac{4.8 \times 10^{18}}{\rho} T_9^3 \exp\left(-\frac{2T_0}{T_9}\right) \qquad T_9 \ll 6 \qquad \text{nondegenerate} \tag{17}$$

where $T_9 = (T/10^9)\,°\text{K}$. Equation (16) gives a value too high by 50% at $T_9 = 3$, and by 10% at $T_9 = 6$, while equation (17) gives a value too low by 25% at $T_9 = 3$.

The pair-annihilation process is, strictly speaking, a volume process: In the nondegeneracy limit, it is independent of matter present and depends only on the radiation-energy density, which is a function of temperature only. When the density of matter increases, the electrons that are present in matter will depress the pair-creation process (because part of the electron states are occupied), and the product $n_+ n_-$ will decrease. Thus, the energy-loss rate decreases with increasing density. This is a feature common to all neutrino processes except the plasma-neutrino process.

Plasma Process (Adams, Ruderman, and Woo, 1963)

When the degeneracy effect suppresses the pair-annihilation process, the plasma-neutrino process will become important. This is a quite subtle process. A gas composed of free electrons will have a dielectric constant $\varepsilon < 1$; in the limit of small electron density, ε is given by

$$\varepsilon = 1 - \frac{\omega_0^2}{\omega^2} \tag{18}$$

As a result, the relation between the frequency ω and the wave vector k of electromagnetic waves is no longer linear, as is usually given:

$$\hbar\omega = kc \tag{19}$$

but is a complicated function, which in simple cases is given by

$$\hbar^2\omega^2 = \hbar^2\omega_0^2 + k^2c^2 \tag{20}$$

Thus, photons behave as if they have a rest mass $\hbar\omega_0$. For a dense electron gas, the correct expression for the plasma frequency ω_0 is given by

$$\omega_0^2 = 4\pi\rho^2 \int \frac{f(E_p)}{E_p}\left(1 - \frac{1}{3}\frac{p^2c^2}{E_p}\right)^2 d^3\mathbf{p} \tag{21}$$

For a degenerate Fermi sea, this becomes

$$\omega_0^2 = \frac{4e^2 p_F^3}{3\pi E_F \hbar^3} \tag{22}$$

In general, two modes of plasma oscillations, longitudinal and transverse, exist. The dielectric constant is no longer given by the simple expression (18) valid for small electron density, but corresponding to each mode is a dielectric constant denoted by ε^t and ε^l, given by

$$\varepsilon^t = 1 - \frac{4\pi e^2}{\omega^2}\int d^3\mathbf{p}f(E_p)\left[\frac{1}{E_p}\left(1 - \frac{1}{3}\frac{p^2}{E_p^2}\right) + \frac{\omega^2}{4E_p^3}\left(1 - \frac{1}{3}\frac{p^2}{E_p^2}\right)\right.$$
$$\left. + \frac{k^2}{2E_p^3}\left(\frac{p^2}{E_p^2} - 1 - \frac{1}{3}\frac{p^4}{E_p^4}\right) + \frac{k^2 p^2}{\omega^2 E_p^2}\left(\frac{1}{3} - \frac{1}{5}\frac{p^2}{E_p^2}\right)\right] \tag{23}$$

$$e^l = 1 - \frac{4\pi e^2}{\omega^2}\int d^3\mathbf{p}f(E_p)\left[\frac{1}{E_p}\left(1 - \frac{1}{3}\frac{p^2}{E_p^2}\right) + \frac{\omega^2}{4E_p^3}\left(1 - \frac{1}{3}\frac{p^2}{E_p^2}\right)\right.$$
$$\left. + \frac{k^2}{2E_p^3}\left(\frac{2p^2}{E_p^2} - 1 - \frac{p^4}{E_p^4}\right) + \frac{k^2 p^2}{\omega^2 E_p^3}\left(1 - \frac{3}{5}\frac{p^2}{E_p^2}\right)\right] \tag{24}$$

and the relations between ω and ω_0 and k for transverse and longitudinal plasmons are given by

$$(\hbar\omega^t)^2 = (\hbar\omega_0)^2 + k^2c^2 \qquad \text{transverse} \tag{25}$$

$$(\hbar\omega^l)^2 = (\hbar\omega_0)^2 + \alpha k^2c^2 \qquad \text{longitudinal} \tag{26}$$

where α is a constant $\ll 1$.

In general, a particle of zero rest mass cannot decay into two particles, because in such a process both energy and momentum conservation laws cannot be satisfied simultaneously. This is the reason why a free photon cannot decay into two neutrinos

$$\gamma \to v + v \tag{27}$$

Inside an electron gas, because of the relations (25) and (26), a photon behaves as if it has a rest mass $\hbar\omega_0$ and, consequently, the decay of such plasmons into neutrino pairs by the $(ee)(v_e v_e)$ interaction [which is the same as the $(ev_e)(ev_e)$ interaction] is possible. The probability of decay depends on the value of ω_0 (which depends on the Fermi momentum) and the temperature. The plasma-neutrino energy-loss rates (in ergs/g-sec) for longitudinal and transverse plasmons are given by

$$Q_t = \frac{1}{\rho} \frac{g^2 \omega_0^6}{12\pi^4 e^2 c^5} \int_0^\infty \frac{k \, dk}{\exp\left[(\hbar\omega/kT) - 1\right]} \tag{28}$$

$$Q_l = \frac{1}{\rho} \frac{g^2 \omega_0^9}{24\pi^4 e^2} \left(\frac{m_e c^2}{\hbar}\right)^3 \frac{16}{315} \frac{1}{\exp\left[(\hbar\omega_0/kT) - 1\right]}$$

$$= \frac{3.15 \times 10^{20}}{\rho} \left(\frac{\hbar\omega_0}{kT}\right)^9 \frac{1}{\exp\left[(\hbar\omega_0/kT) - 1\right]} \text{ ergs/g-sec} \tag{29}$$

The integrals have been expressed in terms of a modified Bessel function of the second kind (the so-called K_n function, occasionally called the "McDonald function").* In general, Q_l is small compared to Q_t, and we shall neglect it. Q_t can be written in a more convenient form involving only one parameter:

$$Q_t = \frac{1.314 \times 10^{15}}{\rho} T_9^9 x^9 F(x) \text{ ergs/g-sec} \tag{30}$$

where

$$F(x) = \int_0^\infty \frac{\sinh^2(\xi) \cosh(\xi)}{\exp\left[x \cosh(\xi)\right] - 1} \, d\xi \tag{31}$$

and

$$x = \frac{\hbar\omega_0}{kT}$$

$$= 2.366 \times 10^{-4} [1 + 6.413 \times 10^{-5} \rho^{\frac{2}{3}}]^{-\frac{1}{4}} \frac{\rho^{\frac{1}{2}}}{T_9} \tag{32}$$

Table I gives values for $F(x)$ and $x^9 F(x)$. Inman and Ruderman (1964) have given asymptotic formulas for $F(x)$:

---

* In some previous papers by the author, these functions were called the "modified Hankel functions." Fowler (1964) informed the author that Gray (1922) had used the name "modified Bessel functions of the second kind," and that he thought that Hankel might protest that his functions were being modified without his consent.

$$F(x) = \frac{1}{x^3}\left[2\xi(3) + \tfrac{1}{2}x^2 \ln x - \tfrac{1}{4}(2 \ln 2 + 1)x^2 + \tfrac{1}{96}x^4 \ln x\right.$$

$$\left. - \tfrac{1}{96}\left\{\ln 2 - \tfrac{1}{4} + \ln 2\pi - \frac{\xi'(2)}{\xi(2)}\right\}x^4\right]$$

$$= \frac{1}{x^3}[2.40412 + x^2 \ln x(0.5 + 0.01042x^2)$$

$$- x^2(0.59658 + 0.02971)] \qquad x \ll 1 \qquad (33)$$

Table I. $x^9 F(x)$ and $F(x)$

How to find q_t: $\log x^9 F(x)$ and $\log F(x)$ are listed as functions of $\log x$ to five decimal places. In general, interpolations from $\log F(x)$ give a more accurate result if higher-order interpolation formulas are used. The plasma-energy loss rate q_t (ergs/g-sec) is given by

$$\log q_t = 9.118595 - \log \rho_6 + 9 \log T_9 + \log x^9 F(x) \qquad (T\text{-}1)$$

The quantity x is given by

$$\log x = -0.625985 - \log T_9 + 0.5 \log \rho_6 - 0.5 \log (\mu_e/2)$$

$$-0.25 \log [1 + 0.6413\rho_6^{\frac{2}{3}}(\mu_e/2)^{\frac{2}{3}}] \qquad (T\text{-}2)$$

$$x = (\hbar\omega_0/kT) \qquad \rho_6 = (\rho/10^6)\,\text{g/cm}^3 \qquad T_9 = (T/10^9)\,^{\circ}\text{K} \qquad (T\text{-}3)$$

$$\mu_e = \frac{(\text{average mass number})}{(\text{average atomic number})} \qquad (T\text{-}4)$$

Given ρ_6 and T_9, one can find x from equation (T-2). Conversely, given ρ_6 and x, one can also find T_9 from equation (T-2).

log x	log F(x)	log x⁹F(x)	log x	log F(x)	log x⁹F(x)
−2.0	6.380 899	−11.619 101	0	0.249 932	0.249 932
−1.9	6.080 878	−11.019 122	0.1	−0.103 225	0.796 775
−1.8	5.780 835	−10.419 165	0.2	−0.476 423	1.323 577
−1.7	5.480 768	−9.819 232	0.3	−0.876 671	1.823 329
−1.6	5.180 671	−9.219 329	0.4	−1.313 168	2.286 832
−1.5	4.880 534	−8.619 466	0.5	−1.798 057	2.701 943
−1.4	4.580 321	−8.019 679	0.6	−2.347 028	3.052 972
−1.3	4.280 009	−7.419 991	0.7	−2.980 094	3.319 906
−1.2	3.979 530	−6.820 470	0.8	−3.722 414	3.477 586
−1.1	3.678 827	−6.221 173	0.9	−4.605 233	3.494 767
−1.0	3.377 779	−5.622 221	1.0	−5.667 360	3.332 640
−0.9	3.076 239	−5.023 761	1.1	−6.957 185	2.942 812
−0.8	2.774 006	−4.425 994	1.2	−8.535 242	2.264 758
−0.7	2.470 733	−3.829 267	1.3	−10.477 321	1.222 679
−0.6	2.165 987	−3.234 013	1.4	−12.879 131	−0.279 131
−0.5	1.859 168	−2.640 832	1.5	−15.860 183	−2.360 183
−0.4	1.549 396	−2.050 604	1.6	−19.571 314	−5.171 314
−0.3	1.235 504	−1.464 496			
−0.2	0.915 864	−0.884 136			
−0.1	0.588 328	−0.311 672			

and for $x \gg 1$, we have

$$Q_t = \frac{1.64 \times 10^{15}}{\rho} T_9^9 x^{7.5} e^{-x} \qquad x \gg 1 \qquad (34)$$

Equation (33) gives a value of $F(x)$ too small by 20% at $x = 0.5$ and is not suitable for $x > 0.5$. For $x \approx 1$, equation (34) gives a value too low by a factor of 3.

As we have said, the decay of a free photon into two neutrinos is forbidden by conservation laws. In the limit of zero electron density $\omega_0 = 0$, the rate of energy loss by neutrino processes is zero. As the density increases, ω_0 increases and we expect the rate of the plasma-neutrino process to be an increasing function of density until eventually complete degeneracy causes it to decrease again. The maximum occurs at $x = 7.1$ with $x^9 F(x) = 3.1 \times 10^3$, and the corresponding density is considerably higher than that given by the criterion $E_F = kT$ (the usual criterion for degeneracy to take place) where E_F is the Fermi energy. Hence, the plasma process can remove stellar energy rapidly from degenerate matter.

Figure 2 divides the ρT plane into two regions, in which $\hbar\omega_0 > kT$ and $\hbar\omega_0 < kT$, respectively. Plasma-neutrino processes are expected to be important in the neighborhood of this line $\hbar\omega_0 \approx kT$. The interiors of red giants, white dwarfs, and neutron stars are known to fall along this line.

Photoneutrino Process (Chiu and Stabler, 1961, and Ritus, 1961)

When a photon is scattered by an electron, there is a certain probability that the energy of the photon will be converted into a pair of neutrinos

$$\gamma + e^- \rightarrow e^- + \nu_e + \bar{\nu}_e \qquad (35)$$

Innocent as this process looks, the calculation of the energy-loss rate is difficult, and so far only four limiting cases have been calculated. A complication exists in the exact calculation of the energy-loss rate when an electron is present in the final state, because it is necessary to take into account the states that were already occupied by other electrons. In evaluating the integrals analytically, certain simplifications have to be made, and this leads to the four limiting cases: when electrons are relativistic and degenerate, relativistic and nondegenerate, nonrelativistic and degenerate, and nonrelativistic and nondegenerate.

Since the calculations are quite lengthy, we give only the expressions for the limiting cases below; we refer interested readers to the original paper (Chiu and Stabler, 1961, and Ritus, 1961) for detailed calculations:*

$$\frac{dU}{dt} = \frac{10^8}{\mu_e} T_8^9 \qquad \text{(nonrelativistic and nondegenerate)} \qquad (36)$$

$$\frac{dU}{dt} = \frac{2.5 \times 10^{14}}{\mu_e} T_{10}^8 (\log T_{10} + 1.6) \qquad \begin{array}{l} \text{(extremely relativistic and} \\ \text{nondegenerate)} \end{array} \qquad (37)$$

$$\frac{dU}{dt} = 1.5 \times 10^2 T_8 \left(\frac{1}{\mu_e}\right)^{\frac{1}{3}} \frac{1}{\rho^{\frac{1}{3}}} \qquad \begin{array}{l} \text{(nonrelativistic and extremely} \\ \text{degenerate)} \end{array} \qquad (38)$$

$$\frac{dU}{dt} = \frac{6.3}{\mu_e}(1 + 5T_9^2)T_9^7\frac{1}{\rho} \qquad \begin{array}{l} \text{(extremely relativistic and extremely} \\ \text{degenerate)} \end{array} \qquad (39)$$

---

* In Chiu and Stabler's original calculation, a factor of 4π was missing in the numerical expressions. This was pointed out by Ritus (1961), and later this error was corrected (Chiu and Stabler, 1963).

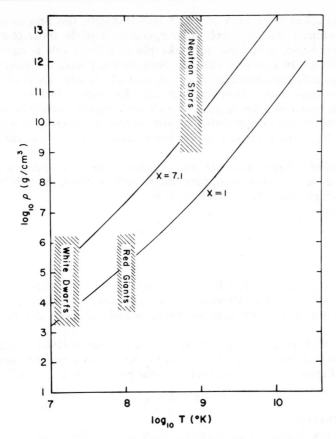

Figure 2. The relation of plasma frequency ω_0 to temperature. The two curves $x = 1$ and $x = 7.1$ are shown and $x = (\hbar\omega_0/kT)$. The plasma neutrino rate is a maximum along the line $x = 7.1$. The regions occupied by neutron stars, white dwarfs, and red giants are also indicated.

Other Neutrino Processes

Since Pontecorvo's (1959) original suggestion that $(ev_e)(ev_e)$ interaction might play an important role in stars, many neutrino reactions have been considered. The most important three processes (in ordinary stars) have been described. Here, we summarize the results of other calculations.

The Original URCA Process. Historically, this was the first neutrino process ever to have been discussed. Gamow and Schönberg (1941) suggested that a cyclic beta and inverse beta process might dissipate stellar energy invisibly:

$$e^- + (Z, A) \rightarrow (Z - 1, A) + v_e \tag{40}$$

$$(Z - 1, A) \rightarrow (Z, A) + e^- + \bar{v}_e \tag{41}$$

Because of the similarity of gamblers losing their money invisibly through a rotating disk (most stars have rotations) in the Casino URCA in Rio de Janeiro to stars losing their energy invisibly through the cyclic process of neutrino emission, the process was called the URCA process.* Unfortunately, this process cannot be important in ordinary stars in which the temperature is of the order of a few keV (at most, 1 meV before a collapse will take place; $kT = 1$ keV is equivalent to $T = 1.16 \times 10^7 \,^\circ$K). In equation (40), the electron must have an energy greater than the energy difference between $(Z - 1, Z)$ and (Z, A), and this is of the order of a few meV for most nuclei. Although there are a few nuclei which have an energy difference of the order of a few tenths of 1 meV with respect to their beta-unstable isobars, these nuclei are also unstable against nuclear reactions at a temperature at which the URCA rate is large. This was investigated by Chiu in an earlier paper (Chiu, 1961b).

During stellar collapse, however, the situation is different. From a temperature of 2 meV up to a temperature of around 50 meV or more, the URCA process becomes important again. For a neutron, it is

$$n + e^+ \to p + \nu \tag{42a}$$

$$n \to p + \bar{\nu} + e^- \tag{42b}$$

$$e^- + p \to n + \nu \tag{42c}$$

Equation (42b) is more important than equation (42c). At still higher temperatures, the production of μ-neutrinos (through π-decay) is more important. The physical condition of a star during the collapse phase is still, however, uncertain (Chiu, 1964).

Another URCA process suggested by Chiu and Salpeter (1964) is expected to be important in neutron stars. This is the scattering of two degenerate neutrons in the top of the Fermi sea, into a final state of four particles, n, p, e^-, and ν_e:

$$n + n \to n + p + e^- + \bar{\nu}_e \tag{43}$$

The inverse process is

$$n + p \to n + n + e^+ + \nu_e \tag{44}$$

Early estimates by Bahcall (1964b) show that this process is probably more important than the plasma process in the core of neutron stars. The calculation is being effected by Bahcall, and results should be available soon.

Photo-Beta Process. Cameron (1959) suggested that the URCA process can also occur among the excited states of nuclei. Among certain states, the transition rate may be large, and also for heavier nuclei there may be many excited states. His result indicates that this process, which he called the "photo-beta process," should be more important than the original URCA process by a few orders of magnitude. However, it is still relatively unimportant compared to the three most important processes, although it may be important in the approach toward equilibrium of elements at high density and lower temperature.

Bremsstrahlung Process. Pontecorvo (1959) suggested that in a bremsstrahlung process neutrino pairs may also be released:

$$e^- + (Z, A) \to e^- + (Z, A) + \nu_e + \bar{\nu}_e \tag{45}$$

However, Gandel'man and Pinaev (1959) calculated the energy-loss rate and found it to be quite small at all temperature and density ranges of a star.

* The reader must be reminded that the Monte Carlo process is *not* a neutrino process.

Photon–Photon Process. Originally, Chiu and Morrison (1960) suggested two processes involving the direct conversion of photons to neutrinos which may be of astrophysical importance:

$$\gamma + \gamma \rightarrow v_e + \bar{v}_e \tag{47}$$

$$\gamma + \gamma \rightarrow \gamma + v_e + \bar{v}_e \tag{48}$$

Gell-Mann (1961) subsequently demonstrated that the cross-section for equation (47) in the first-order approximation to their theory is identically zero. Although, if one includes the intermediate boson in a more sophisticated field-theory calculation, the cross-section might not vanish, the magnitude is still expected to be quite small. Shabalin (1963) investigated the cross-section for the process (48) and found

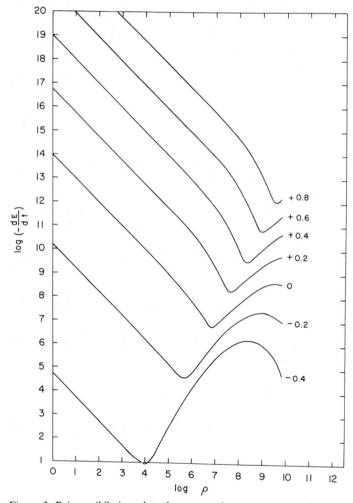

Figure 3. Pair annihilation plus plasma-neutrino energy conversion rates in ergs/g-sec plotted as a function of density. Numbers in the figure refer to log T_9.

Table II. Pair Annihilation Plus Plasma-Neutrino Energy Conversion Rates*

log ρ \ log T₉	−0.5	−0.4	−0.3	−0.2	−0.1	0	0.1	0.2	0.3	0.4	0.5	0.6	0.7	0.8	0.9
0	1.0556	4.7421	7.7496	10.2226	12.2771	14.0067	15.4873	16.7807	17.9376	18.9990	19.9971	20.9556	21.8901	22.8107	23.7231
0.2	0.8556	4.5421	7.5496	10.0226	12.0771	13.8067	15.2873	16.5807	17.7376	18.7990	19.7971	20.7556	21.6901	22.6107	23.5231
0.4	0.6556	4.3421	7.3496	9.8226	11.8771	13.6067	15.0873	16.3807	17.5376	18.5990	19.5971	20.5556	21.4901	22.4107	23.3231
0.6	0.4556	4.1421	7.1496	9.6226	11.6771	13.4067	14.8873	16.1807	17.3376	18.3990	19.3971	20.3556	21.2901	22.2107	23.1231
0.8	0.2556	3.9421	6.9496	9.4226	11.4771	13.2067	14.6873	15.9807	17.1376	18.1990	19.1971	20.1556	21.0901	22.0107	22.9231
1.0	0.0556	3.7421	6.7496	9.2226	11.2771	13.0067	14.4873	15.7807	16.9376	17.9990	18.9971	19.9556	20.8901	21.8107	22.7231
1.2	−0.1444	3.5421	6.5496	9.0026	11.0771	12.8067	14.2873	15.5807	16.7376	17.7990	18.7971	19.7556	20.6901	21.6107	22.5231
1.4	−0.3444	3.3421	6.3496	8.8226	10.8771	12.6067	14.0873	15.3807	16.5376	17.5990	18.5971	19.5556	20.4901	21.4107	22.3231
1.6	−0.5444	3.1421	6.1496	8.6226	10.6771	12.4067	13.8873	15.1807	16.3376	17.3990	18.3971	19.3556	20.2901	21.2107	22.1231
1.8	−0.7443	2.9421	5.9496	8.4226	10.4771	12.2067	13.6873	14.9807	16.1376	17.1990	18.1971	19.1556	20.0901	21.0107	21.9231
2.0	−0.9438	2.7421	5.7496	8.2226	10.2771	12.0067	13.4873	14.7807	15.9376	16.9990	17.9971	18.9556	19.8901	20.8107	21.7231
2.2	−1.1419	2.5420	5.5495	8.0026	10.0771	11.8067	13.2873	14.5807	15.7376	16.7990	17.7971	18.7556	19.6901	20.6107	21.5231
2.4	−1.3343	2.3419	5.3495	7.8225	9.8771	11.6067	13.0873	14.3807	15.5376	16.5990	17.5971	18.5556	19.4901	20.4107	21.3231
2.6	−1.5048	2.1418	5.1494	7.6225	9.6771	11.4067	12.8873	14.1807	15.3376	16.3990	17.3971	18.3556	19.2901	20.2107	21.1231
2.8	−1.6038	1.9416	4.9492	7.4224	9.4770	11.2067	12.6873	13.9807	15.1376	16.1990	17.1971	18.1556	19.0901	20.0107	20.9231
3.0	−1.5425	1.7415	4.7490	7.2222	9.2769	11.0067	12.4873	13.7807	14.9376	15.9990	16.9971	17.9556	18.8901	19.8107	20.7231
3.2	−1.2964	1.5417	4.5486	7.0220	9.0768	10.8067	12.2873	13.5807	14.7376	15.7990	16.7971	17.7556	18.6901	19.6107	20.5231
3.4	−0.9471	1.3440	4.3481	6.8216	8.8765	10.6066	12.0873	13.3807	14.5376	15.5990	16.5971	17.5556	18.4901	19.4107	20.3231
3.6	−0.5632	1.1551	4.1471	6.6210	8.6762	10.4064	11.8873	13.1807	14.3376	15.3990	16.3971	17.3556	18.2901	19.2107	20.1231
3.8	−0.1707	0.9990	3.9457	6.4200	8.4755	10.2061	11.6872	12.9807	14.1376	15.1990	16.1971	17.1556	18.0901	19.0107	19.9231
4.0	0.2228	0.9414	3.7438	6.2184	8.2745	10.0055	11.4871	12.7807	13.9376	14.9990	15.9971	16.9556	17.8901	18.8107	19.7231
4.2	0.6149	1.0645	3.5416	6.0160	8.0729	9.8046	11.2867	12.5806	13.7376	14.7990	15.7971	16.7556	17.6901	18.6107	19.5231
4.4	1.0044	1.3483	3.3417	5.8121	7.8704	9.6030	11.0860	12.3805	13.5375	14.5990	15.5971	16.5556	17.4901	18.4107	19.3231
4.6	1.3904	1.7034	3.1553	5.6061	7.6664	9.4005	10.8847	12.1801	13.3375	14.3990	15.3971	16.3556	17.2901	18.2107	19.1231
4.8	1.7714	2.0771	3.0212	5.3971	7.4602	9.1965	10.6824	11.9793	13.1373	14.1989	15.1971	16.1556	17.0901	18.0107	18.9231

5.0	2.1462	2.4510	3.0222	5.1842	7.2503	8.9903	10.4786	11.7776	12.9369	13.9988	14.9971	15.9556	16.8901	17.8107	18.7231
5.2	2.5128	2.8188	3.2065	4.9697	7.0350	8.7804	10.2726	11.5745	12.7360	13.7987	14.7971	15.7556	16.6901	17.6107	18.5231
5.4	2.8690	3.1772	3.5047	4.7680	6.8114	8.5650	10.0631	11.3691	12.5339	13.5982	14.5970	15.5555	16.4901	17.4107	18.3231
5.6	3.2124	3.5238	3.8369	4.6311	6.5762	8.3411	9.8482	11.1604	12.3298	13.3970	14.3967	15.3555	16.2901	17.2107	18.1231
5.8	3.5404	3.8560	4.1681	4.6518	6.3281	8.1043	9.6250	10.9466	12.1226	13.1945	14.1961	15.1553	16.0901	17.0107	17.9231
6.0	3.8505	4.1715	4.4867	4.8433	6.0780	7.8491	9.3895	10.7250	11.9105	12.9893	13.9946	14.9549	15.8900	16.8107	17.7231
6.2	4.1402	4.4683	4.7885	5.1124	5.8819	7.5694	9.1359	10.4919	11.6913	12.7797	13.7914	14.7540	15.6897	16.6106	17.5231
6.4	4.4077	4.7448	5.0714	5.3921	5.8537	7.2652	8.8568	10.2421	11.4617	12.5637	13.5846	14.5523	15.4891	16.4104	17.3229
6.6	4.6516	4.9999	5.3347	5.6599	6.0095	6.9676	8.5446	9.9685	11.2169	12.3382	13.3723	14.3479	15.2876	16.2099	17.1227
6.8	4.8709	5.2330	5.5780	5.9104	6.2388	6.7924	8.1968	9.6632	10.9508	12.0993	13.1515	14.1392	15.0856	16.0085	16.9220
7.0	5.0649	5.4440	5.8015	6.1431	6.4737	6.8464	7.8414	9.3182	10.6561	11.8417	12.9190	13.9233	14.8798	15.8078	16.7204
7.2	5.2329	5.6326	6.0055	6.3584	6.6967	7.0317	7.6019	8.9329	10.3243	11.5588	12.6701	13.6971	14.6683	15.6043	16.5206
7.4	5.3736	5.7985	6.1900	6.5568	6.9052	7.2410	7.6166	8.5451	9.9484	11.2426	12.3990	13.4566	14.4480	15.3964	16.3191
7.6	5.4854	5.9407	6.3549	6.7386	7.0995	7.4436	7.7820	8.3059	9.5331	10.8851	12.0986	13.1968	14.2153	15.1813	16.1139
7.8	5.5657	6.0575	6.4992	6.9034	7.2795	7.6348	7.9755	8.3392	9.1409	10.4812	11.7609	12.9113	13.9658	14.9556	15.9031
8.0	5.6107	6.1465	6.6214	7.0504	7.4450	7.8140	8.1642	8.5043	8.9589	10.0469	11.3781	12.5926	13.6939	14.7154	15.6835
8.2	5.6157	6.2041	6.7189	7.1779	7.5950	7.9808	8.3433	8.6888	9.0400	9.6897	10.9496	12.2324	13.3926	14.4555	15.4513
8.4	5.5747	6.2256	6.7883	7.2836	7.7280	8.1341	8.5117	8.8681	9.2107	9.6069	10.5159	11.8245	13.0540	14.1698	15.2021
8.6	5.4803	6.2055	6.8254	7.3643	7.8416	8.2724	8.6683	9.0384	9.3894	9.7340	10.2456	11.3786	12.6702	13.8510	14.9303
8.8	5.3237	6.1367	6.8247	7.4158	7.9328	8.3934	8.8118	9.1985	9.5616	9.9080	10.2699	10.9818	12.2387	13.4907	14.6291
9.0	5.0945	6.0110	6.7796	7.4330	7.9976	8.4944	8.9399	9.3468	9.7250	10.0819	10.4259	10.8427	11.7923	13.0822	14.2907
9.2	4.7807	5.8186	6.6823	7.4098	8.0314	8.5716	9.0499	9.4814	9.8779	10.2484	10.5997	10.9474	11.4875	12.6329	13.9068
9.4	4.3680	5.5481	6.5238	7.3388	8.0283	8.6206	9.1384	9.5998	10.0186	10.4056	10.7691	11.1159	11.4839	12.2221	13.4746
9.6	3.8397	5.1864	6.2935	7.2116	7.9815	8.6360	9.2014	9.6987	10.1446	10.5519	10.9303	11.2874	11.6322	12.0591	13.0240
9.8	3.1765	4.7180	5.9789	7.0182	7.8831	8.6115	9.2338	9.7745	10.2532	10.6850	11.0818	11.4524	11.8039	12.1531	12.7057

* The photoneutrino rate is omitted because it has not been accurately calculated.

that the cross-section for this process is 10^8 times smaller than that originally estimated by Chiu and Morrison (1960). Consequently, this process cannot play an important role in stars.

Photon–Coulomb Field Interaction. Matinyan and Tsilosani (1961) and independently Rosenberg (1963) considered the following process

$$\gamma + \text{nucleus} \rightarrow \gamma + \gamma \text{ (virtual Coulomb)} + \text{nucleus} \rightarrow v^- + \bar{v}^- + \text{nucleus} \quad (49)$$

Matinyan and Tsilosani's results are marred by the fact that the amplitude they obtained is not gauge invariant (which is necessary for electromagnetic fields) and presumably is **not** correct. Both their results and Rosenberg's results show that this

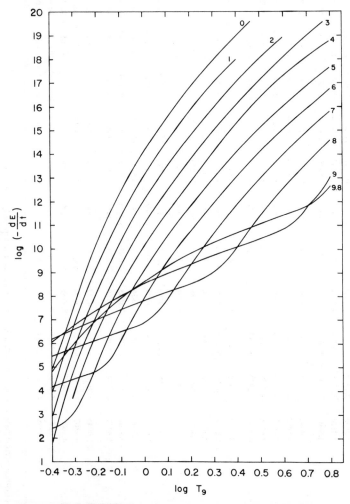

Figure 4. Pair annihilation plus plasma neutrino energy conversion rates in ergs/g-sec plotted as a function of temperature. Numbers in the figure refer to $\log \rho$ (g/cm^3).

process is not important. Rosenberg's result is

$$-\frac{dU}{dt} = 2.3 \times 10^{14} \left(\frac{kT}{mc^2}\right)^{10} \text{ ergs/cm}^3\text{-sec} \tag{50}$$

which also appears to be small compared with other neutrino processes.

Electron–Electron Bremsstrahlung Process. Lomazzo (1964) investigated the process

$$e^- + e^- \rightarrow e^- + e^- + v + \bar{v} \tag{51}$$

and found that it is also unimportant. His result for the relativistic case (which is the more important one) is

$$\frac{dU}{dt} = 2 \times 10^{18} \exp\left(\frac{\rho}{\rho_b}\right) T_{10}^9 \ln^2 (T_9) \text{ ergs/cm}^3\text{-sec} \tag{52}$$

Summary of Neutrino Processes

We have summarized all investigated neutrino processes known to the author. The *combined* rates for a wide temperature and density regime are listed in Table II. For convenience, we have also plotted the temperature dependence of the combined rates for a number of densities and also the density dependence for a number of temperatures (Figures 3 and 4).

REFERENCES

J. B. Adams, M. A. Ruderman, and C.-H. Woo (1963), *Phys. Rev.* **129**: 1383.
H. L. Anderson and C. Lattes (1957), *Nuovo Cimento* **6**: 1356.
J. N. Bahcall (1964a), *Phys. Rev.* **136**: B1164.
J. N. Bahcall (1964b), Private communication.
J. N. Bahcall and S. Frautschi (1964), *Phys. Rev.* **136**: B1547.
A. G. W. Cameron (1959), *Astrophys. J.* **130**: 916.
H.-Y. Chiu (1961a), *Phys. Rev.* **123**: 1040.
H.-Y. Chiu (1961b), *Ann. Phys.* **15**: 1.
H.-Y. Chiu (1961c), *Ann. Phys.* **16**: 321.
H.-Y. Chiu (1963), *Astrophys. J.* **137**: 343.
H.-Y. Chiu (1964), *Ann. Phys.* **26**: 364.
H.-Y. Chiu and P. Morrison (1960), *Phys. Rev. Letters* **5**: 573.
H.-Y. Chiu and E. E. Salpeter (1964), *Phys. Rev. Letters* **12**: 413.
H.-Y. Chiu and R. Stabler (1961), *Phys. Rev.* **122**: 1317.
H.-Y. Chiu and R. Stabler (1963), *Phys. Rev.* **131**: 2839.
R. P. Feynman and M. Gell-Mann (1958), *Phys. Rev.* **109**: 193.
W. A. Fowler (1964), Private communication.
G. Gamow and M. Schönberg (1941), *Phys. Rev.* **59**: 539.
G. M. Gandel'man and V. S. Pinaev (1959), *J. Exper. Theoret. Phys. (USSR)* **37**: 1072. Translation: (1960) *Soviet Physics–JETP* **10**: 764.
A. Gray and T. M. Macrobert (1922), *A Treatise on Bessel Functions and Their Applications to Physics*, MacMillan and Co., London.
M. Gell-Mann (1961), *Phys. Rev. Letters* **6**: 70.
N. van Hieu and E. P. Shabalin (1963), *J. Exper. Theoret. Phys. (USSR)* **44**: 1003. Translation: (1964), *Soviet Physics–JETP* **17**: 681.
C. Inman and M. A. Ruderman (1964), *Astrophys. J.* **140**: 1025.
T. D. Lee and C. N. Yang (1960), *Phys. Rev.* **119**: 1410.
Y. K. Lee, L. W. Mo, and C. S. Wu (1963), *Phys. Rev. Letters* **10**: 253.
A. Lomazzo (1964), Ph.D. Thesis, Columbia University.

S. G. Matinyan and N. N. Tsilosani (1961), *J. Exper. Theoret. Phys.* (*USSR*) **41**: 1681. Translation: (1962), *Soviet Physics–JETP* **14**: 1195.

B. M. Pontecorvo (1959), *J. Exper. Theoret. Phys.* (*USSR*) **36**: 1615. Translation: (1960), *Soviet Physics–JETP* **9**: 1148.

V. I. Ritus (1961), *J. Exper. Theoret. Phys.* (*USSR*) **41**: 1285. Translation: (1962), *Soviet Physics–JETP* **14**: 915.

L. Rosenberg (1963), *Phys. Rev.* **129**: 2786.

Part III. Stellar Evolution
A. Pre-Main-Sequence Contraction

ON CONTRACTING STARS

Chushiro Hayashi

INTRODUCTION

In the pre-main-sequence contraction stage, the surface temperature of a star is low. The star thus has both a hydrogen ionization zone and a hydrogen molecule-dissociation zone. In these regions of ionization and dissociation, the specific heat of the gas is very large, so that the adiabatic temperature gradient is small and the regions are quite likely to be convective. Further, proceeding inward from the ionization zone, the opacity stays high because of the small value of $d \ln T/d \ln P$, and this high opacity extends the convective zone further inward.

The existence of the outer convective region (with polytropic index 1.5) reduces the stellar radius as compared to that of a wholly radiative star (polytropic index about 3) of the same central density. In this way, a star of given mass and luminosity which is in quasi-static equilibrium has a maximum radius or minimum surface temperature, as shown in Figure 1. To the right of the critical curve, which corresponds to a structure that is wholly convective, there is no solution in hydrostatic equilibrium. To the left, of course, there are equilibrium solutions which are composed of outer convective zones and inner radiative cores. In the shaded region, the star must contract dynamically. The star is probably born somewhere in the lower right region of low effective temperature and luminosity and then moves very quickly on the Kelvin–Helmholtz time scale, ending up somewhere on the critical curve. The star then follows a path as shown down to the main sequence and finally begins consuming hydrogen. After the formation of a helium core, the star will become a red giant, but it can never move into the right of the critical curve in Figure 1.

SURFACE CONDITION

If the mass, luminosity, surface temperature, and chemical composition of a star are given, then the pressure at the bottom of the photosphere can be calculated from the theory of stellar atmospheres. At a given effective temperature, the pressure is some function of the opacity and the luminosity. Proceeding from the photosphere, we can calculate the condition for the onset of convective instability. The very outermost regions of the stars are in radiative equilibrium; somewhere just below them the convective region begins. This boundary will be determined by two conditions. The first is the condition of instability: that the radiative temperature gradient is steeper than the adiabatic temperature gradient,

$$(n + 1)_{\text{rad}} \leq (n + 1)_{\text{ad}} \qquad \left(n + 1 = \frac{d \ln P}{d \ln T}\right) \tag{1}$$

This is not sufficient for the onset of convective-energy transport. The second condition is that the convective flux is greater than the radiative flux. The convective-energy transport near the surface is approximately $\beta U V_s$, where β denotes the

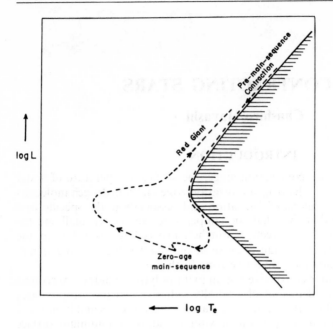

Figure 1. Schematic evolutionary path of a star of 0.8 $M_\odot$.

efficiency of the energy transport, V_s is the velocity of sound, and U is the internal energy density (including the ionization energy). β plays the same role as l/H in the mixing-length theory. The second condition is then

$$\beta U V_s \geq \frac{ac}{4} T_e^4 \qquad (2)$$

If both of these conditions are satisfied, then the inner region becomes convective.

The pressure and temperature diagram for the outer regions of a star is shown in Figure 2. The shaded region is the region of the hydrogen-ionization zone; below it is the photosphere. Curve (2) corresponds to the equal flux condition which depends on the value of β. Curve (1) corresponds to the onset of convective instability, that is, $(n + 1)_{\text{adiabatic}} = (n + 1)_{\text{radiative}}$. In the region of the temperature–pressure diagram above the two curves, the energy transport is predominantly by convection. For high luminosity, the photospheric pressure is p, and stellar structure follows the path pcd; inside point c [condition (2), curve (2)] is the convective zone. But if the luminosity is lower, then the photospheric pressure is higher (p') and the stellar structure follows the path $p'c'd'$; inside the point c' [condition (1), curve (1)] is the convective region. Thus, there are two different cases for the outer boundary of the convective zone: curve (1) and curve (2). The critical stellar luminosity, where the two conditions intersect, is about one solar luminosity.

At point d, the ionization of hydrogen becomes complete. In the internal region of complete hydrogen ionization, the adiabatic relation is

$$P = KT^{2.5} \qquad (3)$$

where K is a function of the surface temperature and luminosity if the mass and composition are given. The value of K is connected with the parameter E, which

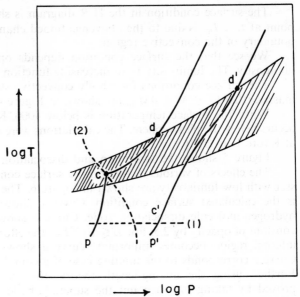

Figure 2.

determines the characteristics of the internal structure,

$$E = 4\pi K G^{\frac{3}{2}}\left(\frac{\mu H}{k}\right)^{\frac{1}{2}} M^{\frac{1}{2}} R^{\frac{3}{2}}$$

The value of $E = 45.5$ corresponds to wholly convective configuration and $E = 0$ to wholly radiative configuration.

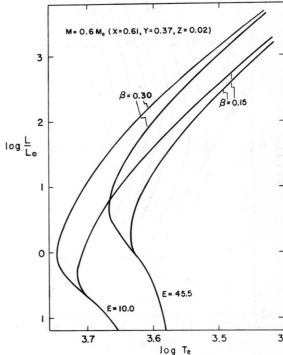

Figure 3.

The surface condition in the H-R diagram is shown in Figure 3. The turning point at $L \simeq L_\odot$ is due to the abovementioned change in conditions for the upper boundary of the convective region.

We see that the surface condition depends on β for high but not for low luminosity. The luminosity is an increasing function of β.

The surface conditions for wholly convective stars, $E = 45.5$, and for various masses, $0.05M_\odot \le M \le 4M_\odot$, are shown in Figure 4.

When the surface temperature is below 4000 °K, the dissociation zone of H^2 occurs below the photosphere. The calculations were made by Nakano and Hayashi at Kyoto University.

Figure 5 shows the ionization and dissociation regions and typical adiabats.

The effects of various factors on the surface condition for a wholly convective star with low luminosity are shown in Figure 6. The mass is $M = 0.2M_\odot$. Curve a is the calculated surface condition. Curve c shows the effect of neglecting the hydrogen-molecule dissociation zone. Curve b shows the effect of increasing the constant of opacity by 2.5. For $L \lesssim 10^{-2}L_\odot$, the effect of electron degeneracy in the internal region becomes important; curve d shows the effect of neglecting this. Curve e corresponds to the limiting case of a completely degenerate configuration. Further, using the mixing-length theory, the above calculation has been improved by taking into account the superadiabatic gradient. This effect is small, $(\nabla - \nabla_{ad} \simeq 0.1\nabla_{ad}$ at the photosphere of $0.2M_\odot)$. Finally, the uncertainty in the photospheric value of pressure is investigated by taking into account the partial energy transport by convection above the photosphere and also the nonisothermal character of the atmosphere. This effect is found to be equivalent to reducing the constant of opacity by a factor of 2.

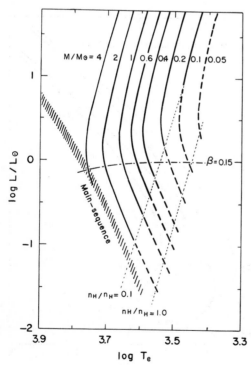

Figure 4.

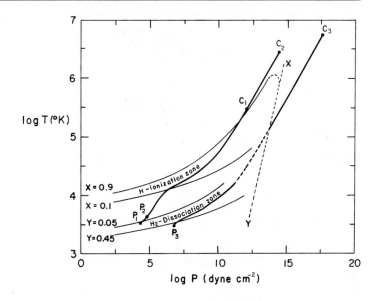

Figure 5.

CONTRACTING STELLAR MODELS

Consider now the interior model instead of the surface condition. A sequence of models of contracting stars were calculated under the simple assumption that $\mathscr{E} \propto T$ and $\varkappa \propto \rho T^{-3.5}$ (Kramers' law). The models consist of a radiative core and a convective envelope with different mass fractions in the radiative core. These models correspond to the different horizontal straight lines shown in Figure 7. The surface condition is represented by the vertical lines. The intersection of the two curves having the same value of E determines the quasi-static solution.

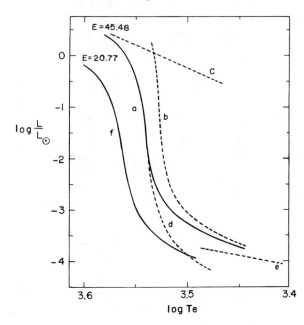

Figure 6.

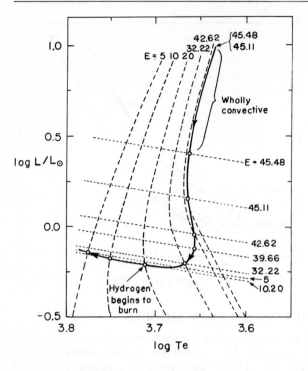

Figure 7.

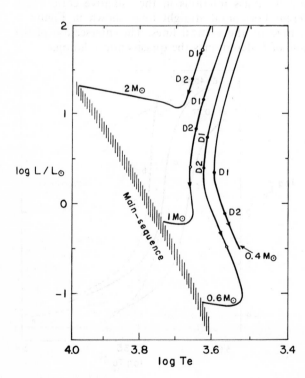

Figure 8.

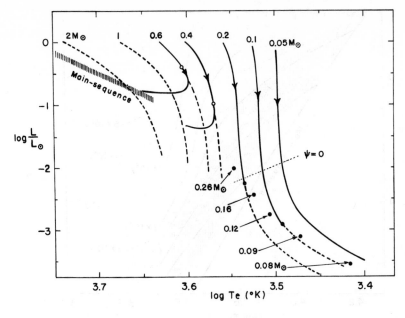

Figure 9.

When the luminosity is sufficiently high, the star is wholly convective and evolves downward along the surface condition curve of $E = 45.5$. Then the radiative core begins to appear, and its mass fraction increases. When this mass fraction becomes about one half, the evolutionary track changes its direction.

Figure 8 shows tracks for $2M_\odot$, $M_\odot$, and $0.6M_\odot$. A radiative core is found to appear before the onset of hydrogen-burning for $M > 0.26M_\odot$, but for stars of $0.26M_\odot > M > 0.08M_\odot$, hydrogen begins to burn during the wholly convective stage and the zero-age main sequence is wholly convective. Thus, the products of the hydrogen-burning are brought up to the surface. For a star of mass $0.26M_\odot < M < 0.08M_\odot$, the p-p chain stops at He^3, so He^3 will be brought up to the surface. For stars of mass $0.26M_\odot$ and luminosity $10^{-2}L_\odot$, the amount of hydrogen burned in 10^{10} years is about 1% of the total mass, so that the amount of He^3 on the surface as compared to He^4 is expected to be about 3%. If the star's mass is less than $0.08M_\odot$, then as the star contracts, the central temperature reaches a maximum and afterward decreases due to electron degeneracy. These stars evolve toward black dwarfs along the surface condition as shown in Figure 9. The critical mass depends on the helium content and changes to $0.12M_\odot$ for a composition $x = 0.90$, $y = 0.08$, and $z = 0.02$.

The locus of lifetime for different stellar masses is shown in Figures 10 and 11. The lifetime is measured from the stage of infinite radius along the curve $E = 45.5$. For stars of $0.08M_\odot$, it takes 8×10^8 years to reach the zero-age main sequence. The solid lines are the present results and the dashed lines the results for wholly radiative models.

Consider now the depletion of deuterium and lithium during the contraction. For stars of $M > 0.26M_\odot$, the temperature at the bottom of convection region increases during the wholly convective stages, attains its maximum value approximately when the radiative core appears, and afterward decreases. Deuterium is

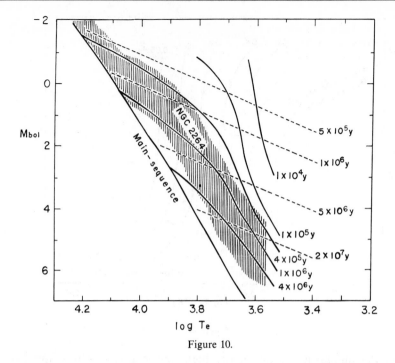

Figure 10.

almost completely destroyed by $D^2(p, \gamma)He^3$ at $T \sim 10^6$ °K for $M < 3.5 M_\odot$. However, for $M > 3.5 M_\odot$, deuterium is expected to remain unburned on the surface. The stages of deuterium burning are indicated by points $D1$ and $D2$ on the evolutionary tracks in Figure 8. Lithium is largely depleted by $Li^7(p, \alpha)He^4$ at

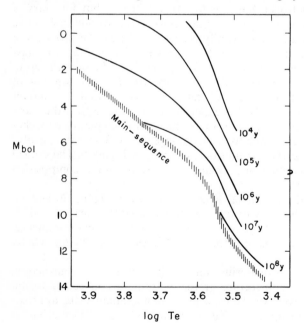

Figure 11.

Table I. Values of log Li/Li$^{(0)}$ on the Main Sequence

Chemical composition	$M = 1.0M_\odot$	$0.6M_\odot$	$0.4M_\odot$
$X = 0.61, Z = 0.02$	-0.02	-0.17	-22
$X = 0.61, Z = 0.04$	-1.0	-1.6	-580
$X = 0.90, Z = 0.02$		-17	
Spectral type	$(\sim G7)$	$(\sim K8)$	$(\sim M2)$

$T \sim 3 \times 10^6\,°K$ for $M < 1M_\odot$, but beryllium is not depleted for $M > 0.4M_\odot$. At the stage when the radiative core appears, the reaction rate for $Li^7(p, \alpha)He^4$ at the center is

$$\frac{1}{\tau_c} = 10^{-7.23}\rho X\left(\frac{T}{3 \times 10^6}\right)^{18.6}$$

The opacity in the relevant ranges of the temperature and density is taken from Morse's table (1940) to be

$$\varkappa = \varkappa_0 \frac{Z(1 + X)}{t/g} \frac{\rho}{T^{3.5}} \qquad \frac{t}{g} = \left(\frac{t}{g}\right)_0 \frac{[\rho(1 + X)]^{0.6}}{T^{0.8}}$$

then

$$\ln\frac{Li}{Li^{(0)}} \propto \frac{1}{\tau_c} \propto X(1 + X)^{2.2}(1 + \tfrac{5}{3}X)^{21.7}(\varkappa_0 Z)^{6.0}M^{-12.4}T_e^{24.1}$$

The results of lithium depletion are shown in Table I. For stars slightly more massive than the sun, lithium remains unburned. However, it is to be noticed that the above values show only the general tendency of the dependence on the mass and the chemical composition, since the reaction rate of lithium-burning depends on the temperature as T^{20} (at $3 \times 10^6\,°K$); a small error in the calculation of the temperature at the bottom of the convective region brings rather large change to the lithium abundance. The important factors are the opacity and the curve of $E = 45$.

REFERENCE

P. M. Morse (1940), *Astrophys. J.* **92**: 27.

THE CONTRACTION PHASE OF SOLAR EVOLUTION

D. Ezer and A. G. W. Cameron

The Henyey method for calculating stellar evolutionary tracks has been used to follow the gravitational contraction of a star of one solar mass. In the normal Henyey method, a stellar model is divided into concentric zones and the equations of stellar structure are solved at the zone boundaries. The external boundary conditions for the outermost zone are obtained from a separate set of atmospheric calculations, wherein a grid of four atmospheres is calculated for different values of radius and luminosity, and physical conditions at the base of these atmospheres form a set within which the external zone-boundary conditions can be interpolated. As Hayashi has pointed out, when the surface temperature of a star is low, the surface conditions play a dominant role in determining the structure of the outer layers. When the external boundary conditions of such a star are handled properly, it is found that there is a deep outer convection region, which may extend all the way to the center. In the transition region where the convection zone gives way to the radiative photosphere, the temperature gradient is extremely superadiabatic. It is customary in the Henyey method to use adiabatic temperature gradients in that part of the interior which is convective. We have adapted the Henyey method to allow use of superadiabatic gradients, but this significantly lengthens the time required to obtain solutions. Consequently, we have found it preferable to extend the atmospheric calculations to cover the outer 2% of the stellar mass. In the atmospheric calculations, we treat convection by the full mixing-length theory. For the interior, we are able to use adiabatic temperature gradients without too great error.

In our first approach to the problem of early solar evolution, we had calculated individual solar models at various stages of contraction, using conventional methods of calculation and assuming homologous contraction in each model. These calculations confirmed Hayashi's ideas about the necessity for deep outer convection zones when the surface temperature is low. In our first calculations with Henyey's method, we started from the point of initial dynamic stability revealed by the previous calculations. At this point, the released gravitational potential energy is just able to supply the internal thermal energy and the dissociation energy of hydrogen molecules and the ionization energy of hydrogen and helium required by the model. In this threshold-stability condition, the solar model had a radius of $57R_\odot$.

The subsequent evolutionary tracks are shown in Figure 1 as the light curves. Two evolutionary sequences are shown, corresponding to the use of two different values for the fundamental parameter which enters into the mixing-length theory. This parameter is the mixing length (l), and we took it as either one times or two times the pressure-scale height (H). We also used a composition for the sun based on measured elemental abundances in meteorites, the sun, and the stars. In this composition the fractional contents (by mass) of hydrogen, helium, and heavy

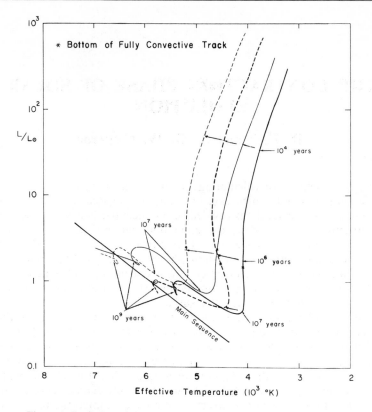

Figure 1. Evolution of the sun toward the main sequence. The light curves
are calculated with initial compositions hydrogen $X = 0.602$; helium
$Y = 0.376$; heavy elements $Z = 0.022$ (by mass). The heavy curves correspond to $X = 0.739$, $Y = 0.240$, and $Z = 0.021$. The solid lines represent
a mixing length of one, and the dashed lines of two, pressure-scale heights.

elements were $X = 0.602$, $Y = 0.376$, and $Z = 0.022$. The hydrogen/helium ratio in
this composition was based on measurements in O and B stars. The opacity of the
material was calculated with the opacity code developed at Los Alamos by A. N.
Cox and his colleagues, with additional crude line corrections applied by us which
were based on calculations of Cox.

The initial stable models of our evolutionary sequences are fully convective,
but their structure differs considerably from that of an $n = 1.5$ polytrope, owing to
the presence of a large region which is not fully ionized. As the protosun shrinks,
it stays fully convective until it has contracted to a radius of $2.5R_\odot$. A radiative
core then starts to develop. Up to this point, the total evolution time is about 10^6
years. When the radiative core has grown to cover more than half of the radius,
the luminosity, which up to then has been steadily decreasing, starts to increase.
Nuclear-energy-generation gradually supersedes the gravitational-contraction
energy release in the central zones as the central temperature increases.

The calculations were carried up to 4.5×10^9 years. At the end of this time,
the star had about the right radius for $l/H = 1$, but much too high a luminosity for
the sun. The luminosity is about twice as large as it should be. Since the luminosity

will change with the mean molecular weight as $\mu^{7.5}$, the most obvious way to decrease the luminosity is to reduce the initial helium content. There are some other indications that the helium content in the sun is less than that in O and B stars.

Recent solar cosmic ray measurements by Biswas, Fichtel, Guss, and Waddington, using data obtained from a rocket flight, have determined the relative abundances of helium, carbon, nitrogen, oxygen, and neon in the sun. The weighted average of the individual determinations, according to Gaustad, is $(\mathrm{He})/(\mathrm{C} + \mathrm{N} + \mathrm{O}) = 54 \pm 6$. The cosmic ray ratio $(\mathrm{Ne}/\mathrm{O}) = 0.1$. These relative abundances and the $\mathrm{H} : \mathrm{O} : \mathrm{Si}$ ratio in the sun quoted by Aller result in a composition for the sun of $X = 0.739$, $Y = 0.240$, and $Z = 0.021$.

Our evolutionary study has been repeated with the above composition. The opacity was calculated, including line corrections, by A. N. Cox and his colleagues at Los Alamos, to whom we are grateful. Results of the calculation are given in Figure 1 by the heavy curves for the two assumed values of the ratio of the mixing length to pressure-scale height. The effective temperatures of the fully convective tracks may be seen here, also. These depend on both the composition and the mixing length. The increase of hydrogen content from 0.602 to 0.739 shifted the tracks to 350 °K lower effective temperature. Doubling the mixing length increased the effective temperature by about 750 °K.

The evolutionary track calculated with the mixing length equal to two pressure-scale heights and with the second composition resulted in bringing the sun nearly to the observed position in the Hertzsprung–Russell diagram at the end of 4.5×10^9 years. The calculated luminosity agreed with that observed for the sun to within 1%, and the radius was 3% smaller than the observed radius. Therefore, the track calculated with the above composition and this mixing-length ratio was taken to be the approximate evolutionary path for the sun from the point of initial stability to its present position. In Figure 2, this evolutionary track and corresponding evolution times are shown.

Assuming that the sun has no angular momentum, it becomes stable against gravitational collapse when it contracts to a radius of $64R_\odot$. Its initial luminosity is about 693 times the luminosity of the present sun. It stays fully convective until it has contracted to a radius of $2R_\odot$. The corresponding luminosity for that radius is $1.5L_\odot$. With further contraction, the radiative core starts to develop. The luminosity keeps decreasing as long as the convective motion of the gases is mostly responsible for the transfer of the energy from the center to the surface layers. When about 58% of the radius and 75% of the mass of the sun have been covered by the radiative core, the luminosity starts to increase, and the evolutionary track turns upward. The evolution time at that point is 10^7 years. The minimum luminosity is $L = 0.512L_\odot$. With further evolution, the radiative core grows continuously until it covers 99% of the mass. After 1.4×10^7 years, the energy generated by nuclear reactions begins to supplement the contribution of the gravitational contraction to the total energy output of the sun. He^3 is formed through the $\mathrm{H}^1(p, \beta^+ v)\mathrm{H}^2(p, \gamma)\mathrm{He}^3$ reactions, and comes into equilibrium with the other nuclei in the proton–proton chain at the center after 2.9×10^7 years. A small convective core develops at the center of the sun and extends out to 9% of the mass while C^{12} is burning to reach its equilibrium value with N^{14} nuclei. After C^{12} is depleted by a sufficient amount, this convective core diminishes in size, and, at the end of 4.5×10^9 years, it covers only 0.1% of the mass. Figures 3 and 4 show the distribution of the luminosity, energy generation, and He^3 and C^{12} abundances, by mass, throughout the interior

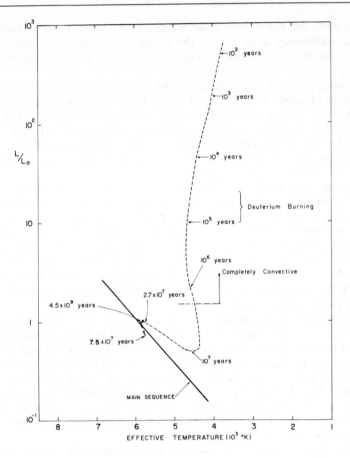

Figure 2. The Hertzsprung–Russell diagram for the sun through its gravitational contraction up to its present main-sequence position, with a mixing length equal to two pressure-scale heights. The evolution times are indicated on the track at several positions.

of the sun. Figure 3 is for the model of age 2.9×10^7 years, where the sun reaches a local maximum luminosity just prior to settling onto the main sequence. Figure 4 represents a model of 7.5×10^7 years, where the luminosity of the sun has its local minimum, i.e. on the zero-age main sequence. Comparing Figures 3 and 4, we see that when nuclear-energy generation superseded the gravitational-contraction energy of the sun, the energy generation concentrated toward the center, so the luminosity of the sun decreased. In the model shown in Figure 3, the gravitational energy became negative, indicating an expansion of the central layers in the process of forming a convective core. In this model, He^3 is still building up toward its equilibrium value throughout the interior of the sun. For the model in Figure 4, He^3 is in equilibrium in the core. Outside the core, it is in local equilibrium up to the mass fraction 0.14, and further out it is still building up toward its equilibrium value. C^{12} has been reduced substantially, but still has not reached its equilibrium value inside the sun.

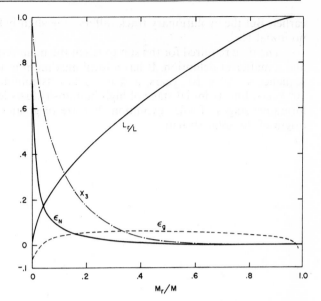

Figure 3. The variation of the luminosity (L_r/L), the nuclear $(\mathcal{E}_N)$ and gravitational $(\mathcal{E}_g)$ energies, and He^3 abundance by mass (X_3) as a function of mass fraction, at the stage where the solar luminosity reaches its local maximum value prior to the main sequence. The scaling factors for the quantities are $L = 3.916 \times 10^{33}$ ergs/sec, $\mathcal{E}_N = \mathcal{E}_g = 25.70$ ergs/g-sec, and $X_3 = 1.724 \times 10^{-4}$.

If the primitive solar material has the same ratio of deuterium to hydrogen as does water on the earth, then an additional 5×10^5 years are added to the contraction time by deuterium-burning, which occurs in the region indicated in Figure 2.

The calculations were also performed with an initial He^3 content of 5.404×10^{-5} by mass. Due to the initial He^3, the core comes into nuclear-burning equilibrium

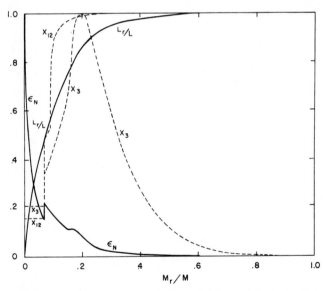

Figure 4. The variation of the luminosity (L_r/L), nuclear energy $(\mathcal{E}_N)$, and the abundances by mass of He^3 (X_3) and C^{12} (X_{12}) inside the sun at the stage for which the luminosity of the sun has its local minimum, i.e., the zero-age main-sequence model. The scaling factors for the quantities are $L = 2.79 \times 10^{33}$ ergs/sec, $\mathcal{E}_N = 29.64$ ergs/g-sec, $X_3 = 6.81 \times 10^{-4}$, and $X_{12} = 4.618 \times 10^{-3}$.

earlier. But the evolutionary track differs only slightly from the one calculated with no initial He^3.

The time required for the sun to reach the main sequence is somewhat arbitrary and a matter of definition. It has a local maximum in the luminosity near the main sequence at 2.9×10^7 years, and it reaches its minimum luminosity thereafter at 10^8 years. But its initial stage of high luminosity lasts less than 10^6 years. It is these formative stages of solar evolution that are of greatest interest for theories of the origin of the solar system.

SCHEMATIC PRE-MAIN-SEQUENCE EVOLUTION

K. von Sengbusch and S. Temesváry

According to Hayashi, the pre-main-sequence evolution of a star should pass through a phase in which the contracting protostar is fully convective and much more luminous than the resulting star on the main sequence. The problem under consideration here is whether this fully convective phase of evolution is inevitable. Many evolutionary problems would be greatly simplified if such a fully convective phase had determined the distribution of chemical elements, angular momentum, and internal magnetic fields, since in such a case the protostar would be independent of the initial conditions.

The pre-main-sequence evolution consists of two phases: first, the collapse of the primordial transparent cloud, which can be described only by full hydro-dynamic equations, including the conservation of angular momentum and the influence of the interstellar magnetic fields; second, a quasi-static state produced when the ionization has become high enough to allow a photospheric surface to be formed. Then, the equation of energy-transport determines the conditions inside the surface, and the further evolution of the system, including also further contraction, is determined by the energy losses through the photospheric surface. Suppose a star reaches hydrostatic equilibrium with a given mass inside a photospheric surface of a given radius. If the surface density is too high, then the density gradient will be less than the adiabatic gradient and the temperature gradient will be very superadiabatic. In such a case, the protostar would flare up and move onto the Hayashi track for fully convective contraction. But the initial photospheric surface need not, of course, contain the entire mass of the system. It is possible that the collapsing protostar becomes optically thick first near the center and that the photospheric surface then moves outward with respect to the infalling mass. The evolution of the star, in this case, would depend on the initial instability. Whether or not the star becomes highly luminous and fully convective depends on the energy content within the photosphere and on the energy added by the infalling mass. To decide this question, the initial phases of pre-main-sequence evolution must be computed. Before embarking on such a difficult project, it is desirable to obtain a general picture of what will occur. For this purpose, a crude model of a contracting protostar with mass accretion was constructed.

The model star consists of a mass M which increases with time according to the quadratic law

$$\frac{dM}{dt} = AM^2 \tag{1}$$

where A is a constant in time. It is assumed to be always in quasi-static equilibrium so that the virial theorem holds.

$$2T + \Omega = 3(\gamma - 1)U + \Omega = 0 \tag{2}$$

where T is the total kinetic energy, Ω is the gravitational energy, and U is the internal energy. The star is assumed to evolve homologously, that is, corresponding always to the same polytrope of index n. The total energy is then

$$E = U + \Omega = \frac{3\gamma - 4}{3(\gamma - 1)}\Omega = -\frac{3\gamma - 4}{(\gamma - 1)(5 - n)}\frac{GM^2}{R} \tag{3}$$

where the ratio of the specific heats $\gamma = \frac{5}{3}$. The luminosity corresponding to an effective temperature T_e of a photospheric surface of radius R is the rate of change of energy

$$L = -\dot{E} = \frac{3}{2(5 - n)}\frac{GM^2}{R}\left(2\frac{\dot{M}}{M} - \frac{\dot{R}}{R}\right) = 4\pi R^2 \sigma T_e^4 \tag{4}$$

where the dot denotes differentiation with respect to time.

The mass-accretion equation (1) can be integrated and gives

$$m = \frac{M}{M_0} = \frac{1}{1 - AM_0 t} \tag{5}$$

where M_0 is the mass at $t = 0$, and the time scale for mass accretion is $\tau_0 = (AM_0)^{-1}$. If the effective temperature were constant in time, then the luminosity equation (4) could be integrated to give the radius as a function of mass and hence of time,

$$r = \frac{R}{R_0} = \frac{m^2}{[1 + c_1(m^3 - 1)]^{\frac{1}{3}}} \tag{6}$$

where

$$c_1 = \frac{1}{2}\frac{L_0 \tau_0}{(-\Omega_0)} \tag{7}$$

is the ratio of the time scale τ_0 of the mass increase to the Kelvin contraction time scale at constant mass. If this ratio c_1 were small, then

$$L = L_0 r^2 \sim M^4 \tag{8}$$

The condition for a fully convective star is that the radiative temperature gradient be greater than the adiabatic temperature gradient throughout the star. For a polytrope with

$$\varkappa = \varkappa_0 \rho T^{-3.5} \qquad \text{and} \qquad \varepsilon = -T\dot{S} \sim T$$

the condition is fulfilled in the center for

$$\frac{L^{\frac{3}{4}}}{M^{\frac{11}{2}}} \geq c_4 f_n T_e \tag{9}$$

with

$$c_4 = \frac{(16\pi)^2 (\pi ac)^{\frac{3}{4}}}{3\varkappa_0}\left(\frac{G\mu_c \beta_c m_H}{k}\right)^{\frac{13}{2}} \nabla_{\text{ad}}$$

$$f_n = \frac{\xi_1^{\frac{3}{4}} \int_0^{\xi_1} \xi^2 \theta^{n+1}\, d\xi}{(n + 1)^{\frac{13}{4}}(-\xi^2 \theta')_{\xi_1}^{\frac{7}{2}}}$$

and

$$\nabla_{ad} = \left(\frac{d \ln T}{d \ln P}\right)_{ad} = 0.4$$

First, the condition for a fully convective model gives for $T_e =$ constant a straight line with $L \sim M^{4.4}$ in an L-M plane with convection for larger L. Then, according to equations (6) and (8), the evolutionary tract is a line with $L \sim M^4$ if the denominator in equation (6) is ≈ 1, or a line with L proportional to an even smaller power of M if c_1 is not negligible. Hence, even if the final point of the evolutionary track lies in the nonconvective region, an earlier part of the track must have been in the convective region.

To get a more realistic picture, a further assumption is needed about T_e as a function of mass and radius. The variation of temperature in the photosphere is given by

$$T^4 = \tfrac{3}{4}(\tau + \tfrac{2}{3})T_e^4$$

Assuming an absorption law

$$\varkappa = \varkappa_0 P^{\alpha_0} T^{\beta_0}$$

the photospheric pressure is

$$P_{ph}^{1+\alpha_0} = \frac{16}{3}\frac{g}{\varkappa_0}\frac{\nabla_{ph}}{T_e^{\beta_0}} \tag{10}$$

with

$$g = \frac{GM}{R^2}$$

and

$$\nabla_{ph} = \left(\frac{d \ln T}{d \ln P}\right)_{ph} = \frac{1 + \alpha_0}{4 - \beta_0}[1 - 2^{-(1-[\beta_0/4])}]$$

$$= 0.28$$

The values of the parameters are given by Hayashi (1961):

$$\alpha_0 = 0.735 \qquad \beta_0 = 3.05$$

$$\varkappa_0 = 10^{-15.58}\frac{X}{0.61}\left[\frac{0.70}{X + (Y/4)}\frac{Z}{0.02}\right]^{\alpha_0}$$

We now wish to eliminate the photospheric pressure. For that purpose, we use a quality of the standard model $n = 3$ and put

$$(1 - \beta)_{ph} \approx \left(\frac{1 - \beta}{\beta^4}\right)_c = BM^2$$

assuming $\beta \approx 1$. We can now integrate equation (4) again and obtain

$$r = \frac{R}{R_0} = \frac{m^2}{[1 + c_1(m^{3.2} - 1)]^{\frac{1}{n}}}$$

$$T_e = c_2 m^{0.05}[1 + c_1(m^{3.2} - 1)]^{\frac{1}{n}} \tag{11}$$

$$L = c_2^4 c_3 \frac{m^{4.2}}{[1 + c_1(m^{3.2} - 1)]^{\frac{6}{n}}}$$

where again

$$c_1 = \frac{8 + 24\alpha_0 + 6\beta_0}{8 + 20\alpha_0 + 3\beta_0}\left(\frac{L_0\tau_0}{-\Omega_0}\right) = 1.38\frac{L_0\tau_0}{(-\Omega_0)}$$

corresponds to the ratio of the time scales for mass accretion and contraction and where

$$c_2 = \left[\frac{16}{3}\frac{GM_0}{\varkappa_0 R_0^2}\nabla_{ph}\left(\frac{3cBM_0^2}{4\sigma}\right)^{1+\alpha_0}\right]^{1/[4(1+\alpha_0)+\beta_0]}$$

$$= (T_e)_0$$

and

$$c_3 = 4\pi\sigma R_0^2$$

The formulas (11) depend only on the mass $m = M/M_0$. We put $M_0 = M_\odot$, $X = 0.75$, and $Y = 0.23$, and adjusted the constant B to obtain Sears' (1959) homogeneous sun. We then calculated the evolutionary track backward in time, as shown in Figure 1. The parameter is

$$A^* = A \times 10^{49.4} = \frac{4 \times 10^8 \text{ years}}{\tau_0} \tag{12}$$

The limiting effective temperature for the fully convective models is roughly a

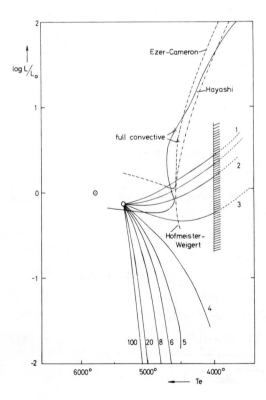

Figure 1. Schematic pre-main-sequence evolutionary tracks. Parameter is $A = 4 \times 10^8$ years/time scale of the mass accretion. The results of Ezer and Cameron (1963), Hayashi et al. (1962), and Hofmeister and Weigert are given for comparison.

vertical line at $T_e = 4000\,°K$ according to equation (9), which is about 500° less than the mean Hayashi track. By eliminating m in equations (9) and (11), we get

$$L \sim T^{300}$$

The formulas (11) are governed by the competition between the terms of equation (4) corresponding to the increase in mass and the decrease in radius, i.e., the contraction of the star. A significant parameter for this ratio is the constant c_1. For sufficiently large c_1 (i.e., a very slow increase in mass) or a small value of A, we approach the case of pure contraction. Accordingly, the fraction of the total mass at the convective limit is large. See the tracks (Figure 1) with $A^* = 1$ to 3. The luminosity decreases and goes through a minimum, e.g., $A^* = 3$. On the other hand, for a sufficiently fast increase in mass (i.e., a large A or a small c_1), the evolutionary tracks in the Hertzsprung–Russell diagram come from below. Slow and fast, here, is with respect to the corresponding Kelvin contraction scale. A time scale for a fast increase in mass, according to equation (12), say, 4×10^6 years, is still long compared to the hydrodynamic time scale. Thus, the assumption made in evaluating equation (4) and in setting ε proportional to the temperature, namely, that the energy gained as kinetic energy of the infalling mass is instantaneously distributed in the whole star, is seen to be justified.

In our investigation, we described the increase in mass inside the photosphere by the accretion formula of Bondi (1952), for a star which is at rest in an interstellar cloud of homogeneous temperature T_a and density ρ_a. Our parameter A then is

$$A = \pi G^2 \left(\frac{5k}{3m_H \mu} \right)^{\frac{3}{2}} \frac{\rho_a}{T_a^{\frac{3}{2}}} \tag{13}$$

Assuming $T_a = 100\,°K$ in our example, then the density is $\rho_a = A^* \times 10^{-20}\,g/cm^3$. We can infer from our model that, if the initial instability of the interstellar cloud is large enough for the total mass to be heated up instantaneously and become confined in an opaque photosphere, the early evolution will go through a fully convective phase. But, if the primordial instability is such that the photospheric layer moves outward with respect to the mass with a time scale appreciably larger than that of sound waves in convection, then the evolutionary track in the Hertzsprung–Russell diagram may come from below and will not be fully convective, except perhaps for the very first nucleus to be formed.

The question of how the pre-main-sequence evolution depends on the initial instability of the primordial interstellar cloud is still open. Although there seems to be only a small margin left to avoid the fully convective phase, it is still possible that the stars do not come down the Hayashi track.

REFERENCES

H. Bondi (1952), *Monthly Notices Roy. Soc.* **112**: 195.
D. Ezer and A. G. W. Cameron (1963), *Icarus* **1**: 422.
C. Hayashi (1961), *Publ. Astron. Soc. Japan* **13**: 450.
C. Hayashi, R. Hoshi and D. Sugimoto (1962), *Progr. Theoret. Phys. Suppl.* **22**: 1.
E. Hofmeister and A. Weigert, Unpublished.
R. L. Sears (1959), *Mem. Soc. Roy. Sci. Liège* **16**: 479.

ROLE OF MAGNETIC ACTIVITY DURING STELLAR FORMATION

E. Schatzman

There are two problems concerned with the effect of electromagnetic activity during the end phase of stellar formation. One is related to the problem of the angular momentum of the stars and the other is related to the problem of surface nuclear reactions.

ROTATION

The main problem is to explain why stars on the lower part of the main sequence are rotating slowly, whereas the stars on the upper part of the main sequence are rotating rapidly. The maximum possible speed of rotation, that is, where the gravitational attraction just balances the centrifugal force, based on the mass–radius relation

$$R \sim M^{\frac{1}{2}}$$

gives an equatorial velocity for a homogeneous star which is

$$V_{\max} = 360 \left(\frac{M}{M_\odot} \right)^{\frac{1}{4}} \text{km/sec} \tag{1}$$

The average observed velocity, taking into account the possible inclination of the axis of rotation, is of the order of

$$V_{\text{obs}} = 270 \left(\frac{M}{M_\odot} \right)^{\frac{1}{4}} \text{km/sec} \tag{2}$$

so the theoretical limit is somewhat higher than the observed values. Thus, there are really two problems, why only stars on the upper part of the main sequence are rotating rapidly and why their equatorial velocity of rotation is slightly smaller than the stability limit.

Electromagnetic activity can explain this. In order to have a dynamo effect which can produce an electromagnetic field, an initial field, rotation, and a convective zone are needed (Parker, 1955). The region of the Hertzsprung–Russell diagram where an outer convective zone exists is bounded by an almost vertical line, i.e., an almost constant effective temperature, where stars with high effective temperature have only a very shallow convective zone (see Figure 1) (Schatzman, 1962). The stars which are following an evolutionary track going through the region where they will possess a convective zone will have electromagnetic activity, while those on the hot side will not have electromagnetic activity. Thus, it may be that the great difference between stars which reach the main sequence at types earlier than F or later than F is the existence of electromagnetic activity during the last part of the contraction stage.

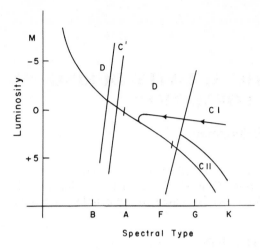

Figure 1. Regions in the Hertzsprung–Russell diagram. Regions C and C′ correspond to stars with a convective zone, D without a convective zone. In region CI, stars present electromagnetic and equatorial mass loss; in region CII, only electromagnetic mass loss occurs.

How is electromagnetic activity important for loss of angular momentum? If angular momentum is lost from the surface of a star when it ejects matter, then the ratio of the specific angular momentum per gram at the surface compared to the average for the star is not favorable. Too small a fraction of the angular momentum of the star is carried away. In order to bring an A star to rest, it must eject 30% of its mass, which is impossible, because it would be seen. If, on the contrary, the ejected mass follows the lines of force, then the angular momentum will be lost at a much greater distance; it will be lost at a distance where the matter is no longer carried by the electromagnetic field. Matter is ejected along the lines of force of the magnetic field. Since the star is rotating, the lines of force become curved due to the Coriolis force acting on the moving matter. When the lines of force become so curved that they are along a roughly circular Keplerian orbit, then the matter is lost from the star. For the case of the sun, this may happen as far out as 100 solar radii. Thus, the angular momentum will be lost at a much larger distance than that from the surface of the star. A factor of 10 increase in the distance reduces the amount of mass ejection needed by a factor of 100, so instead of 30%, only a few thousandths of the mass of the star need to be lost. Thus, in the presence of a magnetic field, a very small mass loss will be sufficient to remove the angular momentum of the star.

If a star reaches the main sequence in the region where there is no convective zone, little electromagnetic activity can be expected, and consequently little loss of angular momentum.

Considering the two sections of the Hertzsprung–Russell diagram, stars which occupy the lower part of the main sequence are expected to have an electromagnetic mass loss and consequently to have been brought, not exactly to rest, but to a small rotational velocity, whereas stars which occupy the upper part of the main sequence have not had much electromagnetic activity and consequently reach the main sequence with a high equatorial velocity.

Why is it that the stars which reach the main sequence with the maximum possible rotational speed are observed to have an equatorial velocity less than the stability limit? This effect has been described by Mestel (1963) as follows: When a star is rotating rapidly, there is a strong meridional circulation producing a strong differential rotation, and this differential rotation can generate a turbulence.

Thus, instead of having a turbulence generated by the convective zone, we may have a turbulence generated by the differential rotation for a fast-rotating star. This turbulence can also generate electromagnetic activity. However, as the star begins to rotate more slowly, these surface currents very quickly sink below the stellar surface and consequently the differential surface rotation, and hence the surface turbulence, quickly become negligible. Thus, for a fast-rotating star, there is an additional effect taking place during the initial pre-main-sequence phase, when the star still has a surface turbulence produced by differential rotation, which tends to decrease the rotation. This disappears after a short time, leaving the star with equatorial velocities of the observed order of magnitude. It is interesting to notice that in the region of the B2 to B5 stars there is also a slightly smaller equatorial velocity. It would be very attractive to relate this to the existence of the helium convective zone.

This completes the first part of this problem: the general scheme of the slowing down of the stars and of the distribution of angular momentum along the main sequence. Further work in greater detail is needed to confirm this outline.

SURFACE NUCLEAR REACTIONS

Now the question of the surface nuclear reactions is considered. At the surface of the sun, during solar flares, fast protons are produced with an energy spectrum which is given by some power law which may be assumed to be

$$n(E) = KE^{-\nu}\,dE \tag{3}$$

In addition, heavy particles will also be accelerated. If the flux of such high-energy particles is large enough, they will produce surface nuclear reactions which will alter the surface composition. The question is how many high-energy protons and alpha-particles, and heavy nuclei, are produced.

The situation is complicated by several problems. First, there are many reactions which must be considered. (p, n) reactions will occur and are important because they produce the neutrons which will alter the surface composition. The reactions

$$\begin{aligned} D + p &\rightarrow He^3 + \gamma \\ He^3 + He^4 &\rightarrow Li^7 + \gamma \end{aligned} \tag{4}$$

must also be considered.

Second, what is the flux of heavy high-energy particles? There is a minimum number of fast particles; these are the fast particles produced by elastic collisions of the fast protons on the nuclei of the chromosphere or the atmosphere of the star. This number is roughly the number of fast protons times the ratio of the abundances of carbon to hydrogen, times a factor which is essentially the mass of the electron over the mass of the proton (due to the fact that this acceleration takes place in flight when the proton is going through the matter):

$$N_{\min} = N_p\left(\frac{X_{\text{heavy}}}{X_H}\right)\frac{m_e}{m_p} \tag{5}$$

This is a small number, but it is the minimum possible number of fast alpha-particles or heavy nuclei. If the accelerating process which takes place at the surface of these stars is more active, there will be more fast particles and their number ratios will approach the chemical composition.

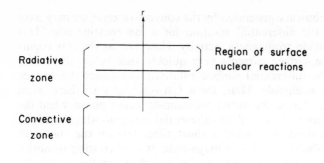

Radiative
zone

Region of surface
nuclear reactions

Convective
zone

Figure 2.

Third, assuming there is a convective zone, a radiative zone, and a region of surface nuclear reaction as shown in Figure 2, the nuclei which are being formed in these regions can be removed by the convective motions. The critical quantity is the time scale of removal of the nuclear reaction products. Consider a layer of 10^{25} hydrogen atoms per square centimeter (neutrons are thermalized in a layer of this thickness, and so this is the layer which is contaminated by the fast neutrons). Then, the surface abundance of He^3 is

$$P_{He^3} = 10^{-22} \frac{(F_p \tau_0)^2 \xi}{1 + 10^{-22} F_p \tau_0}$$

where F_p is the flux of fast protons above threshold, τ_0 is the critical time for dilution, and ξ is the ratio of the rate of neutrons to the protons produced by the p-n reactions. If ξ is of the order of 10^{-3}, i.e., one neutron for 1000 protons, and τ_0 is 10^8 seconds or 5 years, and if the proton flux is 10^{14}, which corresponds to 10^9 ergs cm^{-2} sec^{-1} in the fast-particle spectrum, which is fairly large, then the surface abundance of He^3 is only 10^{21}, which is too small by a factor of 1000 compared to the observation of 3 Centauri A. This difficulty can be surmounted by assuming either a larger flux of protons, which is not impossible, or better, a longer relaxation time for the dilution by convection. Recent investigations by Andouze and Schatzman (1964) have shown that this is possible.

The underabundance of He^4 in 3 Centauri A has not yet been accounted for, because all possible nuclear reactions have not yet been included. Lithium can be formed at the expense of helium, but that is not enough for its destruction.

In addition to convection, mass loss from the surface will remove the products of nuclear reactions from the region of formation. This will be especially true for B stars. For later stars with a convective zone, this may occur by means of the stellar wind.

REFERENCES

J. Andouze and E. Schatzman (1964), *Ann. Astrophys.* (In preparation).
L. Mestel (1963), Preprint.
E. Parker (1955), *Astrophys. J.* **121**: 491.
E. Schatzman (1962), *Ann. Astrophys.* **25**: 18.

Part III. Stellar Evolution
B. Hydrogen-Burning

CALCULATIONS OF MAIN-SEQUENCE STELLAR MODELS

Anders Reiz and Jørgen Otzen Petersen

INTRODUCTION

The fundamental work of Bethe and Weizäcker on nuclear reactions in stellar interiors provided the requisite for a successful attempt to deal with problems of stellar evolution. The advent of fast electronic computers has made possible the practical solution of these problems; in fact, we owe to them the very rapid progress in this field which has taken place over the last decade.

A large group of related problems of growing complexity is gradually becoming accessible, e.g., the advanced evolutionary phases, also including mass reduction, stability problems, star-formation problems, and so on.

We may also expect that the use of computers will mean advances on the technical side, for example in the field of stellar-model computations. In this note, we shall summarize some results we have obtained during the last year at the Copenhagen observatory. An automatic technique for the calculation of stellar models was developed and applied for deriving evolutionary tracks of massive stars on the main sequence.

The solution of the overall problem of stellar structure is achieved by means of numerical integrations of the four classical differential equations, which express the conditions of mechanical and thermal equilibrium in a quasi-stationary state. These equations must be supplemented by the three gas-characteristics relations describing the behavior of the gas, and, further, we have to add the particular conditions for the boundaries.

In the past, the labor of computations had to be greatly reduced; in particular, the outward and inward integrations had to be fitted by labor-saving methods derived on the basis of appropriate transformations. Today, however, the situation has changed completely, thanks to the development of fast computers. Even if one wants to carry through equilibrium solutions for stars of varying masses and over various stages of evolution, this can be done by directly tackling the differential equations in their original form, without applying any of the more sophisticated transformations. Such a direct fitting of inward and outward integrations was followed here.

The procedure developed is straightforward and is perhaps best described by first commenting on the basic differential equations. With the notations $q = M(r)/M$, $f = L(r)/L$, and $x = r/R$, they take the form (see Schwarzschild, 1958):

$$\frac{d \log P}{dx} = -\frac{GM}{R} \frac{q\rho}{x^2 P}$$

$$\frac{dq}{dx} = \frac{4\pi R^3}{M} x^2 \rho$$

$$\frac{df}{dx} = \frac{M}{L}\varepsilon\frac{dq}{dx}$$

$$\frac{d\log T}{dx} = -\frac{3}{16\pi ac}\frac{L}{R}\frac{\varkappa\rho f}{x^2 T^4}$$

or, alternatively, in case of convective equilibrium:

$$\frac{d\log T}{dx} = \frac{\Gamma_2 - 1}{\Gamma_2}\frac{d\log P}{dx}$$

which enters when the condition

$$\left(\frac{d\log P}{d\log T}\right)_{rad} \leqq \frac{\Gamma_2}{\Gamma_2 - 1}$$

is fulfilled. The quantity Γ_2 is defined through the expression

$$\frac{\Gamma_2}{\Gamma_2 - 1} = 4 - 1.5\frac{\beta^2}{4 - 3\beta}$$

where the relative radiation pressure is

$$1 - \beta = \frac{a}{3}\frac{T^4}{P}$$

The equation of state supplies a relation between density, temperature, and pressure. When degeneracy is neglected, this relation is

$$\rho = \frac{H}{k}\mu\beta\frac{P}{T}$$

with

$$\mu = \frac{4}{6X + Y + 2}$$

OPACITY TABLES

Through the kindness of Prof. B. Strömgren, opacity tables have been made available, computed at the Institute for Space Studies, New York, following the procedure of A. N. Cox (1964 and 1965).

The opacity values, including contributions from bound–free and free–free absorption, and electron scattering, are for a specific chemical composition (defined by means of relative amount of hydrogen (X) and heavy elements (Z), given as function of temperature and density). One table may contain as many as 400 opacity values. In a big machine, all these values may be stored in the fast primary memory, which allows fast interpolation. In the present case, however, where the internal memory is comparatively small (1024 words only), subtabulated values for a limited range of $\log T$ and $\log \rho$ are stored in the primary memory, and the machine itself changes the subtabulation as soon as the integration brings us outside the relevant temperature–density range. Second-order interpolation in the actual region yields the opacity values with satisfactory accuracy. In the beginning, we tried to represent the opacity values as function of temperature and density by means of a generalized Kramers' formula; these efforts were later abandoned when it was found that linear or second-order interpolation as described above is preferable. It may also be mentioned that Chebychef representation was tried with success; however, this procedure increased the computing time intolerably and was also abandoned.

ENERGY GENERATION

For the rate of energy generation due to the proton–proton and CNO cycles, $\varepsilon = \varepsilon_{pp} + \varepsilon_{CNO}$, use has been made of the following formulas, which are close approximations to expressions given by H. Reeves (1964).

$$\varepsilon_{pp} = \varepsilon_0(pp)\rho X^2 f_{1,1}\psi(\alpha)$$

$$\varepsilon_{CNO} = \varepsilon_0(CNO)\rho X X_{CNO} \left\langle \frac{X_{14}}{X_{CNO}} \right\rangle f_{14,1}$$

The $\varepsilon_0(pp)$ and $\varepsilon_0(CNO)$ values contained in Tables 9 and 10 of Reeves's review article are approximated with sufficient accuracy by means of a five-point Chebychef polynomial in the temperature interval $8 \leq T_6 \leq 45$:

$$\log \varepsilon_0 = \sum_{k=0}^{5} C_k T_k(\log T)$$

where T_k is the Chebychef polynomial of order k.

The same representation holds with high accuracy for $\log [\alpha/(Y/4X)^2]$, which yields α, by means of which, again, the function $\psi(\alpha)$ is computed:

$$\psi(\alpha) = 1 + 0.958\alpha \left[\left(1 + \frac{2}{\alpha}\right)^{\frac{1}{2}} - 1 \right]$$

The factor $f_{1,1}$ is approximated as follows:

$$f_{1,1} = 1 + 0.25 \left(\frac{\rho}{T_6^3}\right)^{\frac{1}{2}}$$

and for $f_{14,1}$ we write

$$f_{14,1} = 7f_{1,1} - 6$$

for X_{CNO} we take the value $0.6 Z$, and for the mean value $\langle X_{14}/X_{CNO} \rangle$ we adopt 0.96.

In those evolutionary phases where the release of gravitational energy plays a role, the third differential equation has to be modified and is used in the following form (see M. Schwarzschild, 1958)

$$\frac{df}{dx} = \frac{M}{L}\frac{dq}{dx}\left[\varepsilon - \tfrac{3}{2}\rho^{\frac{3}{3}}\frac{\partial}{\partial t}(P\rho^{-\frac{5}{3}}) - \tfrac{3}{2}\frac{\partial}{\partial t}\{P\rho^{-1}(1 - \beta)\} \right]$$

where the time derivatives are approximated as follows:

$$\frac{\partial}{\partial t} = \frac{1}{\Delta t}[(\quad)_t - (\quad)_{t-\Delta t}]$$

Similarly, the hydrogen content in one evolutionary stage $t + \Delta t$ is related to X in the immediately previous stage through the formula

$$X_{t+\Delta t} = X_t + \Delta t \left(\frac{\partial X}{\partial t}\right)_t$$

The change of hydrogen content as function of time $(\partial X/\partial t)$ is given by

$$-\frac{\partial X}{\partial t} = \frac{\varepsilon_{pp}}{E_{pp}^*} + \frac{\varepsilon_{CNO}}{E_{CNO}^*}$$

E_{pp}^* and E_{CNO}^* being the amount of energy released per gram of transmuted matter through the pp and CNO reactions, respectively. The effect of the evolution as set forth through the abovementioned equations will enter in all calculations where the composition parameter appears.

For the hydrogen distribution $X(q)$, no analytical representation has been adopted, but the relevant values are handled by means of an interpolation procedure. This works satisfactorily also in the stages of advanced hydrogen consumption, with gradually shrinking convective core and reduced energy generation in the center, accompanied by increased hydrogen conversion in an outer shell and intensified release of gravitational energy.

INTEGRATIONS

It is well known that divergencies in the numerical integrations are encountered, both when they are carried in the outward direction—essentially caused by the rapidly changing temperature and in the inward direction—due to the occurrence of small radii values.

Hence, a combined use of outward and inward integrations of the four differential equations has to be made. At a conveniently chosen interface, the solutions from both sides have to fulfill the conditions that all physical quantities are continuous at the fitting surface.

The outward-running integration starts from the conditions $q(0) = f(0) = 0$, together with good estimates of the parameters central temperature T_c and pressure P_c. For the inward-running integration, starting from the surface, we assume that pressure and temperature vanish simultaneously; further estimates of the two additional parameters, the luminosity L, and the radius R are needed.

THE FITTING PROCEDURE

The fitting of the inward- and outward-running integrations has been accomplished by means of a variational procedure, which was made completely automatic. Similar techniques have been used by R. L. Sears (1959) and D. Ezer and A. G. W. Cameron (1963). B. Strömgren (1961) has also devised a similar solution of the fitting.

The procedure is based on the assumption that changes in the variables f, q, $\log T$, and $\log P$ are directly proportional to changes in the parameters $L, R, \log P_c$, and $\log T_c$. The essence of the technique is the following:

Assuming that inward and outward integrations, based on reasonably good approximate parameter values and compared at the preassigned fitting point $x = x_f$, yield the residuals Δ_0^i, where the index i refers to the four integration variables q, f, $\log P$, and $\log T$, it is then proposed to make these residuals disappear by imposing the following condition:

$$\frac{\Delta_1^i - \Delta_0^i}{\delta L}\Delta L + \frac{\Delta_2^i - \Delta_0^i}{\delta R}\Delta R + \frac{\Delta_3^i - \Delta_0^i}{\delta \log P_c}\Delta \log P_c + \frac{\Delta_4^i - \Delta_0^i}{\delta \log T_c}\Delta \log T_c = -\Delta_0^i$$

From this linear system, we derive corrections ΔL, ΔR, $\Delta \log P_c$, and $\Delta \log T_c$ to be applied to the initial parameter values, so that the residuals in the improved model are brought to zero. The partial differential quotients at the interface

$$\frac{\Delta_1^i - \Delta_0^i}{\delta L}, \quad \frac{\Delta_2^i - \Delta_0^i}{\delta R}, \ldots$$

are obtained from four sets of inward and outward integrations in which the parameters have been varied one at a time by the amounts δL, δR, $\delta \log P_c$, and $\delta \log T_c$. The residuals at the interface of these integrations are in order $\Delta_1^i, \Delta_2^i, \Delta_3^i, \Delta_4^i$. The procedure is illustrated by means of the scheme shown in Table I.

The procedure can now be repeated, based on the improved parameter values $L + \Delta L$, $R + \Delta R$, $\log P_c + \Delta \log P_c$, $\log T_c + \Delta \log T_c$, and carried until the residuals Δ_0^i are sufficiently small.

The integrations are performed by means of Gill's version of the Runge–Kutta method with integration steps of fixed length h, which can be varied at will, and are generally of the order $0.2\,h$ in the central region. The location of the convective core is found by means of a Newton–Raphson procedure, determining the value x_{conv} in the interval $(x_0, x_0 + h)$ for which the expression $(d \log P)/(d \log T) - (\Gamma_2/\Gamma_2 - 1)$ vanishes. From x_{conv} onward, the integration is continued with a first step equal to $(x_0 + h - x_{conv})$ that brings the integration to the argument $x_0 + h$.

Normally, a limited experience is enough to ensure sufficiently good estimates of the initial parameter values to let the above procedure converge quickly. It has been found, in cases of moderately evolved stars, that the convergence factor is about 10, which means that one integration gives rise to about ten times smaller deviations at the fitting point than the previous one. Another attractive feature of this technique is that the calculation of the corrections ΔL, ΔR, $\Delta \log P_c$, and $\Delta \log T_c$ for not too advanced evolutionary stages can often be based on the original matrix, which means a good deal of reduction of the numerical work. Trial integrations soon indicate a suitable range for choosing the fitting point x_f. For the fitting point outside this region, the numerical instability in the integrations may impair the convergence of the process.

However, inside the range, it seems to be of minor importance for which value of the radius the fitting is performed.

Numerical experiments have also been conducted for a special case of a homogeneous stellar model in order to find how the rate of convergence is influenced by the change of location of the fitting point.

The procedure works satisfactorily with approximately the same efficiency for $x_f < 0.25$. For x_f outside this limit, the rate of convergence is greatly decreased, and

Table I

	Ordinary integration	Integrations for obtaining the partial differential quotients		
Parameter value	L	$L + \delta L$	L	L	L
	R	R	$R + \delta R$	R	R
	$\log P_c$	$\log P_c$	$\log P_c$	$\log P_c + \delta \log P_c$	$\log P_c$
	$\log T_c$	$\log T_c$	$\log T_c$	$\log T_c$	$\log T_c + \delta \log T_c$
Residuals at interface	Δ_0^i	Δ_1^i	Δ_2^i	Δ_3^i	Δ_4^i

for $x_f > 0.28$ the procedure does not work in practice. This is to be expected, since one is here moving out in the region where the numerical stability of outward integrations is impaired.

RESULTS OF CALCULATIONS

The calculations have been performed for population I stars of 6, 10, and 15 solar masses and age-zero composition $X = 0.70$, $Y = 0.27$, and $Z = 0.03$.

We have assumed constancy of mass and complete mixing of elements in the convective core, and further that degeneracy can be neglected. Radiation pressure has been taken into full account. The accurate physical surface conditions have been replaced by the condition that surface temperature and pressure vanish simultaneously. Further, the surface convective zone has been disregarded. The energy generation is by the carbon–nitrogen cycle only; in addition, it is assumed that opacity in the hot central region is due to electron scattering. In no case do the simplifications introduced give rise to serious errors.

The results of the calculations of the evolution in the main-sequence band have been collected in Table II, and the evolutionary tracks in the $(M_{bol}, \log T_e)$ diagram

Table II. Main-Sequence Models

Age (10^6 years)	M_{bol}	$\log T_e$	X_c	q_c	$\log R/R_\odot$	$\log L/L_\odot$	$\log T_c$	$\log \rho_c$
$6M_\odot$								
0.0	−3.05	4.291	0.7000	0.2456	0.4730	3.0617	7.4321	1.1548
12.5	−3.17	4.285	0.5944	0.2174	0.5096	3.1103	7.4414	1.1573
19.0	−3.25	4.280	0.5289	0.2017	0.5350	3.1417	7.4476	1.1609
30.0	−3.40	4.269	0.4011	0.1381	0.5877	3.2020	7.4608	1.1893
35.0	−3.48	4.256	0.3047	0.1245	0.6304	3.2361	7.4721	1.2147
40.0	−3.52	4.241	0.1890	0.1031	0.6673	3.2507	7.4879	1.2638
42.0	−3.55	4.233	0.1314	0.0944	0.6892	3.2618	7.4991	1.2999
44.0	−3.58	4.226	0.0670	0.0850	0.7097	3.2756	7.5181	1.3607
44.8	−3.60	4.227	0.0377	0.0802	0.7128	3.2836	7.5333	1.4076
$10M_\odot$								
0.0	−4.89	4.412	0.7000	0.3269	0.6012	3.8002	7.4777	0.8924
4.0	−5.00	4.408	0.6212	0.3077	0.6307	3.8453	7.4850	0.8910
8.0	−5.14	4.402	0.5285	0.2728	0.6688	3.8986	7.4942	0.8935
12.0	−5.29	4.391	0.4105	0.2158	0.7226	3.9602	7.5067	0.9104
14.0	−5.38	4.380	0.3248	0.1953	0.7608	3.9947	7.5167	0.9293
16.0	−5.47	4.366	0.2223	0.1728	0.8088	4.0305	7.5310	0.9632
17.5	−5.54	4.349	0.1280	0.1534	0.8575	4.0610	7.5492	1.0154
18.0	−5.58	4.342	0.0901	0.1457	0.8765	4.0738	7.5598	1.0468
18.5	−5.60	4.337	0.0490	0.1369	0.8921	4.0853	7.5769	1.0995
$15M_\odot$								
0.0	−6.21	4.494	0.7000	0.4118	0.6997	4.3271	7.5093	0.7109
2.0	−6.32	4.492	0.6296	0.3873	0.7271	4.3731	7.5159	0.7092
4.0	−6.44	4.487	0.5465	0.3576	0.7606	4.4203	7.5239	0.7114
5.0	−6.51	4.483	0.4963	0.3300	0.7829	4.4486	7.5290	0.7154
7.0	−6.66	4.470	0.3806	0.2827	0.8377	4.5088	7.5417	0.7329
9.0	−6.83	4.445	0.2259	0.2444	0.9228	4.5759	7.5624	0.7775
9.7	−6.90	4.429	0.1528	0.2262	0.9683	4.6052	7.5759	0.8137
10.2	−6.96	4.417	0.0925	0.2108	1.0039	4.6280	7.5916	0.8591
10.5	−7.00	4.409	0.0516	0.2003	1.0279	4.6420	7.6084	0.9105

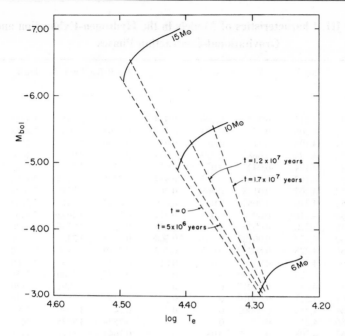

Figure 1. Evolution away from the main sequence for stars of $6M_\odot$, $10M_\odot$, and $15M_\odot$. The dashed lines are the equal-time lines for 0, 5, 12, and 17 million years.

are seen in Figure 1. The initial models define the age-zero line, and curves of ages 5, 12, and 17 million years are displayed.

These results are in reasonable agreement with those obtained by F. Hoyle (1960) and L. G. Henyey, R. LeLevier, and R. D. Levee (1959). The present time scale for the main-sequence evolution is intermediate between these two, apparently reflecting the difference in the assumed hydrogen content.

At the beginning of the computations, we adopted energy-generation rates that were ten times larger than those on which the present computations have been based. In comparing the results, we find very small differences between the models; in particular, the time scales are nearly identical. The effective temperatures of the recent models are higher by an amount corresponding to $\Delta \log T_e = 0.04$, which reflects the difference in radii. Temperatures and densities are slightly higher.

The evolution of the stars of 6 and 10 solar masses has been followed through part of the hydrogen-exhaustion and gravitational-contraction phases, computations being based on the higher energy-generation rates.

So far, the technique described in this paper has worked satisfactorily. There are indications, however, that the greatly increasing release of gravitational energy will violate the procedure; the extrapolation of the parameters R, L, $\log T_c$, and $\log P_c$ to sufficiently accurate values from one stage to the next one is becoming increasingly more difficult.

The results of the calculations have been collected in Table III, and the evolutionary track for 6 solar masses is seen in Figure 2.

Table III. Characteristics of Models in the Hydrogen-Exhaustion and Gravitational-Contraction Phases

Age (10^6 years)	M_{bol}	$\log T_e$	X_c	q_c	f_c	$\log R/R_\odot$	$\log L/L_\odot$	$\log T_c$	$\log \rho_c$
6$M_\odot$									
38.00	−3.38	4.213	0.227	0.100	0.984	0.6971	3.1971	7.4275	1.0986
40.00	−3.41	4.206	0.175	0.093	0.982	0.7165	3.2087	7.4358	1.1247
42.00	−3.44	4.198	0.117	0.085	0.981	0.7369	3.2186	7.4475	1.1638
43.50	−3.46	4.194	0.068	0.079	0.977	0.7509	3.2292	7.4620	1.2108
44.25	−3.48	4.194	0.042	0.075	0.968	0.7547	3.2359	7.4746	1.2509
44.65	−3.50	4.197	0.026	0.072	0.954	0.7526	3.2424	7.4856	1.2848
45.00	−3.52	4.203	0.0128	0.068	0.909	0.7438	3.2516	7.5024	1.3382
45.15	−3.54	4.210	0.0067	0.062	0.829	0.7342	3.2617	7.5160	1.3852
45.22	−3.56	4.216	0.0038	0.055	0.719	0.7264	3.2693	7.5263	1.4289
45.25	−3.58	4.220	0.0026	0.048	0.621	0.7223	3.2766	7.5320	1.4608
45.27	−3.60	4.223	0.00182	0.041	0.508	0.7193	3.2821	7.5359	1.4935
45.295	−3.61	4.226	0.00088	0.023	0.262	0.7159	3.2870	7.5370	1.5691
45.31	−3.60	4.226	0.00047	0.005	0.053	0.7153	3.2839	7.5260	1.6511
45.32	−3.59	4.223	0.00030	0	0	0.7175	3.2800	7.5056	1.7234
45.34	−3.59	4.219	0.00017	0	0	0.7257	3.2801	7.4870	1.8123
45.36	−3.61	4.217	0.00010	0	0	0.7341	3.2877	7.4746	1.8746
45.40	−3.64	4.214	0.00002	0	0	0.7455	3.2980	7.4595	1.9541
45.50	−3.66	4.211	0	0	0	0.7559	3.3075	7.4491	2.0289
45.60	−3.67	4.208	0	0.032	0	0.7649	3.3111	7.4443	2.1075
45.80	−3.69	4.203	0	0.040	0	0.7771	3.3176	7.4430	2.2038
46.00	−3.70	4.198	0	0.048	0.005	0.7896	3.3230	7.4461	2.3022
46.20	−3.71	4.192	0	0.065	0.011	0.8055	3.3268	7.4559	2.4337
10$M_\odot$									
17.50	−5.49	4.308	0.0542	0.1284		0.9303	4.0413	7.5168	0.9359
17.80	−5.52	4.308	0.0281	0.1236		0.9345	4.0502	7.5337	0.9884
17.90	−5.53	4.310	0.0188	0.1215		0.9319	4.0553	7.5437	1.0187
17.97	−5.54	4.314	0.0122	0.1197		0.9265	4.0600	7.5544	1.0512
18.05	−5.57	4.325	0.00458	0.1118		0.9104	4.0700	7.5770	1.122
18.08	−5.60	4.336	0.00171	0.1020		0.8941	4.0824	7.5991	1.198
18.90	−5.61	4.341	0.00075	0.0841		0.8868	4.0876	7.6135	1.263
18.094	−5.62	4.343	0.00042	0.0722		0.8845	4.0923	7.6218	1.310
18.097	−5.63	4.344	0.000199	0.0557		0.8831	4.0936	7.6291	1.365
18.0985	−5.61	4.343	0.000109	0.0399		0.8821	4.0890	7.6322	1.405

It is of interest to compare the results so far obtained for the star of 10 $M_\odot$ with those available from the work of C. Hayashi and R. C. Cameron (1962). Although the assumptions differ in the following respects: mass, age-zero composition, opacity, and energy generation, there is a pronounced qualitative agreement so far as the comparison can be made.

In comparing the stars of 10 and 6 solar masses, the evolution of the less massive is seen to differ from that of the heavier in several respects: it is less violent and more resembles the evolution of a star of 3.89 solar masses computed by Hoyle (1960). When the hydrogen in the central region is nearly exhausted, the temperature drops for a while, due to the rather slow gravitational contraction. This goes on until a substantial helium core has developed. In this phase, the star maintains a stable configuration, in which the energy is generated in a thin shell moving slowly outward.

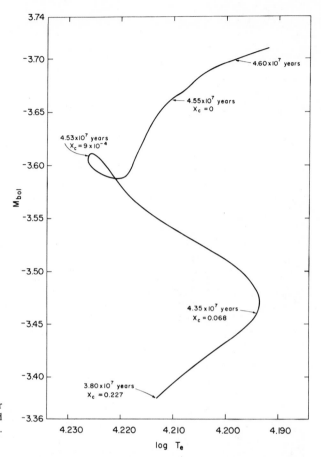

Figure 2. Evolution of a $6M_\odot$ star during hydrogen exhaustion and the beginning of core contraction.

ACKNOWLEDGMENT

The first version of the program, written in ALGOL, was contributed by Mr. Bj. Svejgaard, who also is one of the developers of the GIER computer. We are very grateful to Mr. Svejgaard for this most valuable contribution. We also wish to express our deep appreciation to the Carlsberg Foundation for making available a GIER computer to the Copenhagen observatory.

REFERENCES

A. Cox (1964), in: L. H. Aller (ed.), *Stars and Stellar Systems, Vol. 8*, Chapter 2, to be published.
A. Cox (1965) "Radiative Absorption and Opacity Calculations," this volume, p. 123.
D. Ezer and A. G. W. Cameron (1963), *Icarus* **1**: 422.
C. Hayashi and R. C. Cameron (1962), *Astrophys. J.* **136**: 166.
L. G. Henyey, R. LeLevier, and R. D. Levee (1959), *Astrophys. J.* **129**: 2.
F. Hoyle (1960), *Monthly Notices Roy. Astron. Soc.* **120**: 22.
H. Reeves (1964), in: L. H. Aller (ed.), *Stars and Stellar Systems, Vol. 8*, to be published.
M. Schwarzschild (1958), *Structure and Evolution of the Stars*, Princeton University Press, Princeton, N.J.
R. L. Sears (1959), *Astrophys. J.* **129**: 489.
B. Strömgren (1961), Private communication.

ACKNOWLEDGEMENT

REFERENCES

EVOLUTION OF STARS NEAR ONE SOLAR MASS

P. Demarque

This paper presents series of evolutionary models for stars near one solar mass recently constructed at the University of Toronto using the general method of computation developed by Henyey and his collaborators (1959). In applying Henyey's method to late-type stars, which have extensive outer convection zones, it was found convenient to use new variables. In particular, the parameter in the adiabatic relation $K = PT^{2.5}$ was adopted as one of the variables (Larson and Demarque, 1964). The structure of the hydrogen-convection zone was calculated with the help of the mixing-length theory in the manner described by Demarque and Geisler (1963). The value of K thus obtained from the hydrogen-convection zone was then used as one of the surface boundary conditions. Opacities were from Keller and Meyerott (1955); nuclear-energy generation rates from the work of Reeves (1964).

A number of solar models will first be described, then evolutionary tracks for masses around one solar mass with high and low metal contents will be discussed in an attempt to determine the ages of NGC 188 and the globular clusters.

Models of the sun were calculated for two ages, 4.5×10^9 and 5×10^9 years, and heavy-element abundances ranging from $Z = 0.02$ to $Z = 0.04$. Figure 1 shows the luminosity of the models as a function of their hydrogen content. A relationship is thus found for the two ages between the value of Z and the value of X yielding the present solar luminosity. Figure 2 shows this relation. Some of the results of observational studies are also indicated in the figure: A1 Aller (1953); A2 Aller (1961); F–M Faulkner and Mugglestone (1962); and O–R Osterbrock and Rogerson (1961).

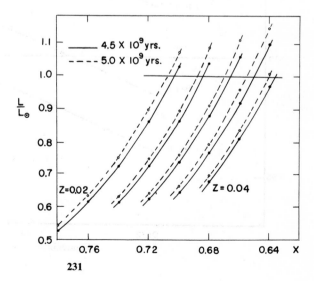

Figure 1.

231

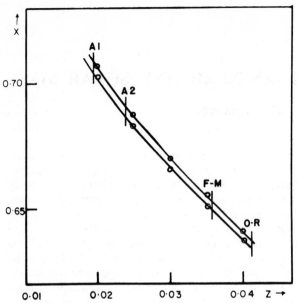

Figure 2.

From Figure 2, one can read possible pairs of X and Z values giving the correct present luminosity for the sun.

The effect of the choice of the effective mixing length in the hydrogen-convection zone in terms of the pressure-scale height is illustrated in Figure 3. The correct

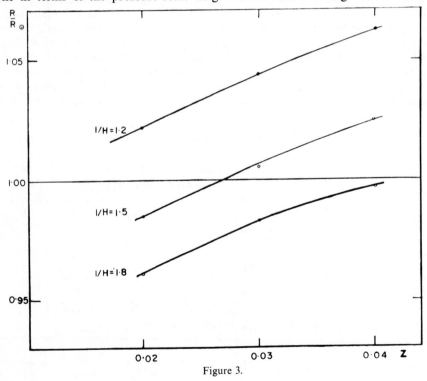

Figure 3.

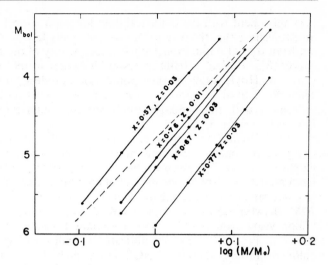

Figure 4.

solar radius is obtained for mixing lengths between 1.4 and 1.8 pressure-scale heights, depending on the composition. In view of the uncertainties in opacities in this region, such a determination of the mixing length may have no close relation to the actual detailed structure of the solar hydrogen-convection zone.

Now consider NGC 188. Unfortunately, the chemical composition of NGC 188 is not very well known. The metal content is estimated not to be very different from that of the sun, so that one can reasonably assume that $Z = 0.03$. In order to determine the helium content, main-sequence models were constructed and the corresponding mass–luminosity curves for different X values plotted. A comparison with the empirical mass–luminosity curve then indicated which helium content yielded

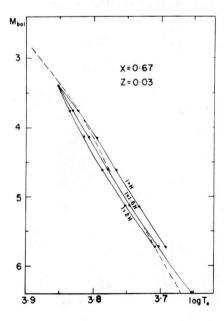

Figure 5.

best agreement with the empirical data for population I stars. It is perhaps reasonable to believe that the original mass–luminosity law of NGC 188 was not too different from that of population I stars in the vicinity of the sun. In Figure 4, one of the curves ($X = 0.77$, $Z = 0.01$) corresponds closely to the composition used by Haselgrove and Hoyle (1959) for their population I models, for the purpose of comparison. Then, for a fixed composition, the value of the mixing length which provides best agreement with the Sandage (1962) zero-age main sequence is determined to be $l = 1.6\,H$ (Figure 5).

Evolutionary sequences of stars with different masses were then calculated using the parameters just determined (chemical composition and effective mixing-length estimate). The masses chosen were $1.03M_\odot$, $1M_\odot$, $0.95M_\odot$, and $0.90M_\odot$. Each model took about one minute of computer time: The iteration process usually required six iterations at the most, each iteration requiring about 10 sec on the IBM 7090. Between 25 and 40 time steps (25 to 40 minutes) are required to cover 10 billion years around $1M_\odot$, with automatic time-step adjustments. The results are shown in Figure 6. Ages are indicated in billions of years. The solid curve is the color-magnitude diagram of NGC 188 as given by Sandage (1962), and the dashed curve is the theoretical main sequence. The mixing-length to scale-height ratio was kept constant throughout the evolution. Notice that the evolutionary track departs from the observations for the $1.03M_\odot$ star after 9.4×10^9 years. There are probably two reasons for this effect: (1) The numerical accuracy began to get poorer when the energy generation approached a shell distribution, and (2) conduction opacity was neglected. This omission is important in the region of electron partial degeneracy near the center. Thus, the age of NGC 188 is found to be between 9 and 10 billion years (assuming $X = 0.67$, $Z = 0.03$).

The same procedure as outlined above was followed for the composition used by Hoyle (1959) for population I stars. A good fit to the zero-age main sequence was found with the mixing length $l = H$ (Figure 7). The evolution of a star with a mass of $0.96M_\odot$ was calculated; the results are shown in Figure 8. Notice that the agreement with observation is also quite satisfactory, although not as good as in the

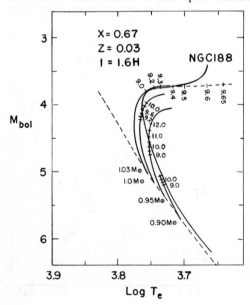

Figure 6.

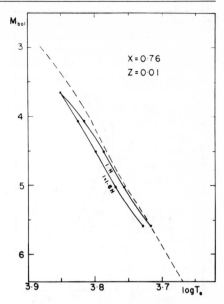

Figure 7.

previous calculation. The corresponding age is a little higher, of the order of 12×10^9 years, but lower than the Hoyle age of 14 to 16×10^9 obtained by Sandage (1962). Of course, this last composition is not as likely since the difference between the metal content of the sun and that of NGC 188 would probably be noticeable if they differed by a factor of 3.*

Next, population II stars were considered. Again the helium content is not known. It is often assumed that population II stars are helium poor for the same

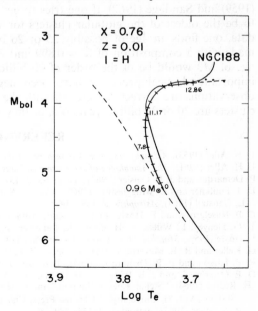

Figure 8.

* This remark may not be valid in view of Gaustad's (1964) recent discussion of the solar composition in which he suggests $Z = 0.02$.

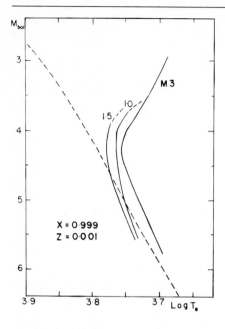

Figure 9.

reason that they are metal poor (thinking in terms of an original galaxy entirely composed of pure hydrogen). The first mixture tried contained no helium: i.e., $X = 0.999$, $Z = 0.001$. The results are shown in Figure 9. Two values of the mixing length ($l/H = 1.0$ and 1.5) were used to show the dependence of the position of the evolutionary track on the mixing-length assumption. The age at the end of both tracks is 22.5 billion years. This agrees fairly well with the results obtained by Hoyle (1959) and Sandage (1962). If one tries to get an age estimate for M3, which seems to be the oldest of the globular clusters for which we possess good observational data, one finds an age of possibly 25 or 26 billion years. Another calculation was made with a composition of $X = 0.899$ and $Z = 0.001$. The curve showed that the age of M3 would be of the order of 20 billion years.* It appears, then, that if no important physical processes have been neglected and if the reductions of the observations are correct, we are left with the following choice: either the globular clusters are 20 or 25 billion years old, or they must have a high helium content.

REFERENCES

L. H. Aller (1953), *Astrophysics: The Atmospheres of the Sun and Stars*, Ronald Press Co., New York.
L. H. Aller (1961), *The Abundance of the Elements*, Interscience Publishers, New York.
P. Demarque and J. E. Geisler (1963), *Astrophys. J.* **137**: 1102.
D. J. Faulkner and D. Mugglestone (1962), *Monthly Notices Roy. Astron. Soc.* **124**: 11.
J. E. Gaustad (1964), *Astrophys. J.* **139**: 406.
C. B. Haselgrove and F. Hoyle (1959), *Monthly Notices Roy. Astron. Soc.* **119**: 112.
L. G. Henyey, L. Wilets, K. H. Böhm, R. LeLevier, and R. D. Levee (1959), *Astrophys. J.* **129**: 628.
F. Hoyle (1959), *Monthly Notices Roy. Astron. Soc.* **119**: 124.
G. Keller and R. E. Meyerott (1955), *Astrophys. J.* **122**: 32.
R. B. Larson and P. R. Demarque (1964), *Astrophys. J.* **140** (in press).
D. E. Osterbrock and J. B. Rogerson, Jr. (1961), Pub. A.S.P. **73**: 129.
 H. Reeves (1964), "Stellar Energy Sources," in: L. H. Aller (ed.), *Stellar Structure* (Stars and Stellar Systems, Vol. 8), University of Chicago Press, Chicago, Illinois.
A. R. Sandage (1962), *Astrophys. J.* **135**: 349.

*More complete calculations, soon to be published, confirm this preliminary result.

THE EARLY EVOLUTION OF STARS BETWEEN ONE AND THREE SOLAR MASSES*

I. Iben, Jr.

The evolution of stars between 1 and $3M_\odot$ has been investigated, with particular attention given to details of the nuclear reactions and to changes in the chemical composition. An evolutionary sequence of models is begun, for convenience, with the fully convective stage of gravitational contraction. The original composition is taken to be

$$X = 0.64 \qquad Y = 0.326 \qquad Z = 0.034$$

Of the heavy elements, O^{16} represents 40%, C^{12} and N^{14} each represent 12.5%. He^3 is taken to be initially zero. The changes in the H^1, He^3, He^4, C^{12}, N^{14}, and O^{16} concentrations are followed in detail. Convection zones are calculated with mixing length equal to one-half the density scale height.

He^3 is created as evolution proceeds and eventually comes into equilibrium with creating and depleting reactions when interior temperatures reach high enough values. If this occurs in a radiative region, the He^3 thereafter maintains its equilibrium abundance; it is locked in equilibrium. Similarly, C^{12} is destroyed until it reaches equilibrium values. It too is locked in equilibrium if the region is radiative. In a convective region, however, the time scale for significant changes in composition is large compared to the mixing time, hence the various elements will be evenly distributed over the entire convective region. If the convective region occurs in a fairly low-mass star where the p–p reactions are dominant, then, since the He^3 local-equilibrium abundance goes up with decreasing temperature, He^3 is pumped from the low-temperature edge to the high-temperature edge of the convective region. The effective energy-generation rate is more steeply dependent on temperature than if local-equilibrium abundances had been chosen—$d \ln \varepsilon / d \ln T$ may increase from 4 to about 5, depending on the size of the convective region. Just the reverse occurs in a convective region within which the C–N cycle reactions dominate energy production. Since the local equilibrium abundance of C^{12} increases with increasing temperature, C^{12} is pumped from the high-temperature edge to the low-temperature edge of the convective region, thereby decreasing the temperature dependence of the CN cycle energy-generation rate.

Consider first the evolution of a $1M_\odot$ star (Figure 1). To evolve from its initial wholly convective state to the point (RC) where a radiative core develops requires 10^6 years. The radiative core grows until it encompasses over 99% of the stellar mass. After 10^7 years, luminosity reaches a minimum and begins to increase. This time of 10^7 years to reach the luminosity minimum agrees with the results of Cameron and Ezer presented at this symposium and is a factor of 10 larger than their previous result based on homologous contraction. At 2.3×10^7 years, nuclear burning begins

* Supported in part by the National Aeronautics and Space Administration and the Office of Naval Research.

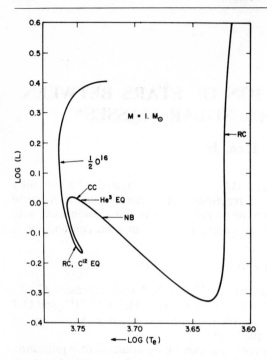

Figure 1. H–R diagram for $1.0M_\odot$—$\log(L/L_\odot)$ versus $\log T_e$.

(NB). Nuclear burning is here defined arbitrarily as the point at which the central energy-generation rate by nucleosynthesis equals the central gravitational-energy-generation rate. At 2.5×10^7 years, He^3 comes into equilibrium (He^3EQ). At 2.6×10^7 years, a convective core appears (CC). The star reaches the main sequence at 3.7×10^7 years, and at 6.20×10^7 years the convective core disappears as C^{12} reaches equilibrium values near the center. The reason for a convective core is the high initial abundance of carbon. The C^{12} (p, γ) N^{13} reaction being relatively fast, carbon burns furiously even before the main sequence is reached. When C^{12} is depleted to equilibrium values near the center, the core comes into radiative equilibrium (RC) and evolution proceeds in the normal fashion. The luminosity now slowly increases, and in about 4.5×10^9 years it reaches the present-day solar luminosity. At this point, the star has 30% by mass of hydrogen at the center. At the point labeled $\frac{1}{2}O^{16}$ in Figure 1, O^{16} at the center has been depleted to half its original value ($t = 6 \times 10^9$ years). During future evolution, the oxygen is not depleted further at the center, since no hydrogen remains to burn it. Evolution is terminated at about 10^{10} years.

The evolutionary track of a $1.5M_\odot$ star is shown in Figure 2. After about 2×10^5 years, a radiative core appears. The luminosity minimum is reached at 3×10^6 years. Nuclear burning sets in at about 8.7×10^6 years. At about 9×10^6 years, He^3 comes into equilibrium in a radiative core. The luminosity then decreases as C^{12} begins to burn rapidly. The star settles on to what might be called a "C^{12} (p, γ)" main sequence ($t = 1.14 \times 10^7$ years). A convective core grows and then decreases as C^{12} is depleted down to equilibrium values. A rapid phase of gravitational contraction carries the star over to the normal main sequence, where the C–N cycle energy generation rate is set by the rate of the N^{14} (p, γ) O^{15} reaction in a convective core.

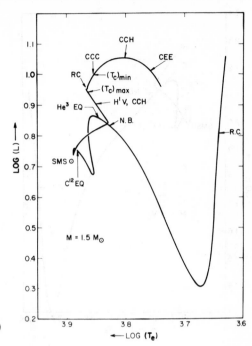

Figure 2. H–R diagram for $1.5M_\odot$—$\log (L/L_\odot)$ versus $\log T_e$.

Also shown in Figure 2 is a standard main-sequence model (computed by the classical fitting procedure) for which it has been assumed that He^3 and C^{12} have local equilibrium abundances in the convective core.

After 1.5×10^9 years, the star depletes the hydrogen in the interior, and the core starts contracting rapidly. Hydrogen is completely depleted in the core at the point labelled H^1 V. The convective core is contracting and heating. At the point where the central temperature reaches a maximum of about 27 million degrees ($t = 1.6 \times 10^9$ years), the flux demanded from the hydrogen-exhausted core by the temperature gradient cannot be supplied by the release of gravitational energy alone. Energy must be supplied from thermal motions, and central temperatures drop. The central temperature gradient becomes radiative (RC) and the core cools as it contracts (CCC) until the central temperature drops to about 21 million degrees ($t = 1.66 \times 10^7$ years). Thereafter the core begins to heat again as it contracts (CCH). The radius of the star begins to expand rapidly and the luminosity decreases as outer layers absorb a considerable fraction of the nuclear energy produced in a thin hydrogen-burning shell. A convective envelope begins to extend into the star at the point labeled CEE ($t = 1.9 \times 10^9$ years). What happens after this point is now being calculated.

Finally, consider the evolution of a $3M_\odot$ star (Figure 3). The star reaches the luminosity minimum in 4.6×10^5 years and begins to burn nuclear fuel at 1.5×10^6 years. A convective core forms, He^3 comes into equilibrium ($t = 1.82 \times 10^6$ years) and is dispersed over the convective core. Carbon starts to burn and the star drops to the "$C^{12} (p, \gamma)$" main sequence ($t = 2.02 \times 10^6$ years). As carbon rapidly burns up and comes into equilibrium, the star contracts briefly and then settles onto the normal hydrogen main sequence. It then moves to the right, having depleted half the hydrogen at its center at about 1.38×10^8 years. At 2×10^8 years, the core

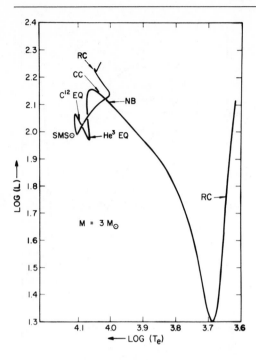

Figure 3. H–R diagram for $3.0 M_\odot$—$\log(L/L_\odot)$ versus $\log T_e$.

starts contracting rapidly. The convective core disappears as central temperatures drop. Further evolution is quite similar to the evolution of the $1.5 M_\odot$ star, except, of course, for the reduced time scale.

The problem of the evolution of smaller mass stars has also been considered. However, the evolution of stars less than $0.5 M_\odot$ occurs on a time scale long compared to the age of the galaxy and its study is beset by several difficulties. (1) In very dense low-temperature regions, partial degeneracy and partial ionization occur simultaneously. This leads to a very complicated equation of state. (2) Screening factors are very important for energy generation. For stars more massive than the sun, it is correct to use the weak screening factor. In dense low-temperature stars, the strong screening factor should be used. However, there is an intermediate region between weak screening at small degeneracy and strong screening at large degeneracy where neither screening expression is valid. (3) For $M < 0.25 M_\odot$ the star is wholly convective, so that surface opacities determine the luminosity and radius; accurate opacities at these very low temperatures, due mainly to H_2^- and Rayleigh scattering, are not yet available.

ON THE PROBLEM OF DETECTING
SOLAR NEUTRINOS

John N. Bahcall† and Raymond Davis, Jr.‡

The evidence supporting the theory of nuclear-energy generation in stars is indirect, based largely upon observations of electromagnetic radiation emitted from the surface of stars and upon theoretical stellar models that have not been subjected to independent experimental tests. It is interesting, therefore, to try to think of a way of *directly* testing the theory of stellar-energy generation in stars. In order to make such a test, one would like to be able to "see" into the deep interior of a star where the nuclear reactions are believed to occur. Thus, an information carrier with a mean free path of the order of 10^{+11} cm ($\sim R_\odot$) is required. In the interior of a star like the sun, light has a mean free path of less than a centimeter. Only neutrinos, which have extremely small interaction cross-sections, can enable us to "see" into the interior of a star. Thus, the observation of solar neutrinos would constitute the most direct test that we can think of for the hypothesis that hydrogen-burning nuclear reactions provide the main energy source for stars like the sun. Moreover, the requirement that a theoretical solar model would have to yield the observed solar neutrino flux would provide an additional, and rather restrictive, condition on acceptable solar models.

Recent theoretical and observational results (Bahcall, 1964a and b, Davis, 1964, and Sears, 1964) have qualitatively changed our ideas concerning the possibility of detecting solar neutrinos. Detailed descriptions of these new results have been given elsewhere; we will summarize here the main conclusions.

The hydrogen-burning fusion reactions in the sun are believed to be initiated by the sequence $H^1(p, e^+ v)H^2(p, \gamma)He^3$ and terminated by the sequences

(1) $He^3(He^3, 2p)He^4$

(2) $He^3(\alpha, \gamma)Be^7(e^-, \gamma)Li^7(p, \alpha)He^4$

and

(3) $He^3(\alpha, \gamma)Be^7(p, \gamma)B^8(e^+ v)Be^8*(\alpha)He^4$

The CNO cycle is believed (Sears, 1964 and 1965) to contribute only a few percent of the energy generation in the sun and a negligible amount of the observable solar neutrino flux. The neutrino fluxes from the hydrogen-burning fusion reactions just listed have been calculated by Sears and others (Sears, 1964 and 1965, Bahcall *et al.*, 1963, and Reeves and Pochoda, 1964) using detailed solar models. When these results are combined with a theoretical discussion of the nuclear physics involved in detecting neutrinos by inverse electron capture, it is found (Bahcall, 1964a and b, and Davis, 1964) that the neutrinos from B^8 decay [sequence (3)] produce 90% of

†Work supported in part by the Office of Naval Research and the National Aeronautics and Space Administration.

‡Work performed under the auspices of the United States Atomic Energy Commission.

the observable reactions, although they constitute less than 0.1 % of the total solar neutrino flux.

The method that will be used to detect solar neutrinos makes use of the inverse electron-capture process $Cl^{37}(v, e^-)Ar^{37}$ and has been extensively discussed by one of us elsewhere (Davis, 1955, 1958, and 1964). On the basis of experience gained in a preliminary experiment (Bahcall, 1964, and Davis, 1964) involving two 500-gal tanks of perchlorethylene, C_2Cl_4, Davis is undertaking an experiment that utilizes a 100,000-gal tank of perchlorethylene as a detector (roughly equivalent to an Olympic-sized swimming pool of cleaning fluid). The most important features of the detection method being employed are that tiny amounts of neutrino-produced Ar^{37} can be removed from the large volume of liquid detector by the simple procedure of sweeping with helium and that the characteristic decay of Ar^{37} can be observed in a counter with essentially zero background. Background effects in an experiment using chemically pure perchlorethylene in a mine 4500 ft deep are expected to be at least a factor of 10 below the predicted rate for solar neutrinos.

Sears (1964 and 1965) has investigated the uncertainties in the predicted neutrino fluxes due to uncertainties in nuclear cross-sections, as well as solar composition, opacity, and age. There are, in addition, uncertainties (Bahcall, 1964) in the predicted neutrino-absorption cross-sections for the reaction $Cl^{37}(v, e^-)Ar^{37}$ due to our incomplete knowledge of the nuclear structure of Cl^{37} and Ar^{37}; experimental studies are currently underway in a number of laboratories in this country that are designed to remove the major gaps in our knowledge of Cl^{37} and Ar^{37}. When the best current estimates of solar and nuclear physics uncertainties were combined, the total predicted number of neutrino captures was found to be $(3.6 \pm 2) \times 10^{-35}$ per Cl^{37} atom per second (Bahcall, 1964a). This rate corresponds to between 4 and 11 predicted solar neutrino captures per day in the 100,000-gal deep-mine experiment.

In order to illustrate how the observation of the solar neutrino flux can be used to determine parameters in the solar interior, we recall that the B^8 neutrino flux is extremely sensitive to the central temperature of the sun (Fowler, 1958). This extreme temperature sensitivity is due to the large Coulomb barrier, compared to solar thermal energies, for the reaction $Be^7(p, \gamma)B^8$ of sequence (3). An experimental upper limit on the central temperature of the sun can therefore be obtained by combining the results of the preliminary experiment (Bahcall, 1964b, and Davis, 1964), which provides an upper limit on the neutrino captures per second per Cl^{37} atom, with the predicted rate and the known temperature dependence of the $Be^7(p, \gamma)B^8$ reaction. Keeping the solar luminosity constant, one finds (Bahcall, 1964b, and Davis, 1964) that the central temperature of the sun must be less than 20 million degrees and that a measurement of the B^8 neutrino flux accurate to $\pm 50\%$ would determine the central temperature of the sun to better than $\pm 10\%$ (Bahcall, 1964b, and Davis, 1964).

It should be noted that a positive result from an experiment of the kind described here would be subject to some ambiguity in interpretation due to the possibility of a galactic source of neutrinos, although present estimates of the probable galactic background indicate it to be negligibly small. A possible method of distinguishing between solar and galactic neutrinos would be to take advantage of the eccentricity of the earth's orbit and measure the 7% difference in solar neutrino intensity between aphelion and perihelion. With a signal as low as 7 captures per day, such an experiment would be marginal, but if a somewhat higher signal is observed a test for the seasonal variation of the neutrino flux will be possible.

POSTSCRIPT

Since the above talk was given, Reines and Kropp (1964) have proposed an experiment to detect solar neutrinos by observing recoil electrons from neutrino–electron scattering. Reines and Kropp point out that such an experiment, which is complementary to the one described in our talk, can in principle give information about the neutrino energy spectrum. In addition, Bahcall (1964c) has shown that under the experimental conditions suggested by Reines and Kropp, all the observed recoil electrons will be confined to a cone of opening angle $\approx 10°$ with respect to the incident neutrino direction. Thus, the observation of electron scattering by neutrinos can in principle enable one to determine the direction (presumably toward the sun) of an extraterrestrial neutrino signal.

REFERENCES

J. N. Bahcall (1964a), *Phys. Rev.* **135**: 137.

J. N. Bahcall (1964b), *Phys. Rev. Letters* **12**: 300.

J. N. Bahcall (1964c), *Phys. Rev.* (to be published).

J. N. Bahcall, W. A. Fowler, I. Iben, Jr., and R. L. Sears (1963), *Astrophys. J.* **137**: 344.

R. Davis, Jr. (1955), *Phys. Rev.* **97**: 766.

R. Davis, Jr. (1958), in: *Radioisotopes in Scientific Research, Vol. 1*, Pergamon Press, New York.

R. Davis, Jr. (1964), *Phys. Rev. Letters* **12**: 302.

W. A. Fowler (1958), *Astrophys. J.* **127**: 551.

H. Reeves and P. Pochoda (1964), To be published.

F. Reines and W. R. Kropp, Jr. (1964), *Phys. Rev. Letters* **12**: 457.

R. L. Sears (1964), *Astrophys. J.* **140**: 477.

R. L. Sears (1965), "Solar Models and Neutrino Fluxes," this volume, p. 245.

SOLAR MODELS AND NEUTRINO FLUXES*

R. L. Sears

The preceding chapter (Bahcall and Davis) described the possibilities of experimentally detecting the neutrino flux from the sun. This paper reports calculations of the expected neutrino flux from the sun, based on models of the solar interior, and attempts to display the range of neutrino fluxes corresponding to the range of uncertainties in solar models.

To build a mathematical solar model, the mass and the distribution of chemical composition throughout that mass must be known (Schwarzschild, 1958, p. 98). The mass of the sun is known, but the distribution of chemical composition is not known from observation. Presumably, the hydrogen/helium ratio is inhomogeneous throughout the sun because of evolution. To find the present chemical-composition distribution, a sequence of evolutionary models must be calculated. First, an initial model with homogeneous composition is constructed. Then, from the theory of stellar evolution, a sequence of models is constructed, stopping at the age of the sun, 4.5×10^9 years. This gives the present model. The reaction rates are then integrated over this model to find the various neutrino fluxes from the various beta decays (Bahcall and Davis, preceding chapter).

The initial composition must be known. It is assumed that the present photosphere reflects the homogeneous initial composition of the sun. Abundances analyses of the photosphere have been made over many years and these can be summarized for this problem in terms of one number—metal-to-hydrogen ratio, or heavy-element-to-hydrogen ratio,

$$\frac{Z}{X} = 0.053 \tag{1}$$

where X is the mass fraction of hydrogen and Z is the mass fraction of all the elements heavier than helium. This result has been obtained from two sources: One source is the oxygen/hydrogen ratio which Osterbrock and Rogerson (1961) derived from forbidden oxygen lines, and the other is the oxygen-to-heavy-elements ratios given, for example, in Aller's (1961) table of the solar system abundances. There is some uncertainty attached to this.

Half the parameters in the initial composition are therefore known; the other parameter needed to construct the solar model is the initial helium content $Y(= 1 - X - Z)$. This is not known from photospheric observation (despite the fact that helium, as its name reminds us, was first discovered in the sun!), but it can be derived by fitting the models to observational data of the sun.

Take a trial value of Y, construct an evolutionary sequence of models, stop at the one for 4.5 billion years, and see whether it agrees with the observations. If it does, the composition is correct: If it does not, it is necessary to pick another trial value of Y and do the problem over again. Since each evolutionary sequence takes

* Supported in part by the National Aeronautics and Space Administration and the Office of Naval Research.

about 10 min on the IBM 7090, this is not too unreasonable a method, although not very elegant. The observational datum to be fitted by a solar model is the luminosity, which is known to within about 1 %. The radius of the sun is also known very accurately, and it would be desirable to fit that, too, but in the present state of convection theory this is impossible, since a calculated solar radius heavily depends on the assumed structure of the convective envelope. Therefore, the radius has not been fitted very accurately in the present work. A very rough theory, namely, an adiabatic convective envelope, has been used to give a very rough radius. Fortunately for this problem, the radius does not affect the central temperature of the sun, nor the luminosity or neutrino fluxes.

The initial helium content which makes the model fit the sun at 4.5 billion years is $Y = 0.32764$, where perhaps only the 0.3 is significant. This number for the helium content depends on several things, of course: it depends on the value assumed for Z/X, it depends on the age assumed, $(4.5 \times 10^9$ years), and it depends on the opacity, on the equation of state, and on the energy generation. For the equation of state, a formula given by Haselgrove and Hoyle (1959) which allows for partial degeneracy has been used; for the opacity the Keller–Meyerott (1955) opacity, by means of the interpolation formula published by Iben and Ehrman (1962), has been used; and for the energy generation all three branches of the proton–proton chain (Fowler, 1960; Parker, Bahcall, and Fowler, 1964) and the carbon–nitrogen cycle (Fowler, 1960) have been included. Since the helium abundance depends on all these uncertain data, the value obtained for it is questionable.

The resulting basic solar model and its dependence on various factors is shown in Table I, which gives the data for the basic model for a fixed $Z/X = 0.053$ and the initial helium content adjusted for the evolutionary sequence to fit the observed luminosity of the sun (3.90×10^{33} ergs/sec). The central temperature is 16.6×10^6 °K, and the neutrino fluxes at the earth are given in the last column. The most significant flux from the observational point of view is that from B^8 which is 4×10^7 neutrinos/ cm^2-sec.

Next, the effect of changes in the convective envelope on the model is shown. K is the constant of proportionality in the adiabatic relation. Taking two different values of K will give two different radii on either side of the observed radius; the observed luminosity is still fitted with the same composition as in the basic model, and the central temperature and the neutrino fluxes are unchanged. Thus, the convective envelope theory used does not particularly affect this problem. The effect of the opacities used, of switching from the Keller–Meyerott to the Los Alamos opacities (Cox, 1964a and b) is shown next. The helium content drops slightly from 0.33 to 0.30. There is a significant effect on the radius because of the line effects on the opacities. The adiabatic proportionality constant K has to be increased in order to get a reasonable radius. The central temperature dropped slightly to 16.3 million degrees, so the B^8 neutrino flux is reduced somewhat (to 3×10^7). This is because the Los Alamos opacities are somewhat lower than the Keller–Meyerott opacities at around 10 million degrees, where the opacity determines the luminosity. The effect of uncertainties in the nuclear reaction rates is shown next. The $He^3 + He^3$ reaction in the proton–proton chain has a very uncertain cross-section (Parker, Bahcall, and Fowler, 1964). A model computed with this cross-section reduced by a factor of 1/6.5 is shown. The resulting central temperature is 16.3 million degrees, but the B^8 neutrino flux goes up somewhat, because with the reduced cross-section the proton–proton chain goes more through the $He^3 + He^4$ branch than through the $He^3 + He^3$ branch. The effect of the possibility that Li^4 is stable, an extreme assumption, is

Table I. Solar Models and Neutrino Fluxes
(Initial $Z/X = 0.053$, Mass $= 1.989 \times 10^{33}$ G)

Identification of model	Initial composition			Surface values			Central quantities			Neutrino fluxes at earth (No. cm^{-2} sec^{-1})			
	X	Y	Z	L (10^{33})	$\log K$	R (10^{11})	T_c (10^6)	ρ_c	X_c	N_{pp} (10^{10})	N_{Be^7} (10^{10})	N_{B^8} (10^7)	$N_{N^{13}} = N_{O^{15}}$ (10^9)
Basic	0.638	0.328	0.034	3.896	−2.6	0.694	16.6	164	0.270	5.2	1.1	4.0	1.8
$\log K$ increase	0.638	0.328	0.034	3.902	−2.4	0.665	16.6	164	0.270	5.2	1.1	4.0	1.8
$\log K$ decrease	0.638	0.328	0.034	3.894	−2.8	0.719	16.6	164	0.270	5.2	1.1	4.0	1.8
Los Alamos opacity	0.663	0.302	0.035	3.897	−2.2	0.690	16.3	176	0.285	5.4	1.0	3.0	1.4
$S_O(3,3) = 200$ keV-b	0.636	0.330	0.034	3.898	−2.6	0.702	16.3	164	0.254	4.7	1.7	4.5	1.3
Li^4 stable	0.655	0.310	0.035	3.890	−2.6	0.730	16.6	182	0.257	4.9	$N_{Li^4} = N_{pp}$		1.9
Age $= 5.5 \times 10^9$ yr	0.646	0.320	0.034	3.902	−2.6	0.708	16.8	183	0.220	5.1	1.2	4.4	2.2
$L_\odot$ variation	0.641	0.325	0.034	3.776	−2.6	0.694	16.4	160	0.284	5.2	1.0	3.4	1.5

Table II. Solar Models and Neutrino Fluxes
(Various Initial Z/X, Mass $= 1.989 \times 10^{33}$ G)

Identification of model			Initial composition			Surface values			Central quantities			Neutrino fluxes at earth (No. cm^{-2} sec^{-1})			
Z/X	O/H	Ne/O	X	Y	Z	L (10^{33})	$\log K$	R (10^{11})	T_c (10^6)	ρ_c	X_c	N_{pp} (10^{10})	N_{Be^7} (10^{10})	N_{B^8} (10^7)	$N_{N^{13}} = N_{O^{15}}$ (10^9)
0.053	0.0014	0.56	0.638	0.328	0.034	3.896	−2.6	0.694	16.6	164	0.270	5.2	1.1	4.0	1.8
0.0364	0.00096	0.56	0.683	0.292	0.025	3.898	−2.7	0.692	16.0	160	0.329	5.6	0.92	2.5	0.79
0.028	0.00096	0.1	0.708	0.272	0.020	3.903	−2.8	0.694	15.7	158	0.359	5.8	0.82	1.9	0.48

shown next. The model does not give very different results. The age of the sun is supposed to be $4\frac{1}{2}$ billion years; if it is $5\frac{1}{2}$ billion years, and a model is constructed to fit that, a somewhat higher central temperature is obtained, and the B^8 neutrino flux is also increased somewhat. So if the sun is any older than the presumed $4\frac{1}{2}$ billion years, the detection of neutrinos will presumably be that much easier. Finally, the effect of the luminosity of the sun is shown in the last row of Table I. A value of 3.90×10^{33} ergs per second has been used (Allen, 1963). The old Kuiper value was about 3% lower, which would lower the central temperature a little and lower the B^8 neutrino flux by approximately 15%.

So much for the variations of the model due to these factors. What about the chemical composition? Recently, rocket observations of solar cosmic rays and flares have established relative number abundances of helium, carbon, nitrogen, oxygen, and neon (Biswas and Fichtel, 1964, and Gaustad, 1964). These five elements, when completely ionized, all have the same charge-to-mass ratio. If they are accelerated by a magnetic process in a solar flare, then perhaps they maintain their relative abundances when they reach our rockets. The carbon/nitrogen/oxygen ratio in these charged particles is found to be 0.6:0.2:1, which is much the same as observed spectroscopically on the surface by Goldberg, Müller, and Aller (1960), 0.6:0.1:1. The neon/oxygen ratio is found to be $Ne/O = 0.1$. It agrees with results by Pottasch (1963) from the emission lines of highly ionized neon in the far ultraviolet and also agrees with Aller's (1961) results for planetary nebulae, but is about one-fifth Aller's value obtained from B stars, which was used for the models in Table I as being the best available.

Another result which has been pointed out by Gaustad (1964) is that the Osterbrock and Rogerson oxygen/hydrogen ratio has slight inaccuracies. Gaustad and Rogerson rediscussed the oxygen/hydrogen ratio of Osterbrock and Rogerson and reduced it by 40% to $O/H = 9.6 \times 10^{-4}$. Both of these new results, the lower oxygen and neon abundances, reduce the ratio of heavy elements Z/X, and so reduce the opacity somewhat. The central temperature must decrease in order for the model to fit the observed sun.

The resulting models for old and new abundances are shown in Table II. The first line is the basic model of the previous table with $Z/X = 0.053$, involving the old oxygen/hydrogen and the old neon/oxygen ratios. The second is for the new oxygen/hydrogen ratio. The helium abundance is 0.29, the central temperature goes down to 16 million degrees, and the B^8 neutrino flux goes down to 2.5 instead of 4.0×10^7. Finally, also adopting the new neon/oxygen ratio, the helium content goes down to 27%, the central temperature drops to 15.7 million degrees, and the B^8 neutrino flux is just about half that in the original model—1.9×10^7. The relative helium abundance found by rocket observations was $He/(C + N + O) = 60$. The helium content from the models, taking the CNO abundances from Aller's tables and the new oxygen and new neon abundances, is about 64, while with the old abundances this ratio is 85.

To summarize: All models predict nearly the same neutrino fluxes, within a factor of 2, so we can have some confidence in the neutrino fluxes predicted from these solar models.

REFERENCES

C. W. Allen (1963), *Astrophysical Quantities*, ed. 2, Athlone Press, London.
L. H. Aller (1961), *The Abundance of the Elements*, Interscience Publishers, New York.
S. Biswas and C. E. Fichtel (1964), *Astrophys. J.* **139**: 941.

A. N. Cox (1964a), this volume, p. 123.
A. N. Cox (1964b), in: L. H. Aller and D. B. McLaughlin (eds.), *Stellar Structure*, University of Chicago Press, Chicago, Illinois.
W. A. Fowler (1960), *Mém. Soc. Roy. Sci. Liège* (Series 5) **3**: 207.
J. E. Gaustad (1964), *Astrophys. J.* **139**: 406.
L. Goldberg, E. A. Müller, and L. H. Aller (1960), *Astrophys. J. Suppl.* **5** (45): 1.
C. B. Haselgrove and F. Hoyle (1959), *Monthly Notices Roy. Astron. Soc.* **119**: 112.
I. Iben, Jr. and J. R. Ehrman (1962), *Astrophys. J.* **135**: 770.
G. Keller and R. E. Meyerott (1955), *Astrophys. J.* **122**: 32.
D. E. Osterbrock and J. B. Rogerson, Jr. (1961), *Publ. Astron. Soc. Pacific* **73**: 129.
P. D. Parker, J. N. Bahcall, and W. A. Fowler (1964), *Astrophys. J.* **139**: 602.
S. R. Pottasch (1963), *Astrophys. J.* **137**: 945.
M. Schwarzschild (1958), *Structure and Evolution of the Stars*, Princeton University Press, Princeton, N.J.

Part III. Stellar Evolution
C. Advanced Stages of Evolution

ADVANCED STAGES OF STELLAR EVOLUTION

Chushiro Hayashi

INTRODUCTION

The nuclear reactions that occur in stars after hydrogen-burning have recently become fairly well known (see the chapter by Reeves). In order to obtain an overall point of view on the advanced phases of stellar evolution, i.e., helium-burning and carbon-burning, calculations for stars of mass $15M_\odot$, $4M_\odot$, and $0.7M_\odot$ were made by Hayashi, Sugimoto, and Hoshi at Kyoto University and by Hayashi and R. C. Cameron at NASA Goddard Space Flight Center, Greenbelt, Maryland.

The basic assumptions used in the calculations are (1) a single star with spherical symmetric distribution of matter, (2) no mixing of chemical composition except in the convective regions, and (3) no mass loss except for possible explosive stages. The essential parameters are the stellar mass and the original chemical composition, which is assumed to be uniform.

GENERAL CHARACTERISTICS OF EVOLUTION

The phases of evolution are shown by the flow diagram given in Figure 1. The characteristics of stellar evolution (i.e., whether the next nuclear burning occurs or not) depend strongly on the mass of the star, but only weakly on the original chemical composition.

The phases are as follows:

1. Gravitational contraction. For stars of less than $0.1M_\odot$, degeneracy occurs during the contraction phase and hydrogen-burning does not occur. The star goes to a black dwarf.

2. Central hydrogen-burning. For stellar masses greater than $0.1M_\odot$, the stars burn hydrogen at their center; this is the main-sequence stage.

3. Shell hydrogen-burning with contracting helium core. For stars of small masses, this stage corresponds to the red-giant branch.

4. Onset of helium-burning at the center. For stars of mass less than $3.5M_\odot$, there is a helium flash. For a star of larger mass, degeneracy does not occur and the star goes straight to the stage of helium-burning in the core. Stars with mass less than $0.5M_\odot$ do not experience the helium flash and go to a white dwarf.

5. Central helium-burning with shell hydrogen-burning.

6. Shell sources of helium- and/or hydrogen-burning with contracting carbon–oxygen core. The core contracts and the central temperature increases.

7. Onset of carbon-burning at the center. For stellar masses greater than $1M_\odot$, carbon-burning occurs; neglecting neutrino emission, the core is nondegenerate for $M > 6M_\odot$, while for the masses between $6M_\odot$ and $1M_\odot$, the electrons are degenerate in the core and a carbon flash occurs. Including neutrino emission, a degenerate core develops for all masses. For stars with mass less than about $1M_\odot$, carbon-burning does not occur and the star contracts to a white dwarf; for $1M_\odot > M > 0.7M_\odot$, another helium flash occurs in the shell, while for $M < 0.7M_\odot$, this second flash does not occur.

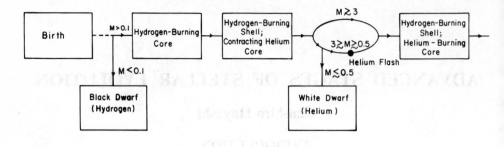

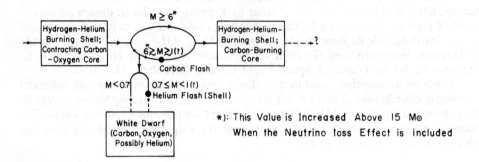

*): This Value is Increased Above 15 Mo
When the Neutrino loss Effect is included

Figure 1.

8. Central carbon-burning with shell helium- and/or hydrogen-burning.

The variation of the central temperature and density during the course of evolution for various stellar masses is shown in Figure 2. The symbols H, He, and C label the stages hydrogen-, helium-, and carbon-burning, respectively. The line $\psi = 0$ is the boundary of the region of incipient degeneracy. The evolutionary tracks for stars of masses $0.7M_\odot$, $4M_\odot$, and $15.6M_\odot$ are shown in Figure 3. Note that during helium-burning the stars of different mass pass through the regions of different types of variable stars.

The qualitative characteristics of the shell structure are as follows: When nuclear shell sources are active, their temperatures are kept nearly constant; these shells remain as nodes of nearly constant radius for the expansion or contraction of the inner and outer regions. For instance, the distance of the shell source from the center for $0.7M_\odot$ and $15M_\odot$ is

$$
\begin{array}{ccc}
 & 0.7M_\odot & 15M_\odot \\
\text{H shell:} & r = \frac{1}{10}R_\odot & \frac{1}{2}R_\odot \\
\text{He shell:} & r = \frac{1}{50}R_\odot & \frac{1}{10}R_\odot \\
\end{array}
$$

In general, shells do not burn simultaneously (some are active and some inactive). A schematic picture of the changes in stellar structure during the phase of helium-burning is shown in Figure 4. The evolutionary tracks were shown in Figure 3. The structural changes that cause the star to move back and forth across the H–R diagram are denoted in Figure 4 by the following letters: (a) onset of helium-burning, (b) $Y_c \cong 0.3$ and contraction begins at the center, (c) shell helium-burning begins,

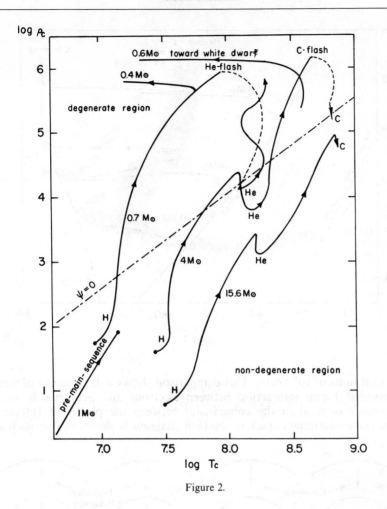

Figure 2.

(d) hydrogen shell source becomes inactive, (e) outer convection zone grows, (f) onset of carbon-burning; and (a′) zero-age helium-burning, (b′) hydrogen shell source becomes inactive, (c′) $Y_c \cong 0.1$ and core begins to contract, (d′) shell helium-burning begins, (e′) outer convection zone grows, (f′) helium shell source begins to be inactive, (g′) hydrogen shell source becomes active again.

If $P_{shell}/P_{center} \ll 1$ (or $T_{shell}/T_c \ll 1$), the core can be treated as a single star. For instance, the luminosity necessary to support a nondegenerate core is proportional to $\varkappa(\mu\langle\beta\rangle)^4 M_{core}^3$ in the case of constant opacity (single star approximation).

EVOLUTION OF A STAR OF $15M_\odot$

The evolutionary track for a star of $15M_\odot$ is shown in Figure 3. The development of the shell structure is schematically shown in Figure 5. At the end of carbon-burning in the core, the star is composed of three zones of mass fractions: hydrogen envelope (73%), helium zone (5%), and metal core, carbon, oxygen, and neon (22%). After carbon-burning, the results are only qualitative. This massive star is a producer of metals rather than helium. The lifetimes in each phase of evolution are shown

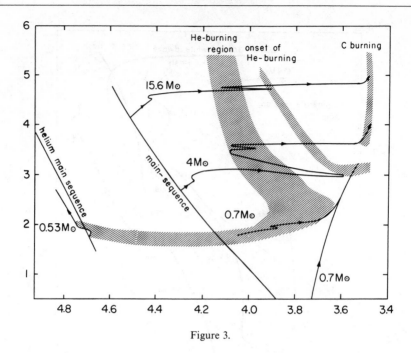

Figure 3.

in Table I (in units of 10^5 years). This comparison shows a discrepancy of factor 10 if the universal Fermi interaction between electrons and neutrinos is assumed. The agreement is good in the comparison between the phases of H (core) and He (core). The evolutionary track in the H–R diagram is shown in Figures 6 and 7.

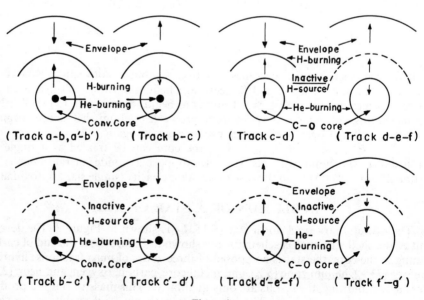

Figure 4.

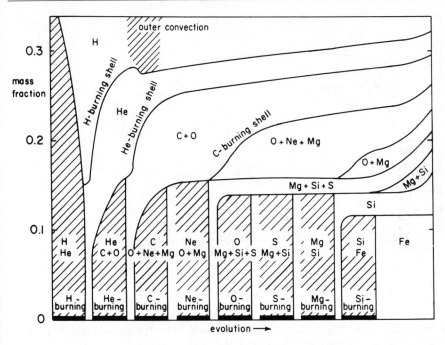

Figure 5.

Table I. Lifetimes (in Units of 10^5 Years)

	Phase of nuclear burning				
	H (core)	He (core)	He (shell)	C (core)	Later
Time (no neutrino loss)	157	12	0.5	2.5	~6
Time (with neutrino loss)	157	12	0.6	~0.2	0
Observations (relative number of stars in in $h + \chi$ Persei)		1.5 ~ 2 (White supergiants)		1 (Red supergiants)	

Figure 6 shows the evolutionary track of a star of $15.6 M_\odot$ superimposed on the color-magnitude diagram of h and χ Persei. Segments of the track correspond to the following phases: (a-b) hydrogen depletion in the core, (b-c) contracting helium core, (c-d) helium depletion in the core, (d-e) contracting carbon–oxygen core, and (e-f) carbon-burning in the core.

The He(core)-burning stage stars are the A supergiants and the later stage stars are the M supergiants. Note that the lifetime of the carbon–oxygen core contraction with He(shell)-burning is extended by the inclusion of neutrinos, because degeneracy will then occur and the central density must rise to a higher value, as will be shown in the following.

The neutrino-loss effect after the formation of a carbon–oxygen core has recently been studied (Hayashi and Cameron) by constructing a sequence of models

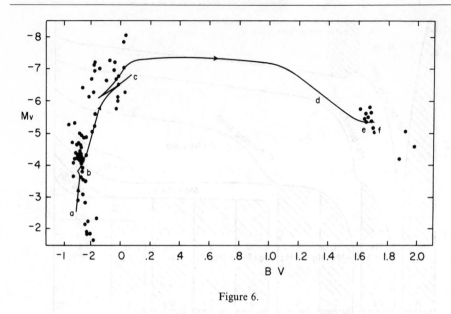

Figure 6.

of the contracting core. The hydrogen envelope and helium zone are neglected for simplicity (single star approximation), and the core is treated as a single star of mass $2.6M_\odot$. As the contraction proceeds, an isothermal core appears at the center because of the rather high temperature dependence of the neutrino-loss rate, the mass of the core grows, and the electrons become degenerate. We have used the newly calculated solutions of a partially degenerate isothermal core in which the relativistic effect is included, since the mass of the isothermal core is large and may be close to the Chandrasekhar limit. In the outer envelope of lower temperature, the energy release (radiation loss) is mainly due to gravitational contraction, and the rate of this release has been taken to be proportional to the temperature. The interface has been determined in such a way that the rate of neutrino loss is equal to the radiation loss, i.e., at the interface, $L_r = 0$ and $\varepsilon_{photon} = \varepsilon_\nu$. The results are shown in Table II. Slightly before the last stage, the carbon flash occurs at the center.

EVOLUTION OF A STAR OF $4M_\odot$

At the onset of hydrogen-burning in the red-giant region, the electrons at the center are marginally degenerate, so that a helium flash does not occur for stars of this mass. The turning point in the H–R diagram in the helium-burning phase is not sensitive to the chemical composition and the nuclear reaction rates. At the

Table II. Results for Star of $15M_\odot$

$M_{core}/M_\odot$	T_{core}	ρ_c	ψ_c	$\psi_{boundary}$	$\log L/L_\odot$
0.00	3.8×10^8	1.8×10^4	-1.8	-1.8	4.23
0.50	4.1×10^8	1.7×10^5	$+0.5$	-1.3	4.42
1.00	5.4×10^8	7.1×10^6	$+9.9$	-1.0	4.72

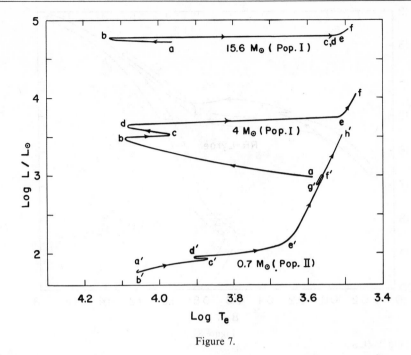

Figure 7.

end of carbon-burning (neutrino loss is neglected), the mass fractions of the different zones are 71% (H), 3% (He), and 26% (metal). This star again produces metals more abundantly than helium. The labels in Figure 7 are the same for the shell structure shown in Figure 4.

EVOLUTION OF A STAR OF $0.7 M_\odot$

In a star of mass $0.7 M_\odot$, there is a helium flash which occurs at the tip of the red-giant branch of population II stars. The conditions just before the helium flash are

$$\frac{M_{He}}{M_\odot} = 0.53 \pm 0.02$$

$$\log \frac{L}{L_\odot} = 3.14 \pm 0.11 \qquad \text{(for } X = 0.9,\ Y = 0.1,\ Z = 0.001\text{)}$$

$$\log \rho_c = 5.94 \qquad \log T_c = 7.95$$

There is some uncertainty due to the opacity and the effect of neutrino losses, but the above values do not depend on the mass in the range $3 M_\odot \gtrsim M \gtrsim 0.55 M_\odot$. After the helium flash, electron degeneracy is dissolved rapidly in the core, the core expands and the envelope shrinks, and the phase of relatively slow central helium-burning in the nondegenerate core begins. This corresponds to the blue end of the horizontal branch of the H–R diagram. The position of the blue end is determined as follows: L is nearly proportional to M_{He}^3, and T_e depends sensitively on M_H/M_{He}. The mass $0.7_\odot$ has been chosen so that the blue end agrees with observation. This mass may be slightly lower than in the stages before the helium flash, in view of the possibility of some mass loss at the time of the helium flash.

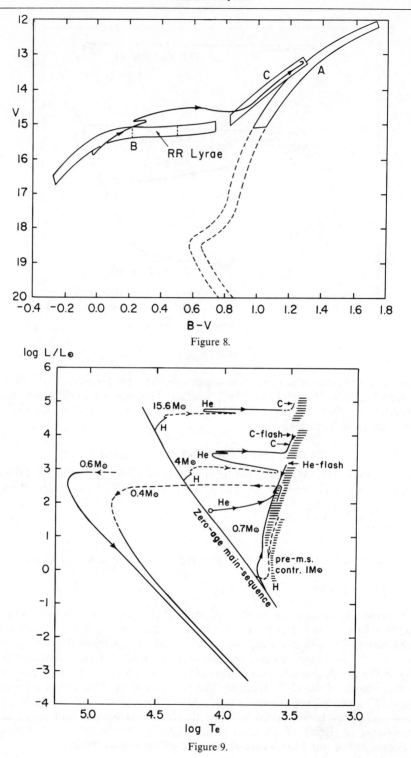

Figure 8.

Figure 9.

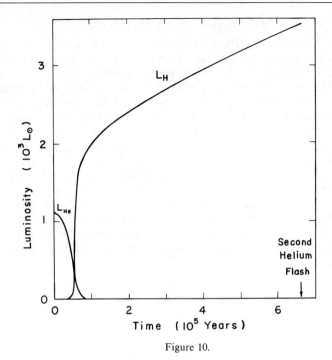

Figure 10.

The evolutionary track in the phases of central helium-burning and shell helium-burning is shown in Figures 7 and 8 (together with the H–R diagram of the globular cluster M5 as observed by Arp) and in Figure 9.

In Figure 7, the first wiggle in the track (at c'd') corresponds to the onset of helium-burning in a shell and the second wiggle (at f'g') corresponds to the re-activation of the hydrogen-burning shell. At the top (f') of the loop, L_{He} begins to decrease, the hydrogen shell source becomes active again, and depletion of hydrogen proceeds to a large extent. The changes in L_{He} and L_H after the stage f' are shown in Figure 10. The mass fractions of the three regions are shown in Table III. Electron degeneracy occurs in the helium shell. When L_H becomes $3 \times 10^3 L_\odot$, the helium zone is heated and there is a second helium flash (at h' in Figure 7), this time in the reactivating shell source. The helium flash occurs in the shell when the mass of the core reaches $M_{C+O} \approx 0.5 M_\odot$. At this stage, $\log \rho_c = 6.2$, $\psi_c = 41$, $\log T_2 = 8.0$, and $\psi_{2e} = 8.5$ (the subscript 2e denotes the bottom of the helium zone). A star of $M < 0.7 M_\odot$ will not experience this second helium flash, but will evolve as a hydrogen-deficient star across the Hertzsprung gap toward the hot subdwarfs, and then white dwarf configurations.

Table III. Mass Fractions

	Top of the loop (%)	Second helium flash (%)
H	24	very small
He	8	29
C + O	68	70

SUPERNOVAE EXPLOSIONS

The above helium and carbon flashes will be candidates for the supernovae explosions which occur at the very last stage of evolution. The other candidate is the implosion of a highly evolved core due to the decomposition of iron into helium and neutrons. At present, it is not clear whether the above flash phenomena generate and propagate to the surface the large amount of energy observed in supernovae. The first helium flash will be more violent than the second helium flash, since the degree of degeneracy is stronger in the former. At any rate, presupernovae are suspected to be red giants (or supergiants) which possess an extensive envelope of low gravity and escape velocity. Ohyama, now at Hiroshima University, made a detailed study on the propagation of shock waves using our envelope solutions (helium zone and extensive hydrogen envelope) of a $15M_\odot$ star in the carbon-burning phase. He found that if a suitable initial shock strength is assumed, the maximum luminosity, the time of duration of this maximum, the expansion velocity, and mass of the ejected material of a type II supernova are all explained in a reasonable way. However, the mechanism of the generation of such a shock wave has not yet been made clear.

THE EVOLUTION OF A STAR OF SEVEN SOLAR MASSES

E. Hofmeister, R. Kippenhahn, and A. Weigert

The Henyey method is the most efficient means of calculating detailed stellar models, especially for phases of the evolution when time-dependent terms in the equations become important. This method has been used to construct a series of evolutionary models for a $7M_\odot$ star. The outer convection zone was calculated using the mixing-length theory with $l/H_p = 1.5$. Four opacity tables were used, one for the outer layers with the original composition ($\varkappa$-values computed with the Los Alamos program) and three for the inner layers with three basic compositions ($\varkappa$-values mainly from the tables of Keller–Meyerott). The equation of state included the perfect gas pressure, the radiation pressure, and the degenerate electron pressure for degeneracy parameters larger than -4. For helium-burning, only the $3\alpha \to C^{12}$ reaction was included; the further burning of carbon into oxygen was neglected. The star was taken to have an extreme population I composition: $X = 0.602$ and $Z = 0.044$.

EVOLUTION

The evolution of the star in the H–R diagram is shown in Figure 1. The evolutionary track confirms the principal features of the tracks of Hayashi for $4M_\odot$. This track is based on more than 200 models, each model containing, as a rough mean, about 30 integration steps above the photosphere and about 200 integration steps below the photosphere.

Some of the physical changes in the interior which give rise to this evolution are shown in Figures 2 and 3. The "cloudy" areas indicate the convective regions, the stripes indicate regions where the energy generation is greater than 10^3 ergs/g-sec,

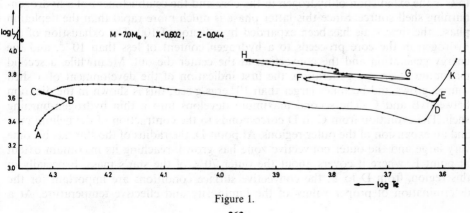

Figure 1.

263

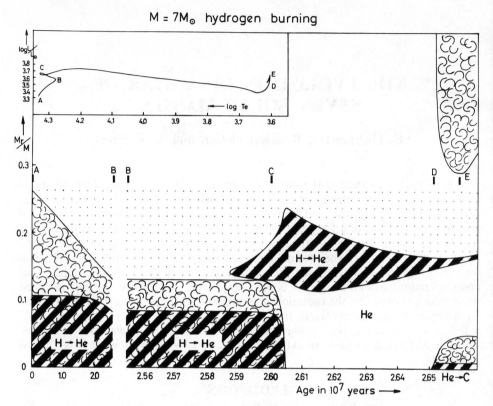

Figure 2.

and the dotted areas indicate the regions where the molecular weight increases inward. The star begins as a homogeneous model on the main sequence. Initially, it has a convective core of about $M_r/M = 0.27$. The evolution from A to B results from the depletion of hydrogen in the core. In approximately the first 25 million years, the hydrogen content of the core decreases to about 2% and the central temperature rises to maintain the energy-generation rate. The convective core shrinks, leaving a region of variable chemical composition. The evolution from B to C is the exhaustion of hydrogen in the core and the establishment of a hydrogen-burning shell source. Since this latter phase is much more rapid than the depletion phase, the time scale has been expanded by a factor of 100. The exhaustion of the hydrogen in the core proceeds to a hydrogen content of less than 10^{-4}, and the energy generation and the convection at the center die out. Meanwhile, a second maximum of energy generation, the first indication of the development of a shell source, grows and becomes larger than 10^3 ergs/g-sec; this is shown in the diagram between B and C. The second maximum develops into a thin hydrogen-burning shell. The evolution from C to D corresponds to the contraction of the helium core and an expansion of the outer regions. At point D, the radius of the star has become very large and the outer convective zone has grown, reaching its maximum extent at point E, where it covers about the outer 70% of the star's mass. Especially in this region, from D to E, the convective surface conditions are important for the determination of proper values of the luminosity and effective temperature. At a

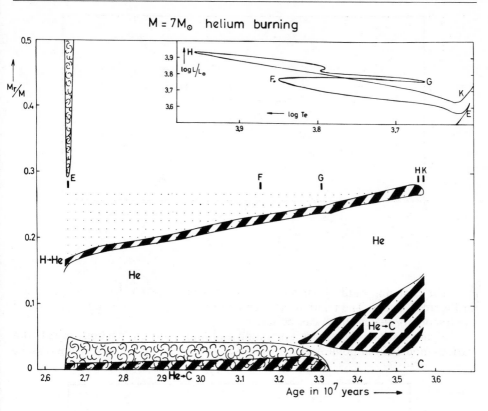

Figure 3.

central temperature of about 10^8 °K (at point D in the diagrams), helium-burning begins at the center, and a new convective core, smaller than that for hydrogen-burning, develops.

At point E, the helium-burning has become efficient enough to stop the rapid contraction of the core, so the outer layers also stop their expansion. During the evolution from E to F, the convective core, in which helium-burning takes place, begins to contract slowly while its mass remains constant. The star's radius decreases. At F, the helium content in the center is about 30%. The evolution from F to G corresponds to the contraction and exhaustion of helium in the core, as the hydrogen shell burns further outward. This hydrogen shell still provides the largest fraction to the star's luminosity, but the helium-burning gradually contributes more and more (about 12% at F, 20% at G, see Figure 9). The helium abundance is 10^{-4} at G. Meanwhile, during the phase F to G, a second maximum of helium-burning is developed from which a helium-burning shell forms. The evolution from G to H corresponds to the contraction of the carbon and oxygen core. But now the stellar radius decreases. At H, the hydrogen shell becomes inactive, finally dying out between H and K. The total energy generation is now due to the helium shell. These last phases are very rapid.

Figures 4, 5, 6 show the temperature, pressure, and density as functions of mass fraction M_r/M through the star for different models. The letters A to K attached to these curves indicate the place in the H–R diagram of the corresponding models.

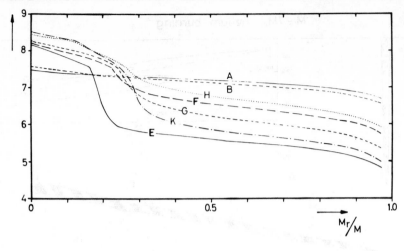

Figure 4.

The changing radii of various mass shells from $M_r/M = 0.005$ to 1 is shown in Figures 7 and 8. The relation between contraction of the core and expansion or contraction of the envelope can be seen.

Figure 9 shows the energy generation and luminosity as functions of mass for various models from point A to just after K. The left-hand graph shows $\log \varepsilon_n$ (solid

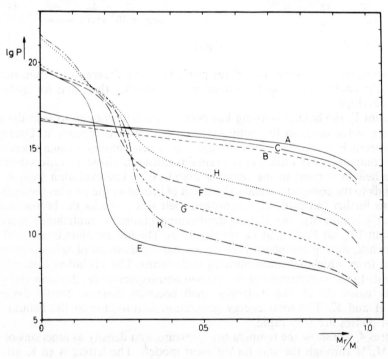

Figure 5.

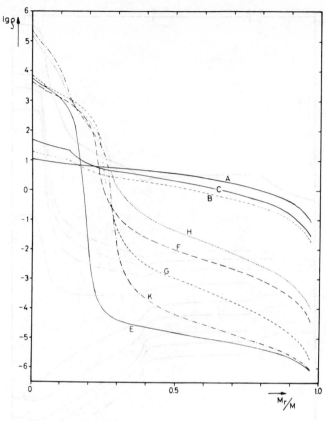

Figure 6.

line), where ε_n is the rate of thermonuclear energy generation in ergs/g-sec, and $\log|\varepsilon_g|$ (dotted line), where $\varepsilon_g = -c_v\dot{T} - P(1/\dot{\rho})$ is the rate of thermodynamic energy release in ergs/g-sec. The right-hand graphs show L_r/L.

During phases of core contraction, the general features to be observed are that a node of ε_g occurs at each shell, which contributes substantially to the luminosity. That is, ε_g abruptly changes sign at the maximum of the active shell sources, and usually the radial velocity changes sign near the shell sources. Thus, there is an accordion effect of expansion–contraction near each active shell source. The evolution of the star is composed of two types of phases: (1) nuclear burning in the core which increasingly depletes the fuel until convection dies out and burning finally stops, and (2) the following contraction of the exhausted core, which raises the temperature and leads to the next nuclear burning stage. For burning and exhaustion of fuel in the core, the corresponding parts on the evolutionary track are: (1) A → B → C, no shell, star moves first to lower than to higher effective temperature, and (2) E → F → G, one shell, star moves first to higher and then to lower effective temperature. For contraction of the core, the corresponding points are: (1) C → D and F → G, one shell, envelope expands; (2) G → H, two shells, envelope contracts and then the hydrogen shell becomes inactive, so that we have (3) H → K, one shell, envelope expands.

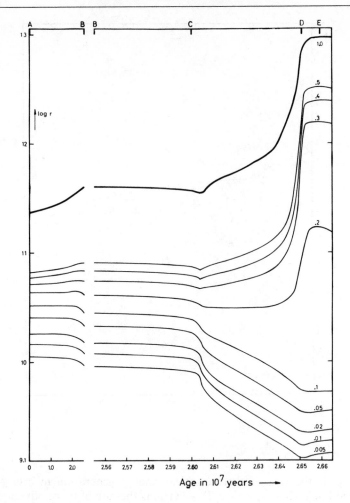

Figure 7.

The lifetimes of the various stages are as follows: Near main sequence (A to C), 26×10^6y; crossing to red-giant region (C to D), 0.5×10^6y; and helium-burning (D to K), 10×10^6y.

CEPHEID VARIABLES

One of the reasons why a $7M_\odot$ star was chosen was that it would pass through the cepheid region. The region of the vibrational instability of the outer layers occurs for about $5200 < T_e < 5500$ °K, and this star makes five passages through this cepheid region. Numbering these five from bottom to top in Figure 1, we see that one goes to the right, two to the left, three to the right, four to the left, and five to the right again. The time intervals the star spends in these five passages are given in Table I.

The first and last are very fast passages, so there is little chance to observe a cepheid in a corresponding phase. The slowest is passage 2, and one would expect

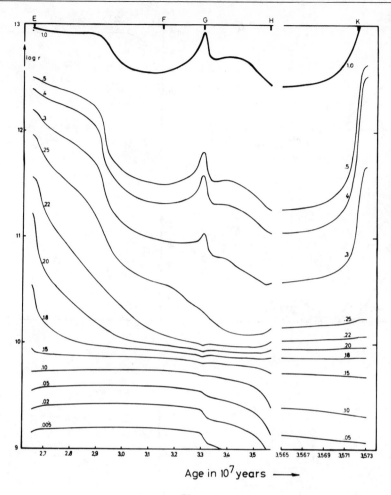

Figure 8.

from Table I that about half of all observed cepheids are in a corresponding phase after the onset of helium-burning.

The next step is to calculate proper eigenfrequencies for these crossings. For crossing 2, this period is 11.4 days (this calculation was done by Temesváry). The evolution of the star as it crosses the cepheid region causes its period to change.

Table I

Passage	Time for crossing (years)
1	2.4×10^3
2	3.5×10^5
3	2.5×10^5
4	1.2×10^5
5	3.6×10^3

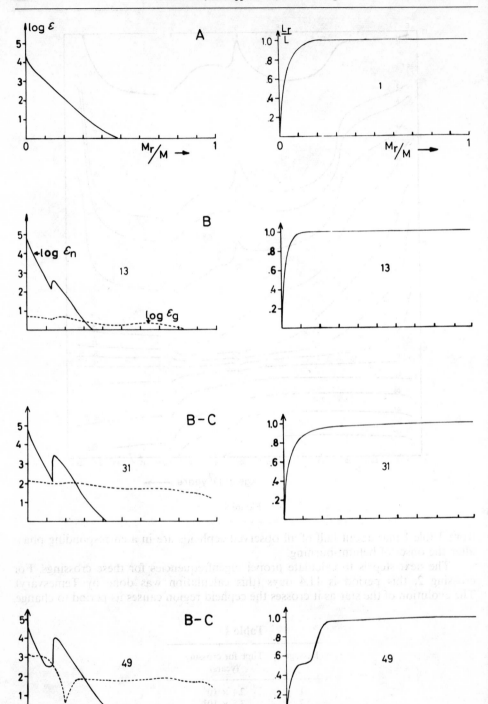

Figure 9(1).

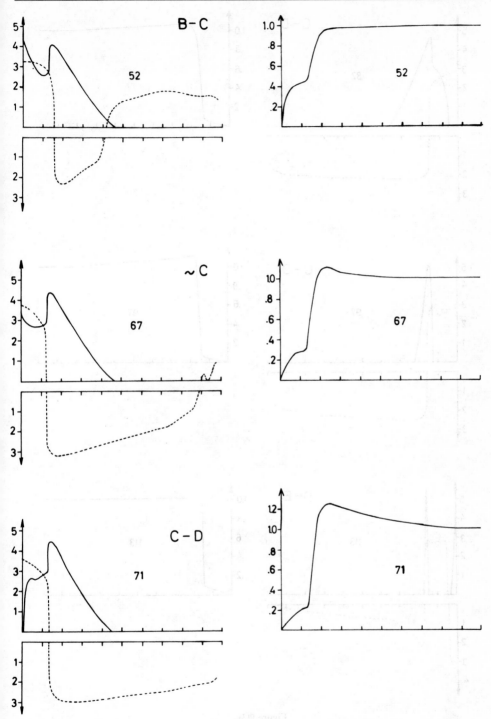

Figure 9(2).

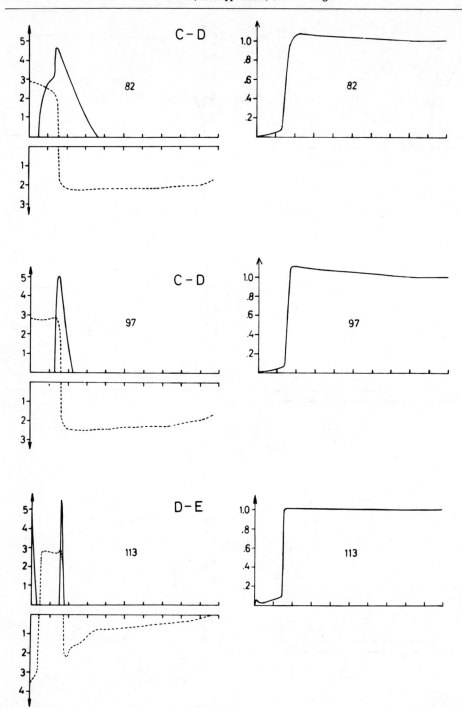

Figure 9(3).

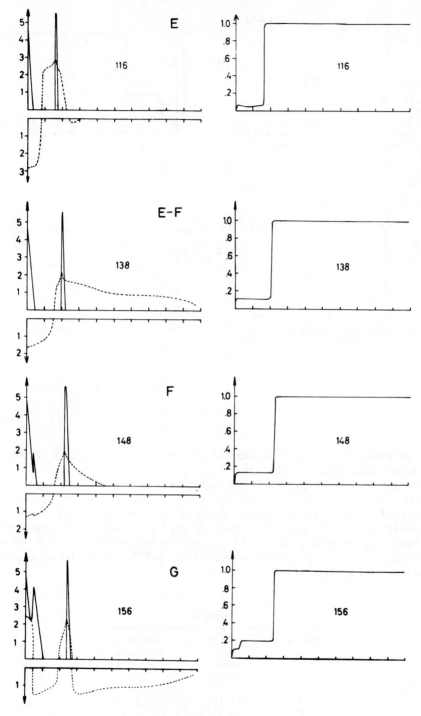

Figure 9(4).

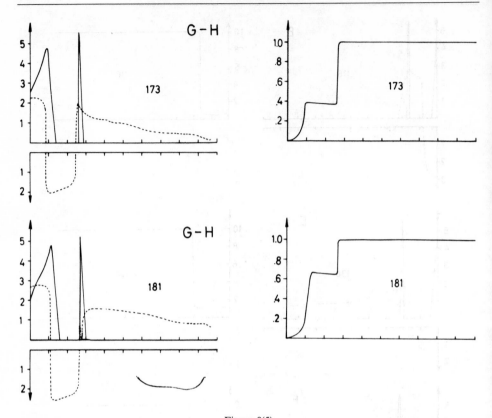

Figure 9(5).

From the period–density relation, the relative changes of the periods in a hundred years have been estimated (Table II).

These values are at least not in contradiction to the observed changes.

The energy input to the oscillations was calculated for various models by N. Baker. The most unstable model of the slowest passage (2) has an effective temperature of 5320 °K and a luminosity of $\log L/L_\odot = 3.663$. This corresponds to a bolometric magnitude of -4^m53 and (with a bolometric correction for this temperature of -0^m18) to a visual magnitude of -4^m35. The fundamental period of this star is 11.42 days.

Table II

Passage	$(\Delta\pi/\pi)$ in 100 years	Increasing ↑ Decreasing ↓
1	5.9×10^{-3}	↑
2	7×10^{-5}	↓
3	1.4×10^{-4}	↑
4	7×10^{-5}	↓
5	8.6×10^{-3}	↑

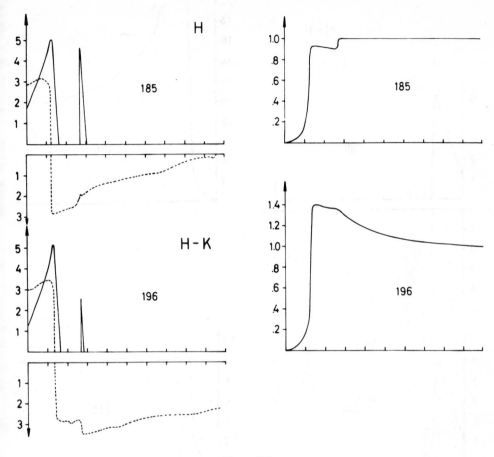

Figure 9(6).

These results may be compared with the observed period–luminosity relation of Kraft, which gives a visual magnitude of -4^m36 for a star of this period. Similar calculations were done for the other four passages. For each of the five passages, Figure 10 shows the visual magnitude of the most unstable model plotted against its $\log \pi^d$-value (π^d = fundamental period in days). The open circles in this figure correspond to observed cepheids and are taken from Kraft's discussion of the period–luminosity relation.

Of course, this good agreement may just be fortuitous. But at least there is no indication from the observations that the cepheids have not come into their region of instability in an evolution similar to that of our models. By more careful and extended calculations, one should try to get especially differential effects, for example, the influence of different chemical compositions on the slope of the period–luminosity relation.

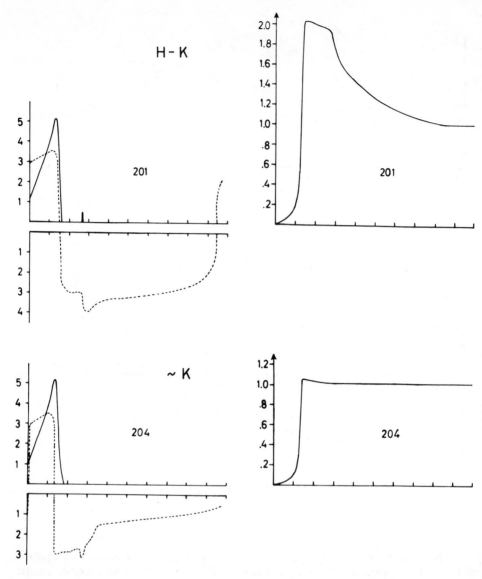

Figure 9(7).

DISCUSSION

W. A. Fowler: You have terminated the helium-burning at carbon and have not included the further burning of the carbon into oxygen. There is, therefore, a pure carbon core in your calculation. I think the details of the evolution will depend considerably on just what the reaction $C^{12} + He^4 \rightarrow O^{16} + \gamma$ does. Your neglecting going on to oxygen might make a major difference.

A. Weigert: I would agree that where the turning points lie will depend on many things which we cannot see.

E. E. Salpeter: Reeves' Figure 9, p. 97, shows the end results of helium-burning. It starts with a pure-helium star, so it would be slightly different for a real one. The figure shows the ratio of carbon,

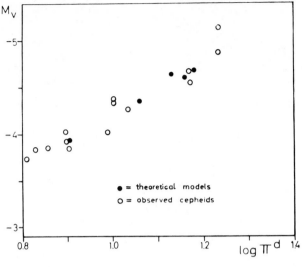

Figure 10.

oxygen, neon, and magnesium as a function of mass. There is still an uncertainty in the ratio of the carbon and oxygen rates, so there are several curves for the final carbon abundance in the core. Only after around $100M_\odot$ does a star begin to produce magnesium, and, as Reeves said, neon is never an end-product at all. The predominant product goes immediately from carbon and oxygen to magnesium.

W. A. Fowler: In low-mass stars at a given temperature, the density is high and the helium-burning to carbon-12 goes as the density cubed, while the other reactions go only as the density squared. Thus there is a very clear-cut, simple separation of the carbon from the oxygen on this basis. In a small-mass star, no matter what the rate of carbon going to oxygen, carbon will always be produced. However, at some mass the density becomes low enough so that appreciable amounts of oxygen will be produced, say, at 10 to $30M_\odot$.

E. E. Salpeter: Just to amplify what Dr. Fowler has said, this question of carbon or oxygen not only affects the mean molecular weight, which is a very small change, but when only 10 or 20% of the helium is left, if you leave out the carbon-to-oxygen stage, you underestimate the actual energy-production rate at a fixed temperature by a factor like 10 or 20, or possibly even more at the very late stages. So the effective energy production rate is quite different when the helium is almost exhausted.

A. Weigert: I would say from these calculations if you change the rate of energy generation in the last stages of helium-burning, the greatest change would come in the luminosity at the peak of the helium-burning, not at the outside, since the outer parts of the stars do not have enough time to follow these changes.

W. A. Fowler: Another point is involved in whether the helium ends up as oxygen or carbon. For example, if the helium-burning ends in oxygen, Hayashi and R. Cameron show that the oxygen-burning is so fast that there are still no red supergiants whether there are neutrino losses or not. Thus many effects depend on the detailed nuclear physics, while, to date, we have only some estimates on which one of these rates may be correct.

A. Weigert: There are, of course, still other points which make it doubtful whether there are giants in the low T_e region. At the onset of helium-burning, the outer 70% of the star is convective, which is very close to the region where the mean molecular weight has changed just a little bit. If there is some penetration, there will be some mixing in the outer layers. What is the effect of this? Will it stop the further growing of the outer convection zone, or increase the outer convection zone to deeper layers?

W. A. Fowler: The ending of the hydrogen-burning is also a very delicate affair. Parker, Bahcall, and I have done what we could to clarify the very complicated processes involved in that stage of the evolution. So the little wiggles in the tracks at the end of hydrogen-burning must also be treated with caution.

PRE-SUPERNOVA EVOLUTION
(NEUTRINO STARS)

Hong-Yee Chiu

NEUTRINOS IN STELLAR STRUCTURE

The later stages of stellar evolution, when the internal temperature is above 10^9 °K, are affected by neutrino processes. Under these conditions, neutrino emission occurs at a rate higher than the photon-emission rate. It is found that the homology relation (density $\sim$ temperature^3) breaks down completely and that, in general, the density rises much faster than the third power of temperature. In regions where nuclear burning commences (C^{12}-O^{16} or O^{16}-O^{16} reactions), the interior stellar structure is determined by the condition (nuclear-energy production rate) + (neutrino-energy dissipation rate) = 0. Such curves have negative power dependence on temperature ($\rho \sim T^{-12}$) or are isothermal. The rapid removal of energy from the core of the star results in a short time scale for these stages, and the star continues to contract. The stars evolve (independently of their mass) to very high central density ($\sim 10^{9-10}$ g/cm^3) at a rather low central temperature (~ 3 to 4×10^9 °K).

The stellar structure equations are

$$\frac{dP}{dr} = - \rho \frac{GM_r}{r^2} \tag{1}$$

$$\frac{dM_r}{dr} = 4\pi r^2 \rho \tag{2}$$

$$\frac{dL_r}{dr} = 4\pi r^2 \rho \mathscr{E} \tag{3}$$

$$\left.\begin{aligned} \frac{dT}{dr} &= - \frac{3}{4ac} \frac{\varkappa\rho}{T^3} \frac{L_r}{4\pi r^2} \quad &\text{(radiative)} \\[2mm] \frac{dT}{dr} &= \left(1 - \frac{1}{\gamma}\right) \frac{T}{P} \frac{dP}{dr} \quad &\text{(convective)} \end{aligned}\right\} \tag{4}$$

The notations used are the standard notations [e.g., those used by Schwarzschild (1958)]. Equation (1) describes the hydrostatic equilibrium, equation (2) is the definition of M_r, equation (3) is the equation of continuity for the energy flux, and equation (4) is the equation of photon energy transport. We refer the reader to the excellent treatise of Schwarzschild (1958) for further details.

We now discuss the influence of neutrino processes on stellar structure. The cross-section for interaction of neutrinos with matter is around 10^{-44} cm^2, in contrast to that for photons, which is around 10^{-20} to 10^{-25} cm^2. The mean free path for neutrinos in lead is around one light year. Only under extreme density conditions ($\rho \gtrsim 10^{11}$ g/cm^3) must one consider the absorption of neutrinos by stellar matter (Chiu, 1964). Neutrino opacities in dense matter have been considered by Bahcall (1964) and Bahcall and Frautschi (1964). Since the star does not absorb neutrinos, they can only enter the stellar-structure equations through equation (3), which describes the energy balance of the star. In equation (3), $\mathscr{E}$ is the energy production term, and we have

$$\mathscr{E} = \mathscr{E}_{gr} + \mathscr{E}_n + \frac{dU}{dt} \tag{5}$$

where $\mathscr{E}_{gr}$ is the rate of gravitational-energy release (in ergs/g-sec), $\mathscr{E}_n$ the rate of nuclear-energy release, and (dU/dt) is the rate energy if dissipated by neutrino emission.

We shall now estimate the rate of photon energy transfer and the rate of neutrino emission inside a star. The photon opacity $\varkappa_{ph}$ of stellar matter is always greater than $\varkappa_e = 0.19$ cm^2/g (the opacity for electron scattering). The mean free path of photons is roughly $(1/\varkappa_e)$ in g/cm^2, and hence is always shorter than 5 g/cm^2. The luminosity of a star, apart from a numerical factor which is not very different from unity, is roughly

$$L \approx \frac{\text{(photon energy content of the star)}}{(R/c)(R/\lambda)} \tag{6}$$

where R is the radius of the star and c is the velocity of light, and λ is the photon mean free path (in centimeters). The relaxation time for a star to be cooled by photon emission τ_{ph} is therefore

$$\tau_{ph} = \frac{\text{(photon energy content of the star)}}{(L)} = \frac{R^2}{c\lambda} \tag{7}$$

By use of a simple model in which the temperature T and the radius R are related by the following condition (which is derivable from the virial theorem)

$$-\text{(gravitational energy)} = \tfrac{1}{2}\frac{GM^2}{R} = \tfrac{3}{2}R_g TM = \text{(thermal energy)} \tag{8}$$

we find

$$\tau_{ph} = 2 \times 10^{11} T_9 \quad \text{[sec]} \tag{9}$$

The corresponding relaxation time via pair-annihilation neutrino emission τ_ν is

$$\tau_\nu = \frac{E}{-(dU/dt)} = 1.5 \times 10^4 \exp\left(\frac{11.9}{T_9}\right) T_9 \left(\frac{\odot}{M}\right)^2 \qquad T_9 \ll 6 \tag{10}$$

at $T_9 = 1$, $\tau_\nu = 2 \times 10^7$ sec, and $\tau_{ph} = 2 \times 10^{11}$ sec. Hence, we can conclude that at $T_9 \gtrsim 1$ the photon-energy transport process cannot be important in the center of stars.

Soon after the temperature exceeds $\sim 10^9$ °K, the rate of energy dissipation by neutrinos will dominate the photon processes. Thus there exist only narrow temperature–density regions where one must consider both photon and neutrino processes simultaneously. Outside this region, one can either completely neglect the photon-energy transfer process (except for luminosity calculations) or completely neglect the neutrino processes. We shall define a star as a "photon star" if the major part of the energy loss of the star is through surface emission, and a "neutrino star" if the major part of the energy loss is through neutrino emission.

The idea of a neutrino star is only an abstraction; in real stars, the surface temperature is between 10^3 to 10^5 °K, and thus in the outer region of a star photon processes must dominate. However, the neutrino-active core will evolve rapidly before the outer region can effectively affect the core's structure (e.g., by the addition of mass to the core, or the addition or removal of energy from the core). Thus, the concept of a neutrino star will be helpful in obtaining the evolutionary properties of a star in its later stages.

Since we neglect L as compared with $-\int (dU/dt)\, dM_r$ (where the integration extends over the whole star), we can also assert that in the neutrino-active region $(dL/dM) \ll -(dU/dt)$. In the first approximation to equation (3), we replace (dL/dM) by zero. Equation (3) then becomes

$$\mathscr{E} = -T\left(\frac{dS}{dt}\right)_M + \mathscr{E}_n + \frac{dU}{dt} = 0 \tag{11}$$

where S is the entropy per unit mass.* Partial differentiation of S with respect to t is taken at a shell of constant mass M_r. (Convection cannot occur in neutrino stars, because the temperature gradient never exceeds the adiabatic temperature gradient.)

Effectively, equation (4) may be ignored in calculating the mechanical structure of the neutrino star in regions where neutrino emission is strong.† The equations for the stellar structure of a neutrino star are given by equations (1), (2), and (11).‡

EVOLUTION OF A NEUTRINO STAR
WITHOUT NUCLEAR ENERGY GENERATION

The energy sources of stars are either gravitational or nuclear in origin. Nuclear reactions occur at well-defined temperatures, and when nuclear energy is not available the star must contract to supply the energy lost through neutrino emission or photon emission.

At lower temperatures ($T \lesssim 10^7$ °K), it was found that stars contract in a simple fashion. The structure of such stars can be described by the so-called polytropes. Under the assumption that the temperature and density relations of a star are such that one can write

$$P = K\rho^{[1 + (1/n)]} \tag{12}$$

where K is a constant depending on the mass of the star and n is the polytropic index, one can simplify equations (1) and (2) by the following substitutions:

$$\rho = \lambda\theta^n \tag{13}$$

$$r = r_0\xi \tag{14}$$

---

* It can easily be demonstrated that $\mathscr{E}_{gr} = -T(dS/dt)$.

† That is to say, the structure of the star is not affected by the inclusion of equation (4). Once the structure is obtained, one can calculate L from equation (4).

‡ Delta-and-epsilonists will be able to prove for us (a) that equations (1), (2), and (11) will yield a solution with ordinary boundary conditions applicable to a star and (b) that the solution is unique.

where r_0 and λ are also constants. The resulting equation is a nonlinear equation

$$\frac{1}{\xi^2}\frac{d}{d\xi}\left(\xi^2\frac{d\theta}{d\xi}\right) = -\theta^n \tag{15}$$

However, a family of solutions is derivable from one existing solution for equation (15): If $\theta(\xi)$ is a solution, then it can be shown that

$$A^{2/(n-1)}\theta(A\xi) \tag{16}$$

is also a solution (this is known as the "homology hypothesis"). Further, when one applies this hypothesis to evolving stars, one finds that inside a star, at a given shell mass M_r, the relation of the density ρ and the temperature T in the course of time is

$$\rho \propto T^3 \tag{17}$$

Hence, if the polytropic index n is a constant in the course of stellar evolution, then equation (17) is valid and the structure of a star can be characterized by the function $\theta(\xi)$; the evolution will introduce only a change in the scale. For proof, see for example, Eddington (1930) or Fowler and Hoyle (1964).

This simple result works remarkably well for many types of stars. One implicit assumption involved in the application of the homology hypothesis to stellar evolution (this is called the "homologous-contraction" hypothesis) is the following: Energy can be redistributed inside stars at a rate much faster than it can be produced by gravitational contraction.

Fowler and Hoyle (1960, 1964) have applied the homologous-contraction hypothesis to stars in their late evolutionary stages. Under certain assumptions, they concluded that a star of mass around $30M_\odot$ will collapse when the center reaches a temperature of about 6×10^9 °K and a density of 2×10^7 g/cm^3. A separate estimate by Chiu (1964) and by Chiu and Fuller (1962) using a similar hypothesis showed that long before this temperature and density is reached a collapse will take place from neutrino processes.

We now examine whether the homologous-contraction hypothesis can be applied to neutrino stars. In the absence of $\mathscr{E}_n$, equation (11) becomes

$$T\left(\frac{dS}{dt}\right)_M = \frac{dU}{dt} \quad (<0)$$

$$= -\left(\frac{\varepsilon_{v_0}}{\rho}\right)T^K \quad \text{for nondegenerate stars} \tag{18}$$

where $T_9 \gtrsim 3$, $\varepsilon_{v_0} = 4.3 \times 10^{15}$, and $K = 9$. For a nondegenerate gas, S is given by the following equation:

$$S = R_g \ln(T^\alpha \rho^{-1}) + \text{constant} \tag{19}$$

where α is a slowly varying constant: $\alpha = 1.5$ for nonrelativistic matter ($T_9 \ll 1$) and $\alpha = 3$ for relativistic matter ($T_9 \approx 6$); at $T_9 = 2$, $\alpha \approx 2.7$.

We now assume that at each shell-mass point, ρ and T evolve according to the following relation:

$$\frac{\rho}{\rho_0} = \left(\frac{T}{T_0}\right)^\beta \tag{20}$$

where ρ_0, T_0, and β are constants. The homologous-contraction hypothesis will require that $\beta = 3$ throughout the star. Substituting equations (19) and (20) into equation (18), one obtains

$$\frac{dT_9}{dt} = \gamma T_9^{K-\beta} \tag{21}$$

where

$$\gamma = \frac{\varepsilon_{v_0} T_0^\beta}{\rho_0 R_g(\beta - \alpha)} \tag{22}$$

Equation (21) can be solved exactly to give

$$T_9 = \frac{T_9^0}{[1 - (t/t_0)]^{1/(K-\beta-1)}} \tag{23}$$

where

$$t_0 = \left(\frac{1}{K - \beta - 1}\right)\left(\frac{\beta - \alpha}{\alpha}\right)\tau_v \tag{24}$$

and τ_v is the relaxation time for the neutrino-energy loss (10). Thus, for a homologously contracting neutrino star, a collapse occurs at a time $t = t_0$. If we use $\beta = 3$, $K = 9$, and $\alpha = 2.7$, then $t_0 = \frac{1}{45}\tau_v$. The free-fall time for a $10M_\odot$ star is reached when $T_9 \sim 3$ (Chiu, 1964).

The collapse can be delayed to higher temperatures if the value of β is close to $K - 1 = 8$. One exact calculation on the evolution of neutrino stars was recently done by Chiu and Salpeter (1964). This calculation confirms our guess that the value of β is close to 8 for the central region of the star. The stellar mass is chosen to be $10M_\odot$, and the initial configuration is chosen to be that for a polytrope of $n = 1.5(\rho \propto T^{1.5})$. No approximation was made in the equation of state, which included contributions from electrons and positrons, radiation, and heavy nuclei (iron). In the $\log \rho$–$\log T$ plane, the structure of the star is initially represented by a straight line of slope 1.5. If the star evolves homologously, then this line will be displaced without distortion along a line of slope 3 (Figure 1). The actual evolution of the star is very complicated, as indicated in Figure 2. Since the neutrino energy-loss rate at high densities is from the plasma-neutrino process, the rate increases with increasing density, and a positive temperature gradient develops throughout the inner one third of the mass of the star.

In this model, we have demonstrated that if nuclear reactions are neglected, then the *homologous-contraction hypothesis cannot be applied to neutrino stars*, for central conditions evolve according to $\rho \sim T^n$, where $6 \leq n \leq 9$.

EVOLUTION OF NEUTRINO STARS WITH NUCLEAR REACTIONS

The time scale τ_v for a star with no energy source other than the gravitational energy to remain at a temperature T is given by

$$\tau_v \approx \frac{E}{(-dU/dT)} \tag{25}$$

In reality, the time scale is shortened by a factor of around 4. On the other hand, nuclear energy (from the O^{16}-O^{16} reaction and the C^{12}-O^{16} reaction) is fairly

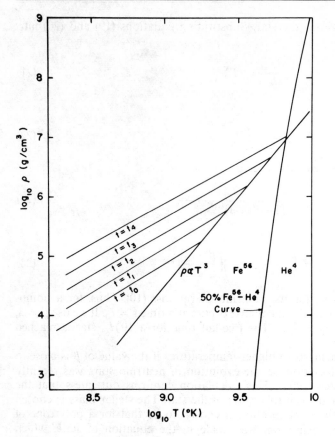

Figure 1. The evolution of homologously contracting polytrope model stars with a polytropic index $n = 1.5$. All parts of the star evolve according to the law $\rho \propto T^3$. For a $10M_\odot$ star, the center will reach the Fe^{56}-He^4 region at a density $\rho \sim 10^7$ and $T \sim 7 \times 10^9$ °K. While the concept of homologous contraction works reasonably well for photon stars, it breaks down completely for neutrino stars, as in Figure 2.

abundant. The time scale τ_n for nuclear burning at a temperature such that

$$\mathscr{E}_n + \frac{dU}{dt} = 0 \tag{26}$$

is several times (~ 10) greater than τ_ν, but still much less than the photon-diffusion time. Because nuclear reactions have much steeper temperature dependence ($\sim T^{30}$) than that of the neutrino rate ($\sim T^9$), a negligible amount of nuclear fuel will be consumed below the temperature defined by equation (26).

Once nuclear burning starts, *in that part of the star*, gravitational contraction can take place only along curves given by the condition

$$\mathscr{E}_{gr} + \mathscr{E}_n + \frac{dU}{dt} = 0 \tag{27}$$

Since, in $\mathscr{E}_{gr}$, ρ and T appear in logarithms [equation (19)], the condition

$$\frac{dU}{dt} + \mathscr{E}_n \cong (\text{constant}) \cdot T \tag{28}$$

determines the temperature and density relation according to which the star can evolve.

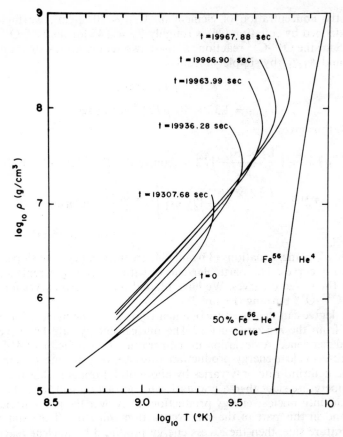

Figure 2. The evolution of neutrino stars. The structures at various times are shown. The highest density point on each curve is the center of the star. Although t is given to 7 decimal places, only the time differences between stages are useful numbers. The evolution is very different from that for a photon star (Figure 1). In this model, the collapse condition is obtained at $\rho \sim 10^9$ g/cm^3, while the star is still composed of iron. This model demonstrates the complete breakdown of the concept of homologous contraction in neutrino stars.

For the C^{12}-O^{16} and O^{16}-O^{16} nuclear reactions, Fowler and Hoyle (1964) have given the following rates:*

$$\log \mathscr{E}_{16\text{-}16} = 55.7 - \log \rho X_{16} X_{16} - (2.3) \log T_9 - \left(\frac{59.04}{T_9^{\frac{1}{3}}}\right)(1 + 0.080T_9)^{\frac{1}{3}} \quad (29)$$

$$\log \mathscr{E}_{12\text{-}16} = 49.7 + \log \rho X_{16} X_{12} - \tfrac{2}{3}\log T_9 - \left(\frac{46.30}{T_9^{\frac{1}{3}}}\right)(1 + 0.080T_9)^{\frac{1}{3}} \quad (30)$$

---

* At present, calculations by Salpeter and Deinzer (1964), Stothers (1964), and Divine (1964) seem to indicate that C^{12}-C^{12} and C^{12}-O^{16} reactions do not take place, since carbon will not be produced by helium-burning in massive stars.

where X_a is the concentration of element a. At $\rho \sim 10^6$ g/cm^3, we find that the temperature defined by equation (26) is roughly $T_9 = 2.45$ for the O^{16}-O^{16} reaction and $T_9 = 1.25$ for the O^{16}-C^{12} reaction; at these two temperatures, one can approximate $\mathscr{E}_{16\text{-}16}$ and $\mathscr{E}_{12\text{-}16}$ by simpler formulas

$$\mathscr{E}_{16\text{-}16} = 8 \times 10^{-5} \rho T_9^{33} \text{ ergs/g-sec} \tag{31}$$

$$\mathscr{E}_{12\text{-}16} = 1.32 \times 10^2 \rho T_9^{31.8} \text{ ergs/g-sec} \tag{32}$$

Equation (28) thus gives

$$8 \times 10^{-5} \rho T_9^{33} + \left(\frac{4.3 \times 10^{15}}{\rho}\right) T_9^9 = (\text{constant}) \cdot T \qquad (O^{16}\text{-}O^{16} \text{ burning}) \tag{33}$$

$$1.32 \times 10^2 \rho T_9^{31.8} + \left(\frac{4.8 \times 10^{18}}{\rho}\right) T_9^3 \exp\left[-2\frac{T_0}{T_9}\right] = (\text{constant}) \cdot T \tag{34}$$

$$(C^{12}\text{-}O^{16} \text{ burning})$$

In the log ρ–log T plane, equations (33) and (34) are lines of negative slope, which we shall call the n-v curves. The particular curves with the constant replaced by zero will be called the n-v-zero curves. We have plotted the n-v-zero curves for O^{16}-O^{16} burning and C^{12}-O^{16} burning (Figure 3).

To what degree can the structure of a neutrino star in the nuclear burning stage itself deviate from the n-v-zero curves? The nuclear-energy rate has a very sharp temperature dependence. A deviation in temperature T by a factor of 4% from the curve will cause nuclear-energy-production rates to vary by more than a factor of 3, while the neutrino rate only varies by about 40%! Imagine that the star is on the lower temperature side; then a gravitational contraction will raise the temperature of the star until nuclear-energy production takes over the gravitational-energy production, and in this part of the star contraction will stop. If the star is on the higher temperature side, then the excess energy produced by nuclear reactions will expand the star until the temperature is low enough so that an energy balance is reached. Thus, the star is always pushed toward the n-v-zero line.

Because of the extremely rapid cooling by neutrino emission, there is a positive temperature gradient in the core of the star, and the maximum temperature, and therefore nuclear burning, occurs in a shell surrounding the core. Thus, the core continues to collapse with $\rho \sim T^8$ until it reaches the n-v-zero line, and nuclear burning begins. The nuclear-burning shell evolves along the n-v-zero line. And, because the time scale of evolution is much shorter than the photon-diffusion time, the energy from the nuclear burning is not transferred to the envelope, which also continues to contract. Thus, the structure of the star approaches the n-v-zero line. When the central density approaches 4×10^7 g/cm^3 for O^{16}-O^{16} burning or 4×10^6 g/cm^3 for C^{12}-O^{16} burning, the plasma-neutrino process becomes important. The plasma-neutrino loss rate Q_t can be approximated to an accuracy of better than 50% at $x \le 1$ by

$$\log Q_t = 6.04 + 3 \log T_9 + \log \rho_6 \tag{35}$$

where, if $\rho_6 \gg 1$, x is given by

$$x = 0.237(1 + 0.6413\rho_6^{\frac{2}{3}})^{-\frac{1}{4}}\rho_6^{\frac{1}{3}} \tag{36}$$

The n-v-zero curve becomes a vertical line, and the temperature is independent of

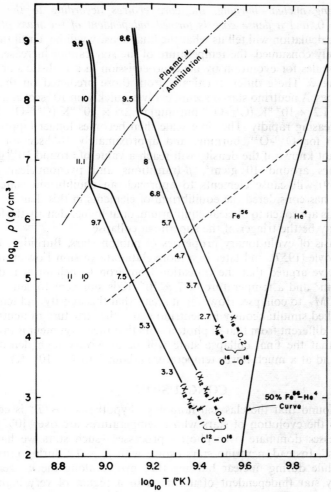

Figure 3. n-ν-zero curves for O^{16}-C^{12} and O^{16}-O^{16} reactions. X_a
is the concentration of the element a, and a curve with $X_a = 0.2$
is also shown for comparison. Along the n-ν-zero curves, the nuclear-
energy release rate is equal to the neutrino rate. The bending at higher
densities occurs because the plasma-neutrino rate becomes important,
and the plasma-neutrino rate has a different density–temperature
dependence from that of annihilation neutrinos. Numbers on the
graphs refer to ($\log \tau_n - \log X_a$). (τ_n is the time scale in seconds for
nuclear burning at that point.) The 50% Fe^{50}-He^4 curve is also
shown. The dotted line (- - - -) is the evolution track of a model star
of $30M_\odot$, studied by Fowler and Hoyle (1964). Actual stars must evolve
along the n-ν-zero lines. The hyphenated line (—·—·—·—) separates regions in which plasma- or annihilation-neutrino processes
dominate as indicated.

the density and is a constant. For the O^{16}-O^{16} burning process, we have

$$\log T_9 = 0.138 - \tfrac{1}{30} \log (X_a X_b) \qquad T_9 \sim 1.38 \qquad (37)$$

$$\text{time scale} \sim 10^8 \text{ sec} \qquad (38)$$

Thus, *during nuclear burning, the core evolves according to the condition* $\mathscr{E}_n + (dU/dt) = 0$, *and a dense core is formed independent of the mass of the star.* Only an exact calculation will tell us what the final density will be. Until the nuclear fuel is completely consumed, the temperature of the star cannot increase.

The time scales for evolution by neutrino emission in a realistic star are indicated on Figure 3. These differ drastically from those predicted on the basis of simple polytropes. A neutrino star, it seems, evolves quickly (in 10^7 sec) at a temperature of around 1.2×10^9 °K (C^{12}-O^{16} burning) or 2.5×10^9 °K (O^{16}-O^{16} burning), its density increasing rapidly. The time scale then becomes longer (approximately 10^8 to 10^9 sec) for O^{16}-O^{16} burning, and approximately 10^{11} sec for C^{12}-O^{16} burning. It is not known if the density will reach a value of around 10^{10} g/cm³.

At densities around 10^7 g/cm³, β-transitions and pycnonuclear reactions (Cameron, 1959) will cause elements to approach an equilibrium state quickly. Tsuruta (1964) has considered the equilibrium of elements in this density–temperature regime. This approach to nuclear equilibrium, or at higher density the formation of neutrons, may be the trigger of the supernova collapse.

On the basis of evolutionary properties of photon stars, Burbidge, Burbidge, Fowler, and Hoyle (1957), and later in a more elaborate version Fowler and Hoyle (1960, 1964), have argued that the transition from iron to helium at a density of around 10^7 g/cm³ and a temperature of 7×10^9 °K is sufficient to cause a star of mass around $30M_\odot$ to collapse. However, it seems that this density and temperature cannot be reached simultaneously in neutrino stars; the structure of neutrino stars is so drastically different from that of photon stars that their argument is invalidated. It is possible that the final collapse state will be at a very high density (about $10^{9\text{-}10}$ g/cm³) and at a much lower temperature (about 3 to 4×10^9 °K).

CONCLUSION

We have found that the classical homology hypothesis $\rho \sim T^3$ is completely inapplicable to the evolution of stars whose temperatures are over 10^9 °K, where neutrino processes dominate over photon processes—such stars we have called "neutrino stars". Instead, neutrino stars evolve with $\rho_c \sim T_c^8$ during gravitational contraction, while during nuclear burning they evolve along the *n-v*-zero curves which lead any star (independent of its mass) to a region of very high density. After the exhaustion of nuclear fuel, the central temperature will increase again. The final collapse of all neutrino stars into supernova will probably occur at a density of around $10^{9\text{-}10}$ g/cm³ and at a temperature of around 3 to 4×10^9 °K. The mechanism for collapse is thought to be the approach to equilibrium of elements (at high density) via β-transitions or pycnonuclear reactions.

ACKNOWLEDGMENT

I would like to thank Dr. L. Adler for his competent programming work on numerical integrations and Dr. A. G. W. Cameron and Dr. E. E. Salpeter for discussions.

REFERENCES

J. N. Bahcall (1964), *Phys. Rev.* **136**: B1164.
J. N. Bahcall and S. Frautschi (1964), *Phys. Rev.* **136**: B1547.
E. M. Burbidge, G. R. Burbidge, W. A. Fowler, and F. Hoyle (1957), *Rev. Mod. Phys.* **29**: 547.
A. G. W. Cameron (1959), *Astrophys. J.* **130**: 884.
H.-Y. Chiu (1964), *Ann. Phys.* **26**: 364.
H.-Y. Chiu and R. Fuller (1962), NASA Goddard Institute for Space Studies preprint.

H.-Y. Chiu and E. E. Salpeter (1964), To be published.

T. N. Divine (1964), Ph.D. thesis, California Institute of Technology.

A. S. Eddington (1930), *Internal Constitution of the Stars*, Cambridge University Press, England.

W. A. Fowler and F. Hoyle (1960), *Astrophys. J.* **132**: 565.

W. A. Fowler and F. Hoyle (1964), "Neutrino Processes and Pair Formation in Massive Stars and Supernovae," to be published in *Astrophys. J. Suppl.* or Cal. Tech. preprint.

E. E. Salpeter and W. Deinzer (1964), *Astrophys. J.* **140**: 499.

M. Schwarzschild (1958), *The Structure and Evolution of the Stars*, Princeton University Press, Princeton, N.J., p. 96.

R. Stothers (1964), Private communication.

S. Tsuruta (1964), Ph.D. thesis, Columbia University.

MECHANISM OF TYPE II SUPERNOVA EXPLOSIONS

Stirling A. Colgate and Richard H. White

The original theory of the origin of supernovae was proposed by Burbidge, Burbidge, Fowler, and Hoyle (1957). It is thought that as a massive star continues to contract after the end of nuclear burning, it will ultimately reach a sufficiently high temperature to cause the endothermic conversion of iron back into helium. This thermal decomposition absorbs so much energy that the resulting decrease in pressure triggers the collapse of the star. Supposedly, the collapsing material eventually becomes so dense that it bounces and generates a shock wave which blows off some of the outer stellar material. It is actually very difficult to propagate a shock "up" the imploding velocity gradient and out of a star, and so another mechanism is needed to transfer the energy of the collapsing core to the outer layers to blow them off. Such a mechanism is found in the very-high-energy thermal

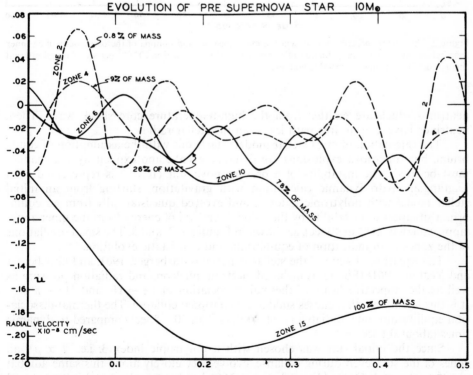

Figure 1. The velocity versus time oscillations of the zone boundaries represent sonic disturbance around the equilibrium condition. The kinetic energy of these oscillations is about 10^{-4} of the internal energy.

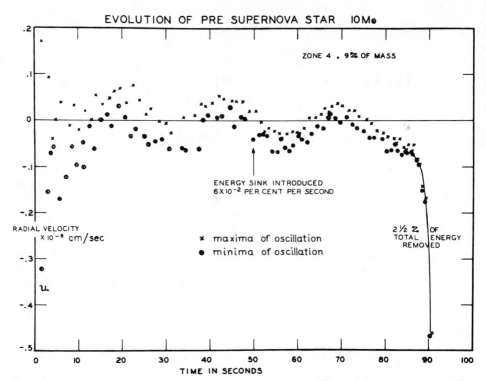

Figure 2. The circles and crosses represent the many maxima and minima of the oscillations of an inner zone before and after the introduction of a slow artificial heat sink $6 \times 10^{-2}\%$ per sec of initial internal energy and $2\frac{1}{2}\%$ at time of unstable collapse.

neutrinos which are emitted from the high-temperature core shock wave. These neutrinos have a mean free path less than the stellar radius.

The calculation of most stellar models starts out with the assumption of equilibrium, but supernova explosions are dynamic events, and explicit hydrodynamics must be used. The instability of a presupernova star $(10M_\odot)$ is represented by a Lagrangian hydrodynamic calculation with gravitation, starting from an initial stellar model with polytropic index 3, and evolved quasistatically from an equilibrium situation to instability by the "slow" removal of energy from the inner zones (approximately $2\frac{1}{2}\%$ in 40 sec), as shown in Figures 1, 2, and 3. The small oscillations of the zones is an indication of equilibrium and quasistatic evolution.

The equation of state of the stellar material (Grasberger, 1961, and Grasberger and Yeaton, 1961) (Figure 4) included electron, nucleon, and radiation pressure as well as the respective heats of thermal dissociation of Fe → He and He → $n + p$. It is this latent heat that causes sudden gravitational collapse. The thermal-dissociation equilibrium time (radiative) is short (less than 10^{-16} sec) compared to dynamic times (about 0.1 sec).

Since the initial star was chosen with a polytropic index 3, i.e., $T \propto \rho^{\frac{1}{3}}$, all zones of the star, even during collapse, evolve very closely along this same adiabat $\gamma = (\frac{4}{3} - \varepsilon)$, $\varepsilon \ll 1$. One additional star of $3M_\odot$ has been evolved and it reproduced the instability.

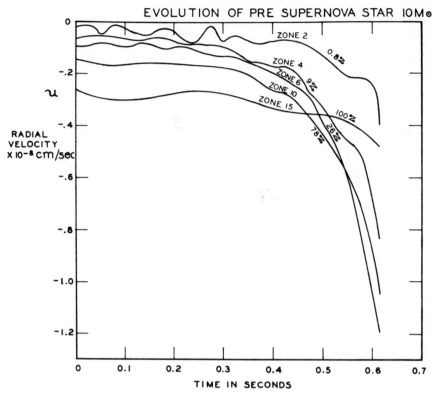

Figure 3. Velocity history of zones at start of collapse. The ordering of the zones in radius is consecutive, but is partially random in velocity space.

A rezoned star, $10M_\odot$, (Figure 5) started at the condition of instability and collapsed inward with no indication of bouncing at a central density of $\rho_{max} = 3 \times 10^{11}$ g/cm^3. Further calculation with the initial equation of state becomes invalid because the rate of inverse β-decay of protons to neutrons becomes comparable to the dynamic collapse time, so that a neutron star equation of state must be used.

The free fall at high density and lack of reflected energy (bounce) (Figure 5) is to be expected from the equation of state, and so, to confirm our understanding of the requirement for a new equilibrium, or "bounce," the equation of state was modified to include a small fraction of initial pressure (10^{-3}) of a hypothetical gas of $\gamma = 2$. The dashed curves show the resulting bounce at the expected 10^3-fold compression and the heavy dots the reflected shock wave. This shock wave is probably strong enough to eject 10% of the mass of the star, although a detailed calculation was not completed. The artifice of introducing a component of $\gamma = 2$ gas was thought to represent, possibly, a small component of angular momentum or dipole magnetic field. The first possibility is rejected on the basis of wheel rotation of the star as a whole and the concomitant small angular momentum of the collapsing core. The second possibility must be rejected because, for a spherical contraction of a dipole magnetic field, the equivalent γ becomes $\frac{4}{3}$ [pressure $\sim B^2 \sim r^{-4}$, volume $\sim r^3$, $pV^{\frac{4}{3}} = $ constant]. Therefore, a further dynamic collapse of the star

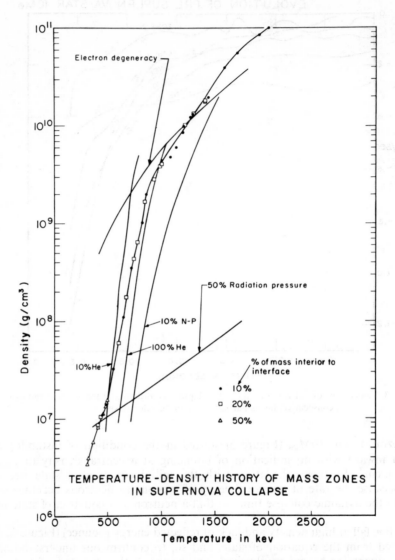

Figure 4. The equation of state of stellar material for a supernova of $10M_\odot$. All zones of the initial star, and subsequently during implosion, fall along the same adiabat.

is expected until a change occurs in the equation of state due to the formation of a neutron star.

An equation of state was synthesized from the cold-matter approximation of Salpeter (1960) shown in Figure 6. The pressure defect (below the $\gamma = \frac{4}{3}$ extrapolation of equilibrium) is much larger in this than in the Wheeler (1964) equation of state, but the difference to the hydrodynamics is negligible because the matter is close to

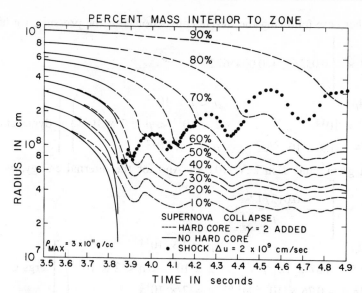

Figure 5. Mass zone interface radial time history during instability collapse of a supernova star. The solid curves describe the case with no hard core, while the dashed curves show the effect of introducing an initial gas of $\gamma = 2$ and fractional energy of 2×10^{-3}. The generation of a shock wave is indicated by the circles showing the emergence of the shock into the outer stellar zones.

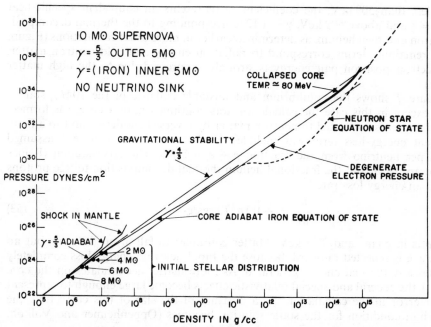

Figure 6. The pressure versus density of stellar matter during a supernova gravitational collapse. When the inverse beta-decay energy sink is used with no neutrino deposition, the stellar material follows the "cold" neutron-star equation of state from the terminal point of the iron equation of state.

free fall in either case following the Fe-He conversion. For the pressure, we have

$$p = 10^{16} \times \left[0.032\rho T + 0.004567 T^4 \right.$$

$$+ \left. \begin{cases} 0.04\rho^{\frac{4}{3}} & \rho < 2 \times 10^{11} \\ 4.67 \times 10^{13} & 2 \times 10^{11} < \rho < 2 \times 10^{12} \\ 23 + 121.9 \times 10^{-22}\rho^{2.6} & \rho > 2 \times 10^{12} \end{cases} \right] \text{dynes/cm}^2 \qquad (51)$$

where ρ is in g/cm^3 and T in keV (10^7 °K). And for the internal energy, we have

$$\varepsilon = 10^{16} \times \left[\varepsilon_0 + \frac{0.0137 T^4}{\rho} \right.$$

$$+ \left. \begin{cases} 0.12\rho^{\frac{4}{3}} & \rho < 2 \times 10^{11} \\ 933.5 - 4.67 \times 10^{13}\rho^{-1} & 2 \times 10^{11} < \rho < 2 \times 10^{12} \\ 23 \ln \rho + 0.76 \times 10^{-20}\rho^{1.6} & \rho > 2 \times 10^{12} \end{cases} \right] \text{ergs/g} \qquad (52)$$

where

$$\varepsilon_0 = \begin{cases} 0.096 T & T < 509 \text{ keV} \\ 0.267 T - 87 & T > 509 \text{ keV} \end{cases}$$

For T less than 509 keV, the coefficient ε_0 represents an equivalent specific-heat ratio $\gamma = \frac{4}{3}$ and above 509 keV, $\gamma = 1.12$, corresponding to the thermal decomposition of iron and then helium, as determined earlier in more exact calculations (Figure 4). The remaining terms correspond to radiation energy and cold neutron matter. (The electron–positron pair energy is strongly suppressed due to the high matter density.)

Figure 7 shows the equilibrium and unstable collapse of the $10M_\odot$ star as before, except in this case the calculation was continued until a core was formed. Normal matter is converted to neutron matter by inverse beta-decay, and so a time-dependent energy-loss term was included in the calculation, with KT assumed emitted per neutrino from the reaction $p + e \to n + v$. The cross-section (Reines and Cowan, 1959) and the fractional helium thermal decomposition to free protons leads to an energy loss rate

$$\frac{d\varepsilon_v}{dt} = -0.1\rho^{\frac{2}{3}}T \text{ ergs/g-sec} \qquad (53)$$

where ρ is in g/cm^2 and T in keV. Matter continues to fall in on the core, but no shock wave is reflected outward, because the rapid neutrino energy loss completely dissipates any thermal energy generated. As more matter accumulates on the core ($> 2M_\odot$), the general and special relativistic effects become large enough to represent a major error in the calculation. As a measure of the size of this correction, the equilibrium condition for the static solution becomes (Oppenheimer and Volkoff, 1939)

$$\frac{dp(r)}{dr} - \frac{[\rho(p) + c^{-2}p]G[M(r) + 4\pi c^{-2}p(r)r^3]}{[r - 2c^{-2}GM(r)]} \qquad (54)$$

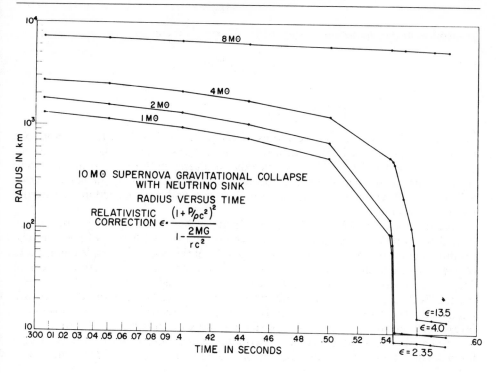

Figure 7. Collapse of a $10M_\odot$ star, including both a neutron-star equation of state and an inverse beta-decay energy sink with the assumption of infinite neutrino mean free path, i.e., no deposition. The shock formed at the surface of the core ($2M_\odot$) does not progress outward faster than the implosion velocity, and so negligible energy is reflected. The combined general and special relativistic correction is large enough by $4M_\odot$ core to cause a major change in the phenomena, presumably in the direction of further implosion.

The factor enhancing the nonrelativistic pressure gradient is shown in Figure 7 as a function of various stages of core collapse and suggests the impossibility of any sizable reflected energy. However, aside from the extreme dubiousness of extrapolating general relativistic static solutions to the dynamic case, a more detailed account of the neutrino energy flux offers the possibility of "exploding" the star before general relativistic effects become of overwhelming importance.

Figure 6 also shows the result of the same calculation, except $(d\varepsilon_v/dt) = 0$. The core temperature of 80 MeV is artificially high due to the neglect of electron-positron and neutrino–antineutrino pairs in the equation of state. However, the core shock wave is demonstrated, and the high temperature focuses attention upon the approximation of stellar-neutrino transparency. The core shock wave is formed by matter falling onto a pressure–equilibrium neutron core. The gravitational potential is approximately 100 MeV per nucleon, so that, following free fall, a shock temperature some fraction of this is expected. Use of a density of 10^{13} g/cm^3 behind the shock from the hydrodynamic calculation results in a temperature of $\simeq 55$ MeV, depending on the number of leptonic degrees of freedom that reach equilibrium. Neutrinos emitted at this energy have a mean free path for inverse beta-decay absorption in the imploding matter of $\rho = 2 \times 10^{12}$ g/cm^3 of

$$\lambda_\nu = \frac{1}{n\sigma} = 10^4 \text{ cm}$$

$$\left[\sigma = \sigma_0 \left(\frac{E}{mc^2} \right)^2 = 1.7 \times 10^{-40} \text{ cm}^2 \right]$$

which is small compared to the radius of the star, $r \sim 5 \times 10^6$ cm. Therefore, we expect a diffusion wave of neutrinos to conduct the core shock-wave heat out through the remainder of the star. Since the mean free path for neutrino scattering is increasing as the neutrino energy decreases, we expect that the neutrino surface will emit such that one half of the neutrino heat flux will be deposited in the external matter of the star.

Figure 8 also shows the calculation of a $10 M_\odot$ star in which 10% of the energy behind the core shock is deposited throughout the external matter of the star. The resulting explosion removes approximately 80% of the original mass, leaving a core of $2 M_\odot$. When no neutrino energy was emitted at any stage of collapse, a core of $5 M_\odot$ was formed and a weak reflected shock removed the outer $5 M_\odot$ of $\gamma = \frac{5}{3}$ matter. Such a stiff γ is unrealistic even for hydrogen, because the electrons are relativistic. The lack of neutrino emission is also unrealistic, since the process of inverse beta-decay has been observed (Reines and Cowan, 1960).

Calculations of the effect of a thermonuclear explosion have also been performed. An equilibrium stellar model was used with an envelope of $3 M_\odot$ with $\gamma = \frac{5}{3}$

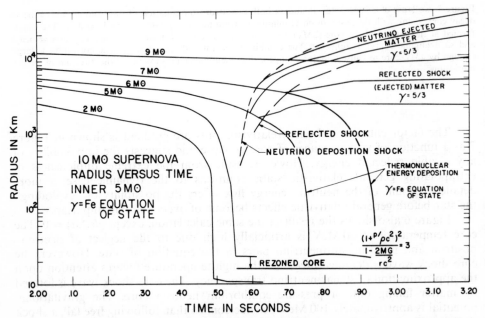

Figure 8. When 10% of the neutrino energy from the collapse of the core and/or from the reflected shock wave is deposited in the envelope $5 M_\odot$, (about 5×10^{19} ergs/g), the resulting explosion removes 70 to 80% of the mass of the star at an initial shock velocity of 4×10^9 cm/sec. When this envelope of $5 M_\odot$ is given a "stiff" $\gamma = \frac{5}{3}$ and molecular weight of hydrogen and no deposited neutrino energy, a weak shock is reflected that removes only a few percent of the star.

and a core of $7M_\odot$ with the iron equation of state. This model was evolved quasi-statically for 74 sec, and then 5×10^{17} ergs/g were injected into the $3M_\odot$ envelope, simulating a thermonuclear explosion corresponding to the synthesis of carbon to iron. It is assumed that there are no neutrino losses, so the star has the maximum possibility of exploding. Immediately following the simulated thermonuclear explosion, the equation of state of the thermonuclear material was shifted from approximately $\gamma = \frac{5}{3}$ to the equation of state of iron, including the potential Fe-He decomposition energy. The injection of this energy and the resulting shift in temperature are shown in Figure 9. The outer few zones that were heated by thermonuclear energy remain essentially static, while the inner zones continue to fall in, as shown in Figure 10. The actual velocities of the outer zones (Figure 11) show that they initially expanded outward but then fell back, because their pressure support had been reduced due to the collapse of the core. Therefore, in order to be blown off, the zones must acquire a kinetic energy comparable to their gravitational potential energy, but their kinetic energy obtained from the thermonuclear explosion is found to be only $\frac{1}{40}$ of their gravitational potential. Therefore, when the inner zones undergo the iron–helium transformation, they fall away from the outer zones so that, regardless of the thermonuclear reaction, the outer zones become unstable and collapse.

Thus, the possibility of blowing off the outer layers of a star by a thermonuclear explosion triggered from the dynamic collapse of the core is exceedingly unlikely. On the other hand, the emission of very-high-energy neutrinos from the collapsing

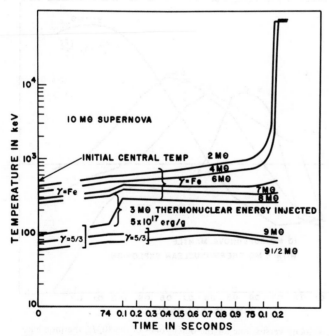

Figure 9. Log temperature versus time for the $10M_\odot$ simulated detonation. The slow initial rise in temperature corresponds to the mock detonation with the deposition of 5×10^{17} ergs/g. The subsequent implosion proceeds unaffected by the detonation.

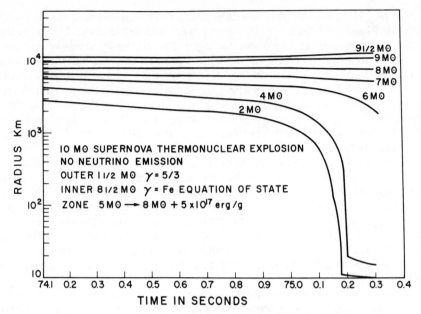

Figure 10. Radius versus time for the $10M_\odot$ detonation.

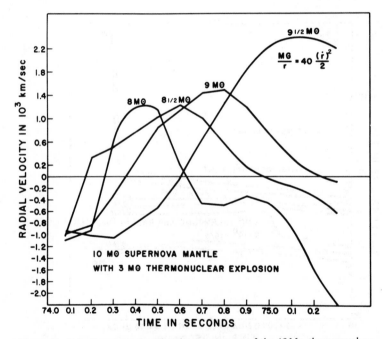

Figure 11. Velocity versus time for the outer zones of the $10M_\odot$ thermonuclear detonation. The reversal in velocity from negative to positive and returning to negative corresponds to (1) initial radial implosion, (2) detonation outward shock, and then (3) reimplosion from the core rarefaction. The kinetic energy of these zones is shown to be $\frac{1}{40}$ of their gravitational potential.

core does provide a mechanism for distributing the dynamic energy of the collapsing core throughout the rest of the mass of the star, and this process, even though inefficient, is still sufficient to blow off most of the outer regions of the star.

REFERENCES

E. M. Burbidge, G. R. Burbidge, W. A. Fowler, and F. Hoyle (1957), *Rev. Mod. Phys.* **29**: 547.
W. H. Grasberger (1961), UCRL–6196.
W. H. Grasberger and J. N. Yeaton (1961), UCRL–6465.
J. R. Oppenheimer and G. M. Volkoff (1939), *Phys. Rev.* **55**: 374.
F. Reines and C. L. Cowan, Jr. (1959), *Phys. Rev.* **113**: 273.
E. E. Salpeter (1960), *Ann. Phys.* **9**: 11.
J. A. Wheeler (1964), in Meeting on "Neutron Stars and Celestial X-Ray Sources" held at the NASA Institute for Space Studies, New York, March 20, 1964 (to be published).
J. A. Wheeler (1964), in: H. Chin and W. F. Hoffman (eds.), *Gravitation and Relativity*, W. A. Benjamin, Inc., New York, chapter 10.

DISCUSSION

E. Schatzman: Povedo has recently discussed the question of the mass ejected from type II supernovae. He considers that at maximum light, if the optical thickness is about 1 (which corresponds to the idea that when the light decays it is because the shell becomes transparent), and assuming a mass per square centimeter of the shell of the order of 1 g (it may be a little larger, but not much), then the development of ejection velocities of about 7000 km/sec gives a mass of the order of $0.2M_\odot$, which is much below what is usually assumed for a type II supernova. He has other evidence concerning this mass based on interpretation of radio sources. It seems that the whole picture makes sense with ejected mass which may not be larger than $\frac{1}{10}$ of a solar mass.

S. A. Colgate: It is certainly crucial whether a supernova has the possibility of ejecting essentially all of its mass, so that it is left with a stable core which is below the Chandrasekhar limit, or whether there is no general mechanism for getting rid of the mass. Unfortunately, optical observations do not have a simple and obvious interpretation, and so, even though a small ejected mass at a given velocity may be observed, it does not exclude the possibility of the existence of a larger mass at a different velocity. Indeed, this velocity gradient of ejected matter is predicted theoretically (Figure 8), and accounts for the discrepancy of the ejected mass with the predictions of Povedo. The difficulty is that the adiabatic expansion of the ejected matter following even the extreme shock heating to 5×10^9 deg causes the ejected matter to cool too rapidly to give rise to the observed (optical) luminosity. An additional mechanism must be sought for heating the matter late in time (weeks), when it has become transparent. I believe this late heating is due to beta-decay of the radioactive debris, but this mass fraction will correspond to the innermost ejected matter which expands the slowest.

W. A. Fowler: The history of the desire to explode a massive star entirely is a rather interesting one, in view of recent developments. We were all motivated in the early days by the feeling that a massive star had to get rid of all of its mass in order to get below the Chandrasekhar limit and become a white dwarf. Clearly, from the recent developments in the application of general relativity to this problem, there are no static solutions in general relativity, but rather dynamic ones which last for an infinite time, and it is not necessary, apparently, for a star to relieve itself of the excess mass over the Chandrasekhar limit.

As pointed out, the question of the ability of thermonuclear reactions to explode a star is very delicate. Even with a competent nuclear reaction such as the burning of helium and carbon on out, the envelope needs something to push against in order to throw itself off. Here, the core has fallen away, but this is for nonrotating cases. Large massive stars have rotation, and rotation acts in just such a way as to give a stiffening of the core. For this reason, we have some hope that one can still eject a fairly substantial fraction, which will now depend in some measure on how much rotation the star has at this stage.

S. A. Colgate: The $\gamma = 2$ fictitious adiabat was a mistaken effort to simulate rotation or magnetic fields. Magnetic fields have a γ of $\frac{4}{3}$, regardless of multipole, and the rotation has an effective γ of $\frac{5}{3}$ in these equations. I would take a fractional energy content of the order of 10^{-2} to 10^{-3} in rotation to seriously perturb these calculations. And there would have to be that fractional energy content within the core. Further, there cannot be solid-body rotation of the star immediately prior to the implosion because a solid-body rotation puts the mantle beyond the stability limit at the time of implosion.

COOLING OF WHITE DWARFS

M. P. Savedoff

Historically, warm white dwarfs have been discussed by Marshak (1940), Mestel (1952), and most recently by Schwarzschild (1958). We shall here consider tests of evolution theories by requiring that the age of Sirius A be consistent with the age of Sirius B, as it should be if our theories are consistent.

The mass of Sirius A is about 2.15 solar masses, and the mass of Sirius B is about 1.05 solar masses (van den Bos, 1960). According to Kushwaha's (1957) stellar model for Sirius A, its maximum age is of the order of 10^9 years; recognizing that Sirius A is more luminous than Kushwaha's model, the age is brought down to 5×10^8 years. From the properties of Eggen's Sirius group (Eggen, 1960)—of which Sirius is naturally a member—namely, that the brightest stars are of approximately absolute magnitude zero, the maximum age is about 1.6×10^8 years (Table I).

These ages can now be compared with the theoretical ages of white dwarfs. First, notice that presumably Sirius B was originally more massive. Possibly, the original mass was about $2.95 M_\odot$, because before rapid mass loss (less than 1 year) with that initial mass, the orbit would have been circular at about four-tenths of the current semimajor axis, or 8 astronomical units. This star evolved, ran out of fuel, and became a white dwarf. The basic assumption is that white dwarfs, when they are formed, are warm. This argument comes from the fact that the velocity of escape from the current surface of Sirius B is of the order of 6000 km/sec. Thus, for material to escape from the star as it is now constituted, it must possess a lot of energy. If a white dwarf forms as a large object that collapses, then the equivalent energy per particle would be available and the star would be hot. It is thus probable that in some stages this star was at least twice as hot as it is now.

Schwarzschild's computation of a white dwarf's lifetime can be summarized as follows: The internal energy of Sirius B is

$$U_I = \tfrac{3}{2} \frac{MkT}{\mu_B} \tag{1}$$

where μ_B is the mean molecular weight per atom. Write this as

$$U_I = C_M T \tag{2}$$

Table I. Ages and Compositions

Source	Age (years)	$\mu(L_B)$
Cluster member	1.6×10^8	> 168
Kushwaha model Sirius A	1.1×10^9	> 15
Luminosity Sirius A	$5 \ \times 10^8$	> 56
Schwarzschild model	3.2×10^9	4.44

where C_M is similar to a specific heat at constant mass. Schwarzschild writes the integrated combined hydrostatic equation and radiative transfer equations for the envelope as

$$L = f(\rho, T) \tag{3}$$

which, for Kramers' scattering, we will call

$$L = C_L \frac{T^{6.5}}{\rho^2} \tag{3'}$$

Next, Schwarzschild fits this exterior to an interior model schematically by a relation between degenerate and nondegenerate electron pressures, i.e.,

$$\rho^{\frac{2}{3}} = C_D T_C \tag{4}$$

To calculate the central temperature (T_c) numerical values of the opacity are needed. The chemical composition used was $Z = 0.1$ and $Y = 0.9$. However, Schwarzschild's model is based on the dominance of electron scattering in the transition layer from radiative to degenerate opacity, and this is uncertain. Combining, one obtains

$$L = C_L \frac{T_c^{3.5}}{C_D^3} = -C_M \frac{dT_c}{d\tau} \tag{5}$$

Thus, the age of the white dwarf (since infinite temperature) can be expressed either in terms of the central temperature as

$$\tau = \frac{C_D^3 C_M}{2.5 C_L} T_c^{-2.5} \tag{6}$$

or in terms of the luminosity, as

$$\tau = \frac{C_D^{\frac{6}{7}} C_M}{2.5 C_L^{\frac{2}{7}}} L^{-\frac{5}{7}} \tag{6'}$$

The age that Schwarzschild obtained for the luminosity of Sirius B, 0.003 $L_\odot$ (assuming a mean molecular weight $\mu_B = 4.44$, which is helium plus the Russell mixture with nothing else), is 8×10^8 years. This exceeds the two numbers obtained for the total lifetime of Sirius A and for the age of the cluster. The above situation would be acceptable, but the true situation is worse. The potential energy of a star is (Chandrasekhar, 1939)

$$-\Omega = 3 \int P \, dV \tag{7}$$

where the total pressure includes the electron pressure and the nuclei pressure. Thus, for a warm white dwarf, some change in the potential energy is expected. A detailed integration using Chandrasekhar's model at $T = 0$ (Savedoff, 1963) and my own model white dwarf at a temperature of 15 million degrees showed the energies to be those shown in Table II. Thus, there is a difference in the energy of the order of 3.7 times the internal energy of the nuclei resulting from a slightly larger radius maintained by the nuclear pressure. Thus, the quantity C_M should be about 3.7 times the value that Schwarzschild calculated. With this new factor for Schwarzschild's mixture, the age of Sirius B is at least 3.2×10^9 years, which is

Table II. Model Binding Energy

Temperature (Kelvin)	E_{total} (ergs)
0	-2.1044×10^{50}
1.5×10^7	-2.0706×10^{50}

incompatible with the age of Sirius A. What molecular weight would make the ages of Sirius A and B consistent? For the maximum age from the Kushwaha model, the molecular weight for Sirius B must be greater than 15; taking the luminosity to give the evolutionary age, the molecular weight must be 56; for Eggen's cluster age, the molecular weight must be greater than 168, which is ridiculous. Notice that there are two other constants in these relations, C_D and C_L; these have not yet been investigated fully, but their dependence on molecular weight has been included. Further, other minor terms have been included.

The failure to achieve agreement with the age of Sirius A of the order of 10^8 years means either that the evolution of white dwarfs is not understood or else that the situation is direct evidence for new energy losses, such as by the production of neutrinos at a temperature of the order of 15×10^6 degrees and a density of the order of 6×10^6 g/cm^3, the conditions found at the center of the white dwarf stars (Chiu, 1965). The dominant mode of neutrino emission under these conditions would be plasmon emission. However, the rate of neutrino emission needed would be more than an order of magnitude greater than the plasmon–neutrino emission under these conditions.

Detailed calculations of evolutionary models are being made by S. Vila and corroborate the preliminary results given here.

REFERENCES

S. Chandrasekhar (1939), *An Introduction to the Theory of Stellar Structure*, University of Chicago Press, Chicago, Illinois, Chapter 3, Th. 2.
H.-Y. Chiu (1965), this volume, p. 175.
O. Eggen (1960), *Monthly Notices Roy. Astron. Soc.* **120**: 563.
R. S. Kushwaha (1957), *Astrophys. J.* **125**: 242.
R. E. Marshak (1940), *Astrophys. J.* **92**: 321.
L. Mestel (1952), *Monthly Notices Roy. Astron. Soc.* **112**: 583.
M. Savedoff (1963), *Astrophys. J.* **138**: 291.
M. Schwarzschild (1958), *Structure and Evolution of Stars*, Princeton University Press, Princeton, N.J.
W. H. van den Bos (1960), *J. Observateurs* **43**: 145.

DISCUSSION

E. Schatzman: The neutrino-production rates cannot be much increased, since they are already close to the maximum possible neutrino flux which is obtained in thermodynamic equilibrium. A source function for neutrinos in thermodynamic equilibrium can be obtained. The emissivity of neutrinos will be the product of the source function and the neutrino absorption coefficient. This gives the maximum possible neutrino emission.

G. Wallerstein: There is another inconsistency with Sirius B. From the mass–radius relation, the large mass leads to the conclusion of a small radius, so from the luminosity, the effective temperature must be high. Weidemann's models show that when the temperature gets above 20,000 degrees, one ought to see helium lines and not hydrogen lines in the white-dwarf spectrum. However, hydrogen lines instead of helium lines are seen.

B. Strömgren: With regard to the cluster age of 1.6×10^8, the position of Sirius A in the H–R diagram is, of course, quite well known; it is very close to the zero-age line. That evidence would confirm the lower age, the cluster age, instead of the full lifetime of the main sequence.

I would like to ask a question in connection with your remark that the distance between the components might change. Is it conceivable that at an earlier phase Sirius A might have gained mass, so that it had a longer life with a smaller mass before its mass became $2.28M_\odot$?

M. P. Savedoff: There is not much cross-section for picking up mass. The current semimajor axis is about 20 astronomical units. The closest approach with an eccentricity of six-tenths is about 8 astronomical units. For a circular orbit undergoing sudden mass loss and becoming an elongated orbit, there is no reason why the system should have been much closer than this. If, instead, there had been a long period of slow mass loss, then the two stars could be brought quite close together. The distances would roughly follow the requirement that total mass times the semimajor axis be constant in the slow phases. However, to get down to essentially stellar radii would require an initial mass of the order of 1000 solar masses.

FORMATION OF HELIUM IN THE GALAXY

J. W. Truran, C. J. Hansen, and A. G. W. Cameron

Modern theories which attribute the origin of the elements to nuclear reactions in stars derive their strength from the observations that the more abundant products of such nuclear reactions are also the more abundant nuclei in nature (Burbidge, Burbidge, Fowler, and Hoyle, 1957; Cameron, 1957). Because all of the nuclei observed in nature can thus be formed, it has usually been assumed that the gas from which the galaxy formed was initially composed entirely of hydrogen.

The product formed in the first stage of nucleosynthesis is helium. The initial composition of the sun apparently contained about 24% of helium (Gaustad, 1964), while the O and B stars formed more recently contain about 35% helium (Aller, 1961). This indicates that the material forming the sun and more recent stars has undergone a large amount of nuclear processing in stellar interiors if the galaxy was initially composed of pure hydrogen.

For some time, it has been argued that the bulk of this nuclear processing would have to take place during the very early history of the galaxy (Burbidge et al., 1957; Taylor and Hoyle, 1964). The essentials of this argument are as follows. One assumes that the sun has a helium content typical of the galaxy as a whole and calculates the energy release in a galactic mass of gas for the conversion of about one quarter of the hydrogen into helium. It is then noted that this energy release is an order of magnitude greater than the present energy output of all the stars in the galaxy, multiplied by a reasonable galactic age. However, the present interstellar medium is only a very small fraction of the mass of the galaxy, and, if the galactic age is large, then the mass of stars formed from the interstellar medium in the last 5×10^9 years will also be relatively small. Clearly, the matter requires closer consideration.

Furthermore, recent studies of stellar evolution indicate that large amounts of helium could not have been produced by a postulated early concentration of O and B stars. Hayashi (1965) has followed the evolution of massive stars through the hydrogen- and helium-burning phases. He finds that, in advanced stages of evolution, the more massive stars have only a thin layer of helium between the hydrogen- and helium-burning shell sources (from 3 to 5% of the stellar mass). The helium-exhausted core has a much larger mass. This leads to the expectation that, when such stars undergo supernova explosions, they will eject only small amounts of helium and comparable amounts of heavier elements. Thus, one cannot appeal to the formation of large numbers of O and B stars in the early history of the galaxy to solve the helium problem.

This paper presents the results of a study in which an attempt was made to follow, numerically, the changes in the compositions of stars and the interstellar medium which take place as a result of stellar evolution and the interchange of gas between the stars and the interstellar medium. There are many uncertainties in our knowledge of the stellar parameters which must be assumed in this study, and hence the study should be regarded only as a reconnaissance of the problem using reasonable values of these parameters.

In carrying out this study, it was necessary to make assumptions about the stellar luminosity birth-rate function, stellar evolutionary lifetimes, the composition distribution at the end of the evolutionary lifetime, the amount of material returned to the interstellar medium, and the rates of production of radioactivities. These are functions of stellar mass.

In order to describe the gross effects of stellar evolution on the compositional history of the galaxy, the following assumptions have been made:

A. The galaxy, in all periods of history, is assumed structureless and homogeneous, and any effects of stellar evolution are immediately felt throughout its volume.

B. The fractional mass of the interstellar medium at any time is completely determined by an assumed prescription.

C. Any material given off by stars at the end of their lifetimes is immediately mixed with the interstellar medium and that mixture is then used for future star formation.

D. The initial luminosity function (or the birth-rate function) is independent of time in the sense that for any time interval the ratio of stars formed of mass M in the mass interval dM to the total of stars formed of all masses in the same time interval is independent of the time or the rate of formation.

E. The lifetime of a star of given mass is independent of time. Thus, we ignore the effects on lifetimes of composition differences resulting from stellar evolution.

F. At time zero, the galaxy is completely gaseous and composed of pure hydrogen.

Following the above, we let $m_g(t)$ denote the mass of the interstellar medium remaining at time t, and $m_s(t)$ denote the total mass of stars formed up to time t, so that

$$m_g(t) = m_g(0) - m_s(t) + m_e(t)$$

where $m_e(t)$ is the total mass of gas ejected by all evolved stars up to time t. $m_g(0)$ is the total galactic mass. Since $m_g(t)$ is assumed known, only m_e and m_s remain to be determined in order to specify the evolutionary turnover of gas through galactic history.

The form of the function $m_g(t)$ is the major variable in this model, since it governs, for the most part, the rate of stellar formation as a function of time. Its form will be taken as a decreasing exponential or modified exponential to conform with the deductions of Eggen (1962), Schmidt (1959, 1963), Salpeter (1959), and Wilson (1964) that the average rate of star formation in the past was much greater than the present rate. The first three authors assume that the rate of formation has been decreasing monotonically since time zero, so that by some early epoch in galactic history most of the primordial gas had already been converted into stars. Wilson, on the other hand, on the basis of the intensity distribution of H and K spectral components in stars, finds that the stellar-formation rate started slowly, reached a peak about the time of formation of 61 Cygni, and has been decreasing ever since. To cover this range of possibilities, two forms for $m_g(t)$ have been chosen:

$$m_g(t) = m_g(0)e^{-at}$$
$$m_g(t) = m_g(0)(1 + at + 0.5a^2t^2)e^{-at}$$

These functions are in order of increasingly deferred periods of maximum stellar-formation rates. Variation of the quantity a in the above relations allows examination of a wide range of possibilities.

We have chosen, as a representative and consistent set of relations between stellar mass, lifetime, and initial luminosity function, those given by Limber (1960). The initial luminosity function for stars with absolute visual magnitudes brighter than +5 consists of an equally weighted mean of three sets of values; two by Sandage (1957) and one by van den Bergh (1957). One of those given by Sandage is derived by modifying the observed luminosity function of the solar neighborhood for the effects of evolution. The remaining two sets are based on the luminosity function of young galactic clusters. For stars fainter than +5, the observed solar-neighborhood luminosity function is used, based on the assumption that this neighborhood constitutes a closed system whose faint stars have not yet evolved away from the main sequence. The values used are given in Figure 1.

It is estimated that the total mass of faint stars of magnitude greater than +20 not accounted for in the initial luminosity function is about 5% of the total mass of stars which is actually present.

The stellar masses and lifetimes are derived by Limber from calculations of Schwarzschild and Härm (1958), Henyey, LeLevier and Levee (1959), and others, for stars on or near the main sequence. Not included are times spent during the later stages of evolution, even though these times represent perhaps as much as 30% of the main-sequence lifetime (Woolf, 1962). Furthermore, the effects of differences in initial chemical compositions have not been taken into account, nor has any attempt been made here to evaluate what these effects imply for this work. The masses and stellar lifetimes used in the calculation as functions of visual magnitude are shown in Figures 2 and 3.

A determination of the abundances of the heavy elements demands that certain assumptions be made concerning the effects of advanced stellar evolution. We must

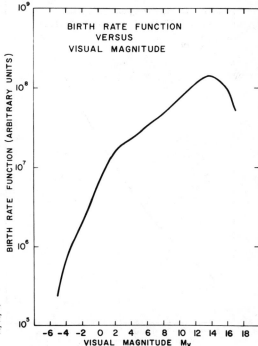

Figure 1. The adopted initial luminosity function, number of stars per interval of magnitude per cubic parsec, as a function of visual magnitude.

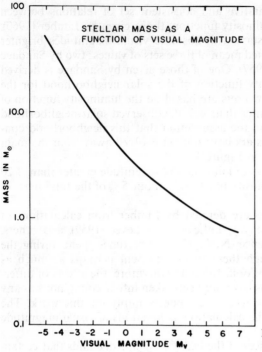

Figure 2. The mass of a star plotted as a function of visual magnitude.

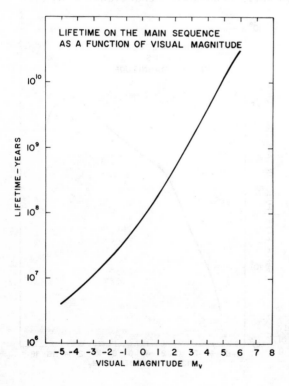

Figure 3. The lifetime of a star on the main sequence presented as a function of visual magnitude.

calculate the mass fractions of the various constituents produced in these advanced stages as a function of stellar mass. The general prescriptions employed in the present work are illustrated in Figure 4 and will be elaborated in the following discussion.

Hayashi, Hoshi, and Sugimoto (1962) (see also Hayashi, 1965) have calculated models for stars of $0.7M_\odot$, $4M_\odot$, and $15.6M_\odot$ through the phase of carbon-burning. We survey briefly the development of inhomogeneous chemical composition due to the depletion of the various nuclear fuels in these stars (see Hayashi, 1965, Figure 5).

The first nuclear-burning stage of stellar evolution is the conversion of hydrogen to helium in a star's core. When hydrogen is exhausted in the core, the core contracts and heats up, and hydrogen burns in a shell source surrounding the core. The core temperature increases until helium in the core ignites. A star of $0.7M_\odot$ will undergo a helium flash due to the degeneracy of electrons in its core when helium begins to burn.

Helium-burning produces predominantly carbon at the temperatures occurring in small-mass stars and predominantly oxygen at the higher temperatures occurring in massive stars. In the initial stage of helium-burning, the burning of hydrogen in the shell source, in $4M_\odot$ and $15.6M_\odot$ stars, results in the growth of the helium core and a contraction of the hydrogen envelope. For a $0.7M_\odot$ star, however, the hydrogen shell source becomes inactive at an early stage of evolution. The depletion of helium in the core results in the contraction of the carbon/oxygen core and the ignition of a helium-burning shell source. The hydrogen-burning shell source also becomes inactive in the massive stars, but toward the end of helium-burning.

Very massive stars have pure-oxygen cores, and the core contracts until the temperature is high enough for oxygen to begin to react with itself. Intermediate mass stars, including $15.6M_\odot$ and $4M_\odot$ stars, have cores of carbon and oxygen, and the cores continue to contract until the temperature is high enough for carbon to react with itself. Electrons become degenerate in a $4M_\odot$ star, and such stars will experience a carbon flash.

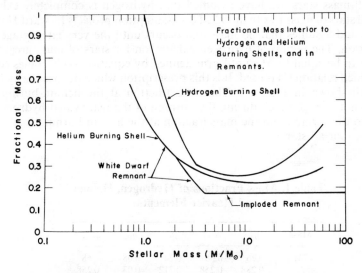

Figure 4. The compositional structure of a star as a function of mass in the final stage of evolution.

For the purpose of our calculations, we have assumed that the compositional structure of a massive star in the final phase of carbon-burning corresponds to the structure at the end of its life. As the central temperature increases still further, neutrino pair emission should speed up the subsequent stages of stellar evolution so much that very little further change in the composition of the outer parts of the star is likely.

A $0.7M_\odot$ star also develops electron degeneracy during contraction of the carbon core, but the temperature never increases to the point at which a carbon flash can occur. The important development for our consideration is the reactivation of the hydrogen-burning shell source. This is found to take place in the later phases of helium-burning as the helium-burning shell approaches the hydrogen envelope (Hayashi, 1965). The helium shell source becomes inactive at this stage. Under these conditions, hydrogen depletion will take place to a large extent.

Table I summarizes the results of Hayashi, Hoshi, and Sugimoto (1962) and Hayashi (1965) for the mass fractions of hydrogen, helium and carbon, oxygen and heavier elements at the end of carbon-burning for the more massive stars, but before the reactivation of the hydrogen shell in the $0.7M_\odot$ star, where q_1 and q_2 are the mass fractions at the hydrogen- and helium-burning shell sources, respectively, and q_H, q_{He}, and q_{heavy} are the mass fractions of these elements. For purposes of computation, we have determined quadratic fits to the numbers for the total mass interior to the hydrogen- and helium-burning shells as a function of stellar mass

$$q_1 m = 0.00527 m^2 + 0.1623 m + 0.413 \tag{1}$$

$$q_2 m = 0.00264 m^2 + 0.1552 m + 0.365 \tag{2}$$

In these and in the ensuing expressions, all masses are in solar mass units. The value of q_1 quoted above for a star of $0.7M_\odot$ does not include the influence of the reactivation of the hydrogen-burning shell source. In an attempt to incorporate this effect for the small-mass stars, we have assumed that hydrogen is completely exhausted ($q_1 = 1$) in all stars of mass less than $M_\odot$. This assumption projects forward Hayashi's calculations and assumes that no mass loss occurs until the very latest stage of the star's evolution. The region of hydrogen exhaustion for stars of mass greater than $M_\odot$ is taken to be equal to $M_\odot$ until $q_1 m$ defined by equation (1) becomes equal to $M_\odot$, after which relation (1) is used. It is this prescription which defines the hydrogen-burning shell shown in Figure 4. The mass fraction at the helium-burning shell source does not change much during the burning of the reactivated hydrogen shell source out to the surface, and the mass fraction at the helium-burning shell is given by relation (2) for all stars.

Table I. Mass Fractions of Hydrogen, Helium, and Heavier Elements

Mass	q_1	q_2	q_H	q_{He}	q_{heavy}
$0.7M_\odot$	0.76	0.68	0.24	0.08	0.68
$4M_\odot$	0.288	0.258	0.712	0.03	0.258
$15.6M_\odot$	0.272	0.22	0.728	0.052	0.22

The stars in our model are divided into two distinct classes: small-mass stars which evolve to white dwarfs and massive stars which are assumed to become supernovae and to leave behind an imploded remnant which, if stable, might become a neutron star. Some assumptions must be made concerning the nature of these evolutionary remnants. In our calculations, we have assumed that all stars of mass greater than $4M_\odot$ will become supernovae. The mass of the imploded remnant is taken to be 0.175 of the mass of the star; a star of $4M_\odot$ then leaves a remnant of $0.7M_\odot$. The end-point of evolution for stars of mass less than $4M_\odot$ is assumed to be a white dwarf. The white-dwarf mass is taken to be $0.7M_\odot$ for all stars satisfying $0.7 \leq q_2 m$. For those stars for which the mass interior to the helium-burning shell is less than $0.7M_\odot$, the mass of the white-dwarf remnant is taken to be equal to $q_2 m$ in order to maximize the helium ejection to the interstellar medium (stars in this range contribute no heavy elements to the interstellar gas). These prescriptions define the appropriate evolutionary end-states as shown in Figure 4.

It is important to notice the implications of our choice of these various prescriptions. In general, Hayashi's results for the more massive stars, $m \geq 4M_\odot$, reveal that they produce little helium. For a $15.6M_\odot$ star, the mass fraction of the helium zone $q_1 - q_2$ is 0.052, while for the $4M_\odot$ case it is only 0.03. Furthermore, these mass fractions are small compared to the mass fraction q_2 of heavy elements in the same stars. Thus, even if the past rate of formation of these massive stars has been much greater than the present rate, it would be difficult to produce large amounts of helium either absolutely or relative to heavy elements. If the galaxy was initially composed of pure hydrogen, it seems apparent that the evolution of small-mass stars must account for the helium content of the interstellar gas. The reactivation of the hydrogen shell source in the late stages of helium-burning for stars of mass $m < M_\odot$ might well provide the necessary helium. The long lifetimes associated with these stars, however, imply high galactic ages.

The size of the evolutionary remnants is the determining factor in the mass of heavy elements produced in these models. Greenstein (1958) has given an average value of $0.55M_\odot$ for a white-dwarf remnant. Our value of $0.7M_\odot$ is chosen to fit smoothly to the mass of the imploded remnant at $4M_\odot$, although there is no physical necessity for continuity. For those stars with $0.7 > q_2 m$, the white-dwarf remnant is assumed to have mass $q_2 m$. By this prescription, we decrease the average white-dwarf mass to be comparable to Greenstein's determination and enhance the production of helium. The choice of the mass of the imploded remnant as $0.175 m$ is governed by the need to produce reasonable heavy-element abundances, although it is difficult to change their production by more than a factor two by using other remnant masses. It is interesting to note that our prescriptions result in the production of approximately equal masses of heavy elements from the two classes of stars.

Having defined the structure of our two classes of stars in the final stages of evolution, the abundances of the various constituents of the interstellar gas can be followed directly. A star formed at time t_1 is formed with mass fractions $X(t_1)$, $Y(t_1)$, and $Z(t_1)$ of hydrogen, helium, and heavy elements. In our calculations, the masses of heavy elements from the two classes of stars are followed individually, Z being the sum of these two contributions. A star of mass m has an associated lifetime $\tau(m)$ determining that time $t_1 + \tau$ at which it will evolve and enrich the stellar gas. The total mass released at the end of the life is simply m, less the mass of the appropriate remnant. The mass fractions of hydrogen, helium, and heavy elements released are determined by the prescriptions established above.

It is further assumed that the hydrogen region, mass $m - q_1 m$, is not composed purely of hydrogen but rather contains the abundances at formation, $X(t_1)$, $Y(t_1)$, and $Z(t_1)$. Similarly, the helium region, mass $q_1 m - q_2 m$, contains a mass fraction $Z(t_1)$ of heavy elements.

At time $t_1 + \tau$, a whole distribution of stars will evolve of varying lifetime, hence varying initial composition. The total mass of the various constituents released by these stars is assumed to be mixed with the interstellar gas. Stars born at this time are formed with the updated mass fractions of hydrogen, helium, and heavy elements.

The results of our investigation are summarized in Figures 5, 6, and 7; each of the prescriptions employed for the rate of star formation has been examined for values of a corresponding to the condition that 5% of the original gas remains in the range 10 to 25 billion years.

In this investigation, we have sought to determine the extent to which the various prescriptions for the rate of star formation result in reasonable values for the composition of the interstellar gas. Assuming that the helium content of the sun is 24% by mass, we can determine the time of formation of the sun from the interstellar gas. The mass fraction of heavy elements at this time should also approximate that observed in the sun, 0.021 (Aller, 1961). The results of all our models agree reasonably well with this value.

The helium content of the gas at the present time can be inferred from a study of the compositions of O and B stars. A reasonable value for $Y(\tau_s + 4.5)$ is 0.35 (Aller, 1961), where τ_s is the time of formation of the sun in billions of years. The values of the present helium content of the gas in our models vary quite noticeably with the prescription for the rate of star formation. The need to satisfy $Y = 0.35$ at the present time leads us toward high galactic ages.

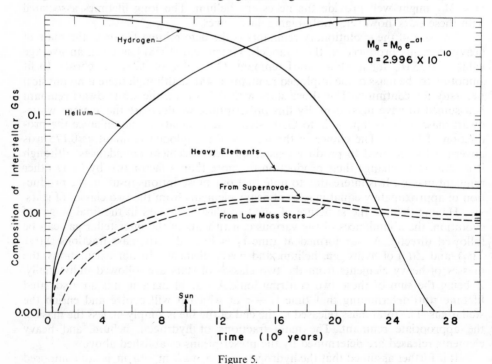

Figure 5.

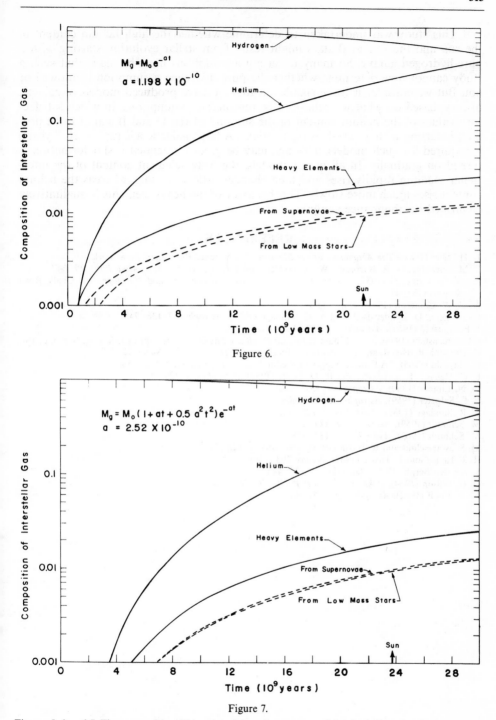

Figure 6.

Figure 7.

Figures 5, 6, and 7. The compositional histories of the galaxy are traced for the indicated prescription for the gas-content function. The time of formation of the sun corresponds to the condition that the helium mass-fraction $Y = 0.24$.

This study was undertaken to determine whether the high helium content of the sun and the O and B stars might result from stellar evolution starting with a pure-hydrogen galaxy. So many uncertain assumptions must be made that such a study cannot now determine whether the pure-hydrogen assumption is allowed or not. But we consider it to be significant that we have produced models of galactic history, based on what we believe to be reasonable assumptions, in which satisfactory values of the helium content of the sun and of the O and B stars can be produced starting with a pure-hydrogen galaxy. A large galactic age ($\gtrsim 2 \times 10^{10}$ years) is required for such models. This age may be greatly extended if star formation is turned on gradually. In all of our models, the heavy-element content of the interstellar medium rapidly rises and then changes little with time, whereas the helium content rises much more slowly. This behavior of the heavy elements is qualitatively confirmed by abundance analyses of stars.

REFERENCES

L. H. Aller (1961), *The Abundance of the Elements*, Interscience Publishers, New York.
E. M. Burbidge, G. R. Burbidge, W. A. Fowler, and F. Hoyle (1957), *Rev. Mod. Phys.* **29**: 547.
A. G. W. Cameron (1957), *Stellar Evolution, Nuclear Astrophysics, and Nucleogenesis*, Chalk River Report CRL-41.
A. G. W. Cameron (1962), *Icarus* **1**: 13.
O. J. Eggen, D. Lynden-Bell, and A. R. Sandage (1962), *Astrophys. J.* **136**: 748.
J. E. Gaustad (1964), *Astrophys. J.* **139**: 406.
J. L. Greenstein (1958), in: S. Flügge (ed.), *Handbuch der Physik, Vol. 50*, Springer-Verlag, Berlin, p. 161.
C. Hayashi, R. Hoshi, and D. Sugimoto (1962), *Progr. Theoret. Phys. Suppl.* **22**: 1.
C. Hayashi (1965), "Advanced Stages of Stellar Evolution," this volume, p. 253.
L. G. Henyey, R. LeLevier, and R. D. Levee (1959), *Astrophys. J.* **129**: 2.
D. N. Limber (1960), *Astrophys. J.* **131**: 168.
E. E. Salpeter (1959), *Astrophys. J.* **129**: 608.
A. R. Sandage (1957), *Astrophys. J.* **125**: 422.
M. Schmidt (1959), *Astrophys. J.* **129**: 243.
M. Schmidt (1963), *Astrophys. J.* **137**: 758.
M. Schwarzschild and R. Härm (1958), *Astrophys. J.* **128**: 348.
R. J. Taylor and F. Hoyle (1964), *Nature* **203**: 1108.
S. van den Bergh (1957), *Astrophys. J.* **125**: 445.
O. C. Wilson (1964), *Publ. Astron. Soc. Pacific* **76**: 28.
N. J. Woolf (1962), *Astrophys. J.* **135**: 644.

Part III. Stellar Evolution
D. Stellar Evolution with Varying G

STELLAR EVOLUTION WITH VARYING G

R. H. Dicke

There is a curious discrepancy among the various ways of dating the galaxy. While these age determinations must be considered with caution, the discrepancy may be significant, for it is to be expected within the framework of the Brans–Dicke cosmology (1961). Various ways of dating the galaxy have been previously discussed by Fowler (1961), and from a different point of view by the author (Dicke, 1962a).

In considering the age of the galaxy, the first and most directly determined value is obtained kinematically from the Hubble constant. The present value of 10 billion years for the Hubble age implies an age of less than 6.6 billion years for a hydrogen-filled closed universe, and of less than 5 billion years for a radiation-filled universe (Sandage, 1961), assuming zero for the cosmological constant.

While the reported Hubble age has increased drastically in the past decade and a half (by two factors of 2), these changes resulted primarily from finding errors in interpretation rather than from new data. There is little reason to expect such large changes in the future, but there is also little reason to believe that the present value is as accurate as the formal 15% probable error suggests.

While a closed universe is slightly favored by the red-shift data, this interpretation is too uncertain to be taken seriously.

From the dynamics of an expanding universe, one would expect galaxies to separate out of the "primordial" expanding hydrogen mass in a relatively short time, less than 10^9 years; hence 10^{10} years can be taken as a reasonable upper limit for the age of the galaxy as determined kinematically from the expansion of the universe.

Physical laws as they are presently understood do not permit the creation of nucleons unaccompanied by a corresponding number of antinucleons. As a great abundance of antinucleons does not seem to exist in the universe, the creation of so much hydrogen 6 to 10 billion years ago seems to present a serious difficulty.

This difficulty does not exist for the closed, oscillating universe, assuming with Lifshitz and Khalatnikov (1963) that the irregularities of the real universe serve to avoid the singularity at the time of maximum contraction. This permits the use over and over again of the same old nucleons, the holocaust of the contraction phase resulting in the decomposition of all the nuclei and providing fresh hydrogen for the next expansion of the universe. For a recent discussion see Dicke, Peebles, Roll, and Wilkinson (1965).

While this is cosmological speculation of the more adventurous type, it avoids the embarrassment of requiring the creation of material that cannot be made, and it also explains why the "primordial" material of the galaxy was free of heavy elements. If this argument is valid, the universe is closed, with an age of less than $(6.6 \pm 2.0) \times 10^9$ years. The error assumed here is double that quoted by Sandage.

It should be noted that during the contraction phase the universe would be a very uncomfortable place, for the galactic red shift would turn to a blue shift, first for the nearby galaxies, later for the more distant ones. The Obers paradox (Bondi,

1960) would gradually give way, the sky eventually filling with radiation brighter than the surface of the sun. The radiation temperature would be expected to rise to over 10^6 °K while the matter density was still very low, perhaps as low as 3×10^{-16} g/cm^3.

Observationally, it is not yet possible to exclude helium from the "primordial" material from which the galaxy was formed. However, it is important to know if helium was present, for the composition of the mix affects the determination of the evolutionary age of population II stars. We shall follow the cosmogonic speculations a bit further in attempting to decide whether helium was present in the "primordial" material.

Our ignorance is too great to give a firm answer to this question. At the time of the conference and the writing of this note, there was little to be said, but in reconsidering (May 1965) I note that we now know a bit more. Dicke, Peebles, Roll, and Wilkinson (1965) have interpreted some observations of Penzias and Wilson as indicating thermal radiation in space at a temperature of about 3.5 °K. Such radiation must have been much hotter in the past, suggesting that the universe expanded from a high-temperature state. Thus, the original conjecture is reasonable. The universe may be closed and oscillating, with the ashes of the previous cycle reprocessed to hydrogen during this high-temperature (10^{10} °K) phase. Peebles has noted that the expected helium content of the protogalaxy depends critically upon the nucleon density at this temperature. Assuming ordinary general relativity and a present matter density of 6×10^{-31} g/cm^3, he finds that the galaxy should have been formed from essentially pure helium. By contrast, Dicke finds that with the Brans–Dicke cosmology, essentially zero helium should have been present. Presumably, a similar result would also be obtained if the universe was filled with some other new field capable of providing enough "rigid energy" of some other type. By "rigid energy" I mean energy for which the pressure is equal to the energy density.

Another kinematically determined age is the time difference between the formation of the first population II and the formation of the first population I stars. It is observed that the dispersion in the velocities of population II stars is quite high, of the order of 75 km/sec perpendicular to the galactic plane, whereas for population I stars it is only about 10 km/sec. All mechanisms for producing these high velocities give a short time scale for their duration.

The high velocity of the population II apparently mirrors a corresponding motion in the gas from which the stars were formed. There have been three reasonable suggestions made for the large velocities. One suggestion is that population II stars were formed during the initial collapse of the galaxy, giving a free-fall time of a few hundred million years for the duration of population II production. Another suggestion involves the existence of a very high turbulence velocity in the initial gas of the galaxy. The decay time for the turbulence is quite short, of the order of 10^8 years (Oort and Spitzer, 1955). The third suggestion is based on the fact that in the initial population II there undoubtedly was a class of star, the massive population II star, which has never been seen by man, and which may have been very supernova prone. The radiation from these many supernovae could have driven the turbulence in the galactic gas. Even without these supernovae the sudden appearance with the initial population of so many O and B stars could have driven the turbulence. As a massive population has a short life, this also leads to a fairly short time scale.

All of the above ways of interpreting these high velocities lead to short time scales. On the basis of this, the onset of population I star formation was a relatively short time (less than 10^9 years) after that of population II.

There is another aspect of the problem which yields a reason for believing that massive population II stars may have provided the energy source for driving turbulence. It also yields an independent reason for believing that the population II stars are less than 10^9 years older than the oldest of the population I. The very oldest population I stars show a large heavy-element abundance (Arp, 1961). Furthermore, the observations indicate that there is little if any dependence of the heavy-element content upon the age of a population I star. This strongly suggests that it was primarily population II stars which were the source of the heavier elements. If the much more numerous population I stars were as prolific as the population II in generating heavy elements, there should exist in the population I a continuous range of Z, an essentially exponential distribution function with low metal content more probable than high content.

An examination of the Salpeter (1961) creation function shows it is the more massive stars, with lifespans of less than 10^9 years, which are capable of returning the bulk of the matter in a population to the interstellar medium through supernova explosions. Assuming that the Salpeter creation function is applicable to the initial stellar population, roughly 33% of the mass of the population could be called "massive," living as stars 10^9 years or less. If it is presumed that a typical star in this massive part of the population II dies as a supernova and that the residual central body (be it a freely falling, continuously imploding "Schwarzschild star," a white dwarf, or a neutron star) has about 1 stellar mass, about 75% of the matter from these massive stars would be thrown back into the gaseous medium. Under this assumption, 25% of the mass of the initial population would be processed through supernova explosions in the first billion years, 20% being processed in the first 5×10^8 years. This would represent about 10% of the mass of the galaxy. This is to be contrasted with a present rate of only about $10^{-11}\%$ per year (based on an assumed mean period of 300 years per supernova and $2M_\odot$ ejected matter per supernova). Assuming that massive population II stars tend to be explosive (implosive?), it is easily seen that the bulk of the heavy elements would have been produced in the first half billion years.

We conclude from the observed approximate constancy of heavy-element abundance in the population I that the population II is at most only 10^9 years older than the oldest of the population I stars. This is in agreement with the kinematic arguments. (Furthermore, the early supernova activity in the population II, that presumably resulted in the creation of the heavy element, may have driven the gaseous turbulence that resulted in the high velocity for the population II stars.)

While the Hubble age of the universe is now believed to be free of large errors, this was also true in the past and it is desirable to have an independent way of dating the galaxy. This is provided by the decay of U^{235} using the technique of Burbidge, Burbidge, Fowler, and Hoyle (1957). The theory of nucleosynthesis provides a production ratio for $U^{235}/U^{238} = 1.64$. The known abundance ratio at the time of formation of the solar system (4.5×10^9 years ago) was 0.283. Assuming an instantaneous production of all the uranium, this event took place 2.1×10^9 years before the formation of the solar system. Because the solar system was separated from the interstellar medium some 4.5×10^9 years ago, the above age is not influenced by uranium production since that time. As a result, if one were to make the assumption that there has been a continuous production of uranium by population I stars, amounting to as much as 40% of the present content of the interstellar medium, this assumption is almost without effect on the computed date.

Assuming that at least 60% of the uranium now found in the interstellar medium was formed by the more massive population II stars in the first 10^9 years, the galaxy is computed to be 7 to 7.5×10^9 years old (Dicke, 1962b).

Inasmuch as this dating technique depends upon the decay of U^{235}, an isotope with a short half life, this age is quite insensitive to the assumed production ratio for U^{235}/U^{238}, a change by a factor of 2 in this ratio amounting to only 8.5×10^8 years.

This is to be contrasted with the elaboration of this dating technique introduced by Fowler and Hoyle (1960), and later by Clayton (1964). In order to avoid the necessity for an assumption of an explicit production distribution curve (prompt, continuous, etc.), Fowler and Hoyle (1960) attempted to make use of the decay of U^{238} relative to Th^{232}, and Clayton used Re^{187} to supplement the information obtained from the ratio U^{235}/U^{238}. This appears to be a questionable procedure. All three of the isotopes U^{238}, Th^{232}, and Re^{187} have long mean lives, requiring an accurate knowledge of both the production ratios and the present abundances before any useful conclusions can be drawn. Unfortunately, the abundance ratios needed are of chemically different substances, and the effects of chemical differentiation on abundance ratios make conclusions uncertain. The variations in the Th/U abundance ratio in chondritic meteorites have recently been discussed by Morgan and Lovering (1964). Variations of as great as a factor of 2 are found among the various types of chondritic meteorites. If one accepts the arguments of Mason, Ringwood, and Morgan and Lovering (1964), the type I carbonaceous meteorites represent nearly primordial solar material. In this case, the present observed ratio Th^{232}/U^{238} is 2.7, in contrast to the value 3.8 assumed by Fowler and Hoyle. If one were to assume that these elements were made promptly 7.0×10^9 years ago, the ratio now would be 3.4 (assuming 1.65 as the production ratio).

To see how difficult it is to draw firm conclusions from the use of the Th/U ratio, it may be noted that the difference in decay constants of U^{238} and Th^{232} is only 1.04×10^{-10} per year, and that a factor of 2 error in either the production ratio or the present abundance ratio would result in an error of 6.6 billion years in the age determination based on this decay scheme.

When Re^{187} is used, the situation is far worse, for in addition to these uncertainties there is also doubt about the decay constant. Assuming a decay constant of 1.7×10^{-11} per year, an uncertainty of 20% in either the production ratio or the present ratio of Re^{187} to radiogenic Os^{187} results in an error of 10.4 billion years in an age determination based on this decay scheme.

It is our conclusion that with the present state of our knowledge, the Re^{187} decay scheme is essentially worthless and the method based on the decay of Th^{232} must be used with extreme care. Apparently, it is necessary to know something about the history of the production of uranium before firm conclusions can be drawn.

It is sometimes stated that the age determinations based on the decay of U^{235} are doubtful because the present content of U^{235} in the solar system could be chiefly the result of a single supernova explosion shortly before the formation of the solar system. It should be noted, however, that the ratio U^{235}/U^{238} at the time of formation of the solar system was 0.28. If it be assumed that the formation ratio was 1.64, this would imply that 0.17 of the total uranium content of the earth arose in this single explosion. If the solar system can be considered typical, this is very unlikely, for 10^8 or more supernovae were probably involved in producing the heavier elements. To have $\frac{1}{5}$ of the heavy elements in a given star produced by a single supernova would imply an unbelievably small amount of mixing of the interstellar medium.

To recapitulate, the observations of metal line strengths suggest strongly that the heavy elements were produced chiefly by some type of star present only in the population II. Also, a consideration to all the possible ways of generating the large velocity dispersion in this population strongly suggests a short period of time (less than 10^9 years) for the formation period. As a result, the conclusion based on the decay of U^{235} is that the galaxy is 7 to 7.5 billion years old, provided that the Hoyle–Fowler formation ratio U^{235}/U^{238} is correct. A liberal allowance for error in the production ratio would give the value $7.25 \pm 0.4 \times 10^9$ years for the age of the galaxy.

This age for the galaxy and the somewhat cruder value $(6.6 \pm 2.5) \times 10^9$ years obtained by combining the Hubble constant with adventurous cosmological speculations are to be compared to a stellar evolutionary age for population II stars of 15 to 25 billion years. The greater age is obtained from the color magnitude diagrams for globular clusters, without an allowance for reddening and assuming little initial helium content. The lower value includes a generous allowance for reddening. Thus, these ages are 2 to 3 times as great as that obtained from the decay of U^{235}.

The oldest population I stars (NGC 188) appear to be over 10^{10} years old. This discrepancy is here not as great, but is also appreciable.

These contradictions could be resolved in a few years in any number of ways; for instance, the stellar models may be found to be wrong. However, it is interesting for the moment to investigate a way of removing these inconsistencies based on the following assumption: There exists in addition to the two known zero-mass boson fields, electromagnetism and gravitation (with particle spins of 1 and 2, respectively), a zero-mass scalar field (zero spin). The presence of this scalar, generated by the matter in the universe, affects the measured value of the gravitational constant. As the universe expands, the strength of gravitation decreases with time, causing stars to burn more and more slowly. This causes their ages as inferred from their present burning rate to be greater than their true ages. As the properties of a relativistic scalar field are not generally known, they will be described briefly here.

The existence of this scalar field is very difficult to demonstrate experimentally, but the laws of relativity enable its general properties to be predicted reliably. The reasons for introducing such a field have been largely philosophical. This was first done by Jordan (1959) to give a theory of Dirac's (1938) cosmology, and later by Brans and myself (1961) to avoid a subtle difficulty in general relativity in relation to Mach's principle.

The chief properties of the scalar field are easily summarized: (1) It gives an attractive force between all matter; (2) the force falls off inversely as the square of the distance; (3) the force is proportional to the mass of the body. (It is, therefore, extremely difficult to distinguish such a force from gravitation because it behaves so much like gravity.) The field also has these properties: (4) It produces a velocity-dependent force proportional to $[1 - (v/c)^2]^{\frac{1}{2}}$. Thus, a rapidly moving body will experience a weak force only. In particular, a light ray is not deflected by this force field. This leads to one of the possibilities for a specific test for a scalar field—investigation of the gravitational deflection of light. (5) Because of velocity dependences, there will also be a different result for the relativistic perihelion rotation of planetary orbits. (6) The scalar interaction is weak, being somewhat weaker than the gravitational attraction. It is essentially impossible to make a strong interaction of this kind (Dicke, 1962c), because the total matter distribution in the universe contributes to the strength of the scalar observed locally. The contribution from the rest of the universe to the value of the scalar is always very large compared to the contribution from a local body. Thus, the local contribution is miniscule and the resulting force is small. (7) Finally, the mass of any particle is equal to some constant times a universal function of the scalar:

$$m = m_0 f(\varphi)$$

Now consider the effect of the existence of such a scalar field on the strength of gravitational interactions. This is most conveniently discussed by comparing the gravitational interaction between any two elementary particles (such as protons) with their electrostatic interaction. This ratio is

$$\frac{Gm_p^2}{e^2}$$

a dimensionless number with the value 8×10^{-37}. The strength of the gravitational interaction is also conveniently measured by the so-called gravitational coupling constant

$$\frac{Gm_p^2}{\hbar c} \sim 6 \times 10^{-39}$$

Because the mass of an elementary particle is a function of the scalar φ, the coupling constant is also a function of φ. In the mass-coupled picture of the Brans–Dicke theory, which is being described here, the mass of a particle varies with φ as $\varphi^{-\frac{1}{2}}$,

$$m = m_0 \varphi^{-\frac{1}{2}}$$

Consequently, the gravitational coupling constant varies as φ^{-1}. The gravitational coupling constant could be said to be small because of the large contribution to φ generated by the large amount of matter in the universe. The Brans–Dicke theory permits a calculation of the coupling "constant." This value turns out to be roughly

$$\frac{m_p}{\rho R^3} \frac{R}{\lambda_p}$$

where ρ is the mean mass density of the universe, R is the Hubble radius, and $\lambda_p = \hbar/m_p c$ is the Compton proton wavelength.

Inasmuch as the general expansion of the universe causes a gradual increase in the value of φ, the gravitational coupling constant decreases gradually with time (by a few parts in 10^{11} per year). The above discussion has been based upon a definition of units of measure such that $\hbar$, c, and G are constant by definition (Dicke, 1961). Alternatively, with a unit of mass based on the atom and $\hbar$ and c still constant by definition, the gravitational "constant" G is no longer constant. The easiest way to see this is to note that the value of the gravitational coupling constant, being dimensionless, is independent of the choice of units and that the new units $\hbar$, c, and m_p are constant by definition. Expressed in this way, the chief effect of the gradual increase with time of the scalar field is a corresponding gradual decrease of the value of the gravitational constant.

Assume that G is decreasing with time: What does this imply for stellar evolution? The cosmological solution for a closed model of the universe is shown in Figure 1, which gives the radius of the universe, the scalar field (which varies inversely as the gravitational constant), and the acceleration parameter (of the cosmological model) as functions of time. The significance for stellar evolution is shown in

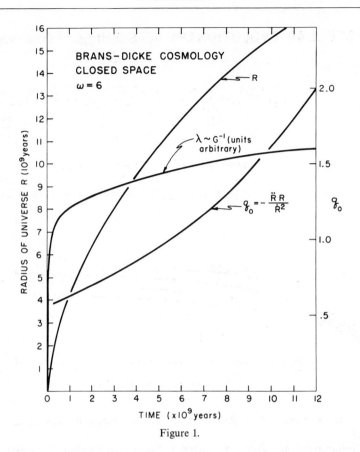

Figure 1.

Figure 2. The stellar ages in Figure 2 are expressed in units of $\frac{2}{3}$ the Hubble age (approximately the present age of the universe). Assuming that the luminosity or burning rate of a star is proportional to G^8, the evolution rate would have varied as φ^{-8} and would have been greater in the past, giving an "apparent" stellar age greater than its "true" age. The curves in Figure 2 are based upon the assumption of a flat cosmological model. The corresponding closed-model curves are more difficult to calculate, but do not differ much from these curves, provided that the acceleration parameter is less than 1.5. The parameter ω is the constant defined in the Brans–Dicke theory, essentially a measure of the strength of the gravitational interaction when compared with the scalar interaction. ($\omega = \infty$ implies no scalar interaction.)

These curves were calculated from the relation $\varphi = t^{2/(4+3\omega)}$ where φ is the value of the scalar and t is time measured from the start of the expansion. The Hubble age is given by

$$T_H = \frac{4 + 3\omega}{2 + 2\omega} t$$

The apparent stellar age is

$$t_2 = \int_{t_0 - t_1}^{t_0} \left(\frac{\varphi_0}{\varphi} \right)^8 dt$$

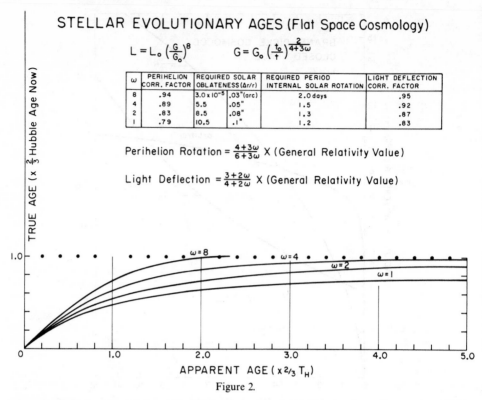

Figure 2.

where t_0 is the present age of the universe, φ_0 is the present value of φ, and t_1 is the star's true age.

To summarize the situation, Table I shows the computed "apparent" age of the sun, and the "true age" of several classes of stars for two different choices of Hubble age T_H and ω.

It is of interest to note that the choice $\omega = 4$ and $T_H = 13$ billion years gives a pattern of ages in agreement with the cosmological speculations discussed above.

Table I. Computed "Apparent" Age of the Sun and "True Ages" of Several Stars

Object	"Apparent" age (× 10^9 years)	"True" age $T_H = 10^{10}$ years, $\omega = 4$ (× 10^9 years)	"True" age $T_H = 10^{10}$ years, $\omega = 8$ (× 10^9 years)	"True" age $T_H = 1.5 \times 10^{10}$ years, $\omega = 4$ (× 10^9 years)	"True" age $T_H = 1.5 \times 10^{10}$ years, $\omega = 8$ (× 10^9 years)
Sun	7.9	4.5	—	—	—
Sun	6.0	—	—	4.5	—
Sun	6.2	—	4.5	—	—
Sun	5.3	—	—	—	4.5
NGC 188	12	5.3	6.3	6.8	8.0
Globular cluster	17	5.83	—	7.9	9.3
Universe	—	6.25	6.45	9.4	9.7

Table II. Age Pattern with Hubble Age
$T_H = 13$ **Billion Years and** $\omega = 4$

Object	Apparent age ($\times 10^9$ years)	True age ($\times 10^9$ years)
Sun	6.6	4.5
NGC 188	12	6.3
Globular cluster	17	7.1
Universe	—	8.1

These ages are listed in Table II. The following points should be noted in connection with this choice:

1. The Hubble age of 13×10^9 years is in satisfactory agreement with the observations of the galactic red shift.
2. The delay of 10^9 years in the formation of the first stars is a reasonable period to permit the collapse of the protogalaxy and the formation of the first stars.
3. The age of the oldest stars is in good agreement with the age of U^{235}.
4. The short period between the formation of the first stars (population II) and NGC 188 is in agreement with the kinematic arguments based on velocity dispersion and with the argument that the heavy elements were formed by the massive population II stars.
5. The helium content of the sun is not well-enough known to permit a choice between an evolutionary age of 4.5 and 6.6 billion years (Pochoda and Schwarzschild, 1964).

Unfortunately, stellar systems are so complicated that it is doubtful that it would ever be possible to base a really convincing argument for the existence of a scalar field upon the facts discussed above. By contrast, if there were a good experimental demonstration of the existence of this field, the age discrepancies would be understandable, being explained by these arguments.

It is fortunate that there are feasible ways of testing for the existence of a scalar field. Perhaps the most interesting and direct would be to compare a gravitational time scale (based on the motion of the moon and planets) with an atomic time scale, hence to look for the secular decrease of G. Another possibility would be to make an accurate determination of the gravitational deflection of light, which would be $[(3 + 2\omega)/(4 + 2\omega)] \times$ (the Einstein value) if a scalar field is present. What may be the quickest way of testing for the field is indirect. One could measure the oblateness of the sun. If the sun could be demonstrated to be free of an appreciable oblateness, hence gravitational quadrupole moment, the observed relativistic part of the perihelion rotation of Mercury's orbit would agree with an accuracy of 1 or 2% with Einstein's value, demonstrating a zero (or at most a small) contribution from a scalar field (Dicke, 1964). The perihelion rotation correction factor $(4 + 3\omega)/(6 + 3\omega)$, the required solar oblateness, and the solar interior rotation period which would produce this much flattening of the sun are tabulated in Figure 2 for several values of ω.

The gravitational light deflection results are plotted in Figure 3, where the result expected in the absence of a scalar field is 1.75″ arc deflection for a grazing light ray. It is clear that much better results for this observation would permit a calculation of the deficiency, if any, from a scalar field.

One further remark is necessary about the age of the galaxy inferred from the decay of U^{235}. The age calculated above, based upon a duration of 10^9 years for the

GRAVITATIONAL DEFLECTION OF LIGHT

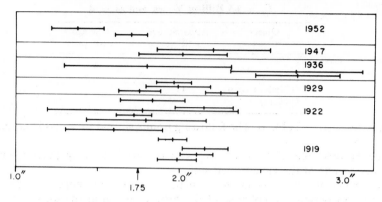

LIGHT   DEFLECTION   (Seconds of arc)

Figure 3.

production of heavy elements, was predicated upon an assumed constant gravitational "constant." Assuming a scalar field with $\omega = 4$, the rapid burning of early stars would shorten this time interval substantially, and the uranium decay age for the galaxy would be reduced to

$$6.9 \pm 0.4 \times 10^9 \text{ years}$$

To summarize: It is possible to make a self-consistent interpretation of the Hubble red-shift data, the high velocities of population II stars, the U^{235}/U^{238} abundance ratio, and the nearly uniform heavy-element abundance in the population I stars. This interpretation yields an age of $(7.25 \pm 0.4) \times 10^9$ years for the oldest stars in the galaxy. This is to be contrasted with an evolutionary age of 15 to 25×10^9 years for the oldest stars. One way in which the discrepancy can be understood is within the framework of the Brans–Dicke cosmology, which is capable of accounting quantitatively for the difference. Assuming the validity of this cosmology, the uranium decay age of the galaxy is reduced to $6.9 \pm 0.4 \times 10^9$ years.

REFERENCES

H. Arp (1961), *Science* **134**: 810.
H. Bondi (1960), *Cosmology*, ed. 2, Cambridge University Press, New York.
C. Brans and R. H. Dicke (1961), *Phys. Rev.* **124**: 925.
E. Burbidge, G. Burbidge, W. Fowler, and F. Hoyle (1957), *Rev. Mod. Phys.* **29**: 547.
D. D. Clayton (1964), *Science* **143**: 1281.
R. H. Dicke (1961), *Phys. Rev.* **125**: 2163.
R. H. Dicke (1962a), *Rev. Mod. Phys.* **34**: 110.
R. H. Dicke (1962b), *Nature* **194**: 329.
R. H. Dicke (1962c), *Phys. Rev.* **126**: 1875.
R. H. Dicke (1964), *Nature* **202**: 432.
R. H. Dicke, P. J. E. Peebles, P. G. Roll, and D. T. Wilkinson (1965), *Astrophys. J.* (in press).
P. A. M. Dirac (1938), *Proc. Roy. Soc. Astron.* **165**: 199.
W. Fowler (1961), *Conference Report Rutherford Jubilee International Conference*.
W. Fowler and F. Hoyle (1960), *Ann. Phys.* **10**: 280.
P. Jordan (1959), *Ztschr. Phys.* **157**: 112.
E. Lifshitz and I. Khalatnikov (1963), *Advan. Phys.* **12**: 185.

J. W. Morgan and J. Lovering (1964), *J. Geophys. Res.* **69**: 1989.
J. H. Oort and L. Spitzer (1955), *Astron J.* **121**: 6.
P. Pochoda and M. Schwarzschild (1964), *Astron. J.* **139**: 587.
E. E. Salpeter (1961), *Astron. J.* **129**: 608.
A. Sandage (1961), *Astron. J.* **133**: 355; *Proc. IAU Symposium No. 15*, Santa Barbara, California.

DISCUSSION

Demarque: These ages are based on clusters being very poor in helium. Isn't it simpler just to accept the fact that they could contain 25% helium or something like that?

Dicke: If you like to start your universe loaded with helium, this is a possibility.

Part IV. Stellar Variability

SIMPLIFIED MODELS FOR CEPHEID INSTABILITY

Norman Baker

INTRODUCTION

Of the various approaches which have been used to study stellar pulsation, the technique which in recent years has received the most attention is the stability analysis of the linearized equations of motion. An equilibrium configuration is assumed to exist, and oscillations of infinitesimal amplitude about this equilibrium are studied. Since the mechanical system is not conservative, the infinitesimal oscillations will, in general, tend to decrease or increase in amplitude. In the former case, the system is stable, in the sense that any small perturbation results in an oscillation which will eventually die out. In the latter case, it is supposed that the amplitude of the oscillation will increase with time until nonlinear effects, which are not discussed, set a limit to the size of the fluctuation. The system is in this case said to be "vibrationally unstable" or "overstable."

The complete equations of the problem in the spherically symmetrical case are a set of coupled nonlinear partial differential equations, the system being of fourth order in the space variable and third order in time. The chief advantage of the approach outlined above is that these equations are made linear, resulting in a very great simplification in the analysis. Furthermore, since the coefficients do not explicitly involve the time, it is possible to perform a harmonic analysis to separate the time. The system is thereby reduced to a fourth-order set of ordinary linear differential equations. This system can be solved numerically with presently available computers, even when the equilibrium models are quite elaborate.

Of course, one does not suppose that such an approach can provide a detailed model of cepheid oscillations, which are, in general, of large amplitude and clearly nonsinusoidal. On the other hand, it has proved to be possible to use the method to investigate the vibrational stability of specific stellar models and thus to begin to identify, from a theoretical viewpoint, the regions of instability in the H–R diagram. However, when the linear-pulsation theory is applied even to quite simple equilibrium models, a complex numerical calculation is necessary, and it is not easy to isolate the essential physics of the problem from the numerical details. It is the purpose of the present discussion to consider a model sufficiently simple to be solved analytically, but which nevertheless illustrates the physical basis of pulsational instability.

GENERAL FORMULATION AND LINEARIZATION OF THE PROBLEM

In our simplified model, we shall consider only the behavior of a thin layer or shell of gas in the star during a pulsation. It is convenient, however, to write general equations applicable to the entire star; the restriction to a small layer is made later. The independent variable is taken to be the mass M_r lying internal to the radius r

in the equilibrium model; this is then a Lagrangian variable which moves with the gas in a radial pulsation. The mechanical behavior of the star is described by the equation of motion

$$\frac{\partial P}{\partial M_r} = -\frac{1}{4\pi r^2}\left(\frac{GM_r}{r^2} + \frac{\partial^2 r}{\partial t^2}\right) \tag{1}$$

supplemented by a continuity equation

$$\frac{\partial r}{\partial M_r} = \frac{1}{4\pi r^2 \rho} \tag{2}$$

The gas is characterized by an equation of state

$$P = \frac{\mathscr{R}}{\beta\mu}\rho T \qquad \text{where } \beta = \frac{P_{\text{gas}}}{P} = 1 - \frac{aT^4}{3P} \tag{3}$$

and by the specific heat at constant pressure C_p, as well as the bulk moduli,

$$\alpha = \left.\frac{\partial \ln \rho}{\partial \ln P}\right|_T \qquad \delta = -\left.\frac{\partial \ln \rho}{\partial \ln T}\right|_p \tag{4}$$

The transport of radiation is described by the usual diffusion equation

$$L_r = -\frac{64\pi^2 acr^4 T^3}{3\varkappa}\frac{\partial T}{\partial M_r} \tag{5}$$

where $\varkappa(P, T)$ is the Rosseland mean of the mass-absorption coefficient and L_r is the total radiation flux through a spherical surface of radius r.

The energy equation is then

$$\frac{\partial L_r}{\partial M_r} = \mathscr{E} - \frac{\partial Q}{\partial t} = \mathscr{E} - C_p\frac{\partial T}{\partial t} + \frac{\delta}{\rho}\frac{\partial P}{\partial t} \tag{6}$$

where $\mathscr{E}$ is the nuclear-energy production per unit mass per unit time and Q is the heat content per unit mass. It should be noticed that equation (1) contains the "dynamic" term $\partial^2 r/\partial t^2$ and equation (6) contains the "nonadiabatic" term $\partial Q/\partial t$; these are the only time-dependent terms, since the composition is taken to be constant over the time scale we are considering. The effects of viscosity in the momentum and energy equations are also neglected.

With the help of equation (3), we eliminate ρ in equations (2) and (6); then equations (1), (2), (5), and (6) constitute a set of four first-order (in M_r) differential equations for the four dependent variables r, P, T, and L_r. We next assume that an "equilibrium" model, i.e., a solution to these four equations when the time derivatives are ignored, is available. This solution is specified by the quantities r_0, P_0, T_0, and L_{r0}; the solution to the full time-dependent problem is then defined to be

$$r(M_r, t) = r_0(M_r)[1 + r'(M_r, t)] \tag{7}$$

Thus, r' is a nondimensional variable describing the fluctuation of r from its equilibrium value. Writing similar expressions for P, T, and L_r, we define corresponding fluctuations p', t', and l'.

By neglecting all terms of higher than first order in the nondimensional quantities r', p', t', and l', which thus assume the meaning of small perturbations to the equilibrium model, we obtain linear differential equations for these quantities. Since we shall be interested only in the outer layers of a star, we assume $\mathscr{E} \equiv 0$. We further define

$$\sigma_0^2 = \frac{g_0}{r_0} = \frac{GM_r}{r_0^3} \qquad C = \frac{C_p \mu \beta}{\mathscr{R} \delta}$$

$$\varkappa_0 = \varkappa(P_0, T_0) \qquad \varkappa_p = \left.\frac{\partial \ln \varkappa}{\partial \ln P}\right|_T \qquad \varkappa_T = \left.\frac{\partial \ln \varkappa}{\partial \ln T}\right|_p \tag{8}$$

The linearized equations thus become

$$\frac{\partial p'}{\partial M_r} = \frac{1}{4\pi r_0 P_0}[\sigma_0^2(4r' + p') - \ddot{r}'] \tag{9}$$

$$\frac{\partial r'}{\partial M_r} = -\frac{1}{4\pi r_0^3 \rho_0}(3r' + \alpha p' - \delta t') \tag{10}$$

$$\frac{\partial t'}{\partial M_r} = \left(\frac{1}{T_0}\frac{\partial T_0}{\partial M_r}\right)[l' - 4r' + \varkappa_p p' + (\varkappa_T - 4)t'] \tag{11}$$

$$\frac{\partial l'}{\partial M_r} = -\frac{P_0 \delta}{L_{r0} \rho_0}(C\dot{t}' - \dot{p}') \tag{12}$$

THE ONE-ZONE MODEL

Since the coefficients of this linear system are not time dependent, it is possible to perform a separation of the time and space variables. There remains, however, a fourth-order system in the space variable M_r, whose coefficients are generally nonconstant. The essential simplification in the one-zone model is to ignore, so far as possible, the space dependence of both the coefficients and the variables. We consider a single layer, a spherical shell of gas having mass m and lying somewhere in the interior of the star. The equilibrium state of the layer is described by a single set of variables r_0, P_0, T_0, and L_{r0}. The fluctuations r', p', and t' are assumed to be constant through the layer at any given time, so that

$$\frac{\partial r'}{\partial M_r} = \frac{\partial p'}{\partial M_r} = \frac{\partial t'}{\partial M_r} = 0 \tag{13}$$

It is not possible, however, to make such a simple assumption for l' without losing essential physical properties of the system. If we set $\partial l'/\partial M_r$ equal to zero, it means that the pulsation is completely adiabatic, and no damping or excitation is possible. In general, however, the flux passing through will be modified by the gas in the shell, and we wish to show that this effect can lead to either a positive or a negative damping of the pulsation. Let the flux variation at the lower boundary of the layer be l'_L, and that at the upper boundary, l'_U. Then we make the following approximations:

$$l' = \frac{l'_U + l'_L}{2} \tag{14}$$

$$\frac{\partial l'}{\partial M_r} = \frac{l'_U - l'_L}{m} \tag{15}$$

The system will now be completely soluble if some assumption is made about l'_L. The simplest approach is to set

$$l'_L = 0 \tag{16}$$

In this case, the flux entering the layer at the bottom is constant; any variation of the emerging flux is due solely to the properties of the layer being considered.

It might be possible to obtain a somewhat more realistic one-zone model by allowing l'_l to be nonzero, since in fact there will be a variable flux impinging on any layer one might select in the star. We have chosen condition (16), however, because we wish our model to be as simple as possible and to be completely independent of the neighboring regions. Our approach will then be to try to understand the effects of interactions between different layers by comparing the one-zone model with the continuous model discussed later. Thus, the final condition defining the one-zone model, obtained by combining equations (14), (15), and (16), is

$$\frac{\partial l'}{\partial M_r} = \frac{2l'}{m} \tag{17}$$

In a formal sense, we may imagine that we are approximating the differential equations by difference equations, having divided up the star into a large number of small zones. In general, the values of the variables in a given zone depend upon the values in the neighboring zones. In the one-zone model, we have made special simplifying assumptions about the quantities in the neighboring zones in order that we may discuss the behavior of a single zone independently. In this way, we may illustrate certain physical properties which we believe to be important, while neglecting other complicating factors. We will discuss some of these neglected effects in the section on the distributed model.

In physical terms, we are considering a bulked mechanical system which is interacting with a thermodynamic system. The gas gives a "spring constant" to the shell; if set into motion, it will oscillate with a frequency of order of magnitude σ_0. If the oscillation is adiabatic, the amplitude will remain constant. The essential nonadiabatic effect to be studied here is the interaction of the gas with the radiation passing through. The gas can modulate the constant flux entering at the bottom, producing a variable outward flux at the top; this can happen in such a way that energy is either removed from or fed into the mechanical modes. In the former case, the oscillation will be damped, much as a sound wave is damped in passing through a medium of finite thermal conductivity. In the latter case, the oscillation is "negatively damped" by a similar but inverse process, and the amplitude increases at the expense of a slight decrease in the mean radiative flux emerging.

Upon applying conditions (13) and (17), which characterize the one-zone model, equations (9) to (12) become

$$\ddot{r}' = \sigma_0^2(4r' + p') \tag{18}$$

$$3r' + \alpha p' - \delta t' = 0 \tag{19}$$

$$l' - 4r' + \varkappa_P p' + (\varkappa_T - 4)t' = 0 \tag{20}$$

$$C\dot{t}' - \dot{p}' = -K\sigma_0 l' \tag{21}$$

where we have defined

$$K = \frac{2L_{r0}\rho_0}{mP_0\delta\sigma_0} \tag{22}$$

If $K = 0$ we have $C\dot{t}' = \dot{p}'$. This is the fully adiabatic case, in which the thermal properties of the oscillations are essentially uncoupled from the mechanical behavior. Thus, the nondimensional parameter K, which couples the mechanical and thermal modes, is a measure of the degree of departure from adiabacy.

Equations (18) to (21) may be combined to yield a single third-order equation

$$\dddot{r}' + K\sigma_0 A\ddot{r}' + \sigma_0^2 B\dot{r}' + K\sigma_0^3 Dr' = 0 \tag{23}$$

The coefficients in this equation are

$$A = -\frac{[(\varkappa_T - 4)\alpha + \varkappa_p\delta]}{(\alpha C - \delta)} \tag{24}$$

$$B = \frac{[3C - 4(\alpha C - \delta)]}{(\alpha C - \delta)} \tag{25}$$

$$D = \frac{[(4\alpha - 3)(\varkappa_T - 4) + 4\delta(\varkappa_P + 1)]}{(\alpha C - \delta)} \tag{26}$$

It may be noted that the quantities α, δ, and C are related to the well-known adiabatic exponents (Chandrasekhar, 1938) by the following identities:

$$\Gamma_1 = \frac{C}{\alpha C - \delta} \qquad \Gamma_2 = \frac{C}{C - 1} \qquad \Gamma_3 - 1 = \frac{1}{\alpha C - \delta} \tag{27}$$

If r' is assumed to have an exponential time dependence

$$r' = \xi(M_r)\,e^{st} \tag{28}$$

then equation (23) becomes a cubic equation in s:

$$s^3 + K\sigma_0 As^2 + \sigma_0^2 Bs + K\sigma_0^3 D = 0 \tag{29}$$

This is the eigenvalue equation for the one-zone model.

In the adiabatic case ($K = 0$), the equation reduces to one of second order, having the solutions

$$s_{ad} = \pm i\sqrt{B}\,\sigma_0 = \pm i\sqrt{3\Gamma_1 - 4}\,\sigma_0 \tag{30}$$

This is the period–mean density relation for the one-zone model. If Π is the period, and if we define the mean density of the sphere interior to r_0 by

$$\bar{\rho}_r = \frac{M_r}{\frac{4}{3}\pi r_0^3} \tag{31}$$

then equation (30) may be written

$$\Pi\bar{\rho}_r^{\frac{1}{2}} = \left(\frac{3\pi}{(3\Gamma_1 - 4)G}\right)^{\frac{1}{2}} \tag{32}$$

If the entire star is considered, the mean density

$$\bar{\rho} = \frac{M}{\frac{4}{3}\pi R^3} \qquad (33)$$

appears in place of $\bar{\rho}_n$, and equation (32) is simply one form of the usual period–mean density relation. If $3\Gamma_1 - 4 < 0$, s_{ad} in equation (30) is real, corresponding to an exponential increase or decrease in the perturbation. This is the well-known case of dynamic instability.

THE STABILITY CRITERIA

If we consider equation (29) with $K \neq 0$, the criterion that all three roots have negative real parts (stability) is

$$\sigma_0^2 B > 0 \qquad (34a)$$

$$K\sigma_0^3 D > 0 \qquad (34b)$$

$$K\sigma_0^3 (AB - D) > 0 \qquad (34c)$$

We have seen above that condition (34a) is necessary for *dynamic* stability. It will be shown below that (34b) is the condition for *secular* stability. We now assume that both (a) and (b) are fulfilled; then condition (34c), necessary for *vibrational* stability, may be written

$$\frac{4}{C} - \left(\frac{\varkappa_T}{C} + \varkappa_P\right) - \frac{4}{3\Gamma_1} \equiv \frac{\Lambda}{C} > 0 \qquad \text{(stability)} \qquad (35)$$

which also defines the quantity Λ.

By an inspection of equation (35), we can immediately see several factors which influence the vibrational stability. The first term is $4/C$, which is always positive. This term reflects the fact that mechanical energy is lost through radiation when the gas heats during compression and cools during expansion. It is the analog of the conductive damping of a sound wave. The quantity C, which may be written as $\Gamma_2/(\Gamma_2 - 1)$ according to equation (27), is a "dimensionless specific heat," having the value $\frac{5}{2}$ for an ideal monatomic gas. In an ionization zone, C may become very large because the mechanical energy of compression can go into ionization energy as well as thermal energy. Thus, the temperature fluctuation is much smaller and the stabilizing effect of this term is diminished. In the limit $C \to \infty$ ($\Gamma_2 \to 1$), the oscillation is isothermal and no energy is lost.

The second term in equation (35) is $-(\varkappa_T/C + \varkappa_P)$; it shows the effect of the change in opacity during a pulsation. Since both pressure and temperature are greater during contraction than during expansion, positive values of $\varkappa_T$ and $\varkappa_P$ would imply that the opacity is greater in the contracted state and less when the shell is expanded. This means that the change in opacity is such that more energy is removed from the flux passing through during contraction than during expansion; energy is fed into the mechanical modes during contraction, which is the proper phase to provide a driving effect. This is the so-called $\varkappa$-mechanism (Baker and Kippenhahn, 1962); the destabilizing effect of positive $\varkappa_T$ and $\varkappa_P$ is obvious from equation (35). In fact, $\varkappa_P$ is positive and $\varkappa_T$ negative throughout most of a stellar envelope (for Kramer's opacity, $\varkappa_P = 1$, $\varkappa_T = -\frac{9}{2}$), and in general it turns out that $-(\varkappa_T/C + \varkappa_P) > 0$. On the other hand, this term can have a strongly *destabilizing* effect in an ionization zone where C is large. Thus, even though $\varkappa_T < 0$, the specific heat may be so high that there is actually a net increase in opacity during compression because of the very small temperature increase. In addition, there are regions in the H$^-$ and H ionization zones where $\varkappa_T$ becomes positive (see Baker and Kippenhahn, 1962, Figures 1 and 2).

The third term in equation (35), $-4/3\Gamma_1$, which is always destabilizing, is an effect of the spherical geometry. If one constructs a plane one-zone model by ignoring curvature effects in the basic equations, this term does not appear. It reflects the change in density of a spherical shell during pulsation. The thermal conductivity of the gas depends upon the density as well as upon $\varkappa$. In the pulsation of a plane layer, the density also increases, but this increase is exactly canceled by the decrease in the total thickness of the layer; the mass per unit area remains constant. Thus, the total impedance to radiation of the layer is unchanged if $\varkappa$ is constant. In the compression of a spherical layer, however, the increase in density more than compensates for the decrease in thickness; in this way, the spherical geometry exerts a "throttling effect."

THE SATURATION EFFECT

It was seen above that a large value of C, which occurs in a region where an abundant element (hydrogen or helium) is partially ionized, can lead to instability in the sense of an exponentially increasing amplitude of infinitesimal oscillations. It is also apparent that the amplitude cannot increase indefinitely. In the linear model, C is determined by the equilibrium values of P and T and is assumed to be constant in time. However, when the amplitude becomes so large that the element in question is fully ionized in the contracted state or fully recombined in the expanded state, the effective value of C will decrease and the damping may become positive at the extremes of the fluctuation. This "saturation effect" may then limit the amplitude.

Thus, we might imagine a simple nonlinear one-zone model obtained by allowing the coefficients in equation (23) to be functions of the amplitude. A zone in which infinitesimal oscillations are negatively damped would then behave somewhat like a van der Pol oscillator; the positive damping at large amplitudes will set a limit to the size of the fluctuation. The nonlinear oscillations of such a model would then execute a sort of limit cycle. The analytic treatment of a third-order nonlinear equation such as equation (23) is very difficult. In a few specific cases it has been possible to show (see Rauch, 1950) that something like a limit cycle exists, in that the motion must be confined to a certain small region of the three-dimensional phase space. The present model seems reasonable, however, from a physical point of view, and equation (23) could easily be solved numerically. The variation of the coefficients A, B, and D would presumably be principally due to the change in C.

Clearly, there will also be other layers, lying just above or below the ionization regions in the equilibrium model, which have quite the opposite behavior. Consider such a region, which might be positively damped in the linear theory. As the amplitude increases to the extent that the gas becomes partially ionized during part of the cycle, the damping might well decrease and even become negative. There might then be a range of amplitudes for which such a layer is negatively damped. As the amplitude increases still further, however, the saturation effect would take over and the damping must again become positive.

The rough physical picture of nonlinear cepheid oscillations suggested by the above considerations may be summarized as follows: We consider a layer of gas in the star lying between M_r and $M_r + dM_r$. Such a layer undergoes oscillations whose frequency and amplitude are determined by the star as a whole. If the gas in the layer is fully ionized or fully un-ionized throughout the cycle, the layer will probably provide a positive damping—a drag on the rest of the star. A layer in which the gas is partially ionized during all or part of the cycle may produce a net negative damping over an entire cycle. This will depend, among other things, upon the mean pressure and temperature of the layer, the ionization potential of the atom or ion in question, and the amplitude of the oscillation. If the amplitude becomes sufficiently large, the saturation effect will assure that any layer will ultimately be positively damped.

It should be kept in mind that the saturation effect is not necessarily the only or even the principal mechanism which limits the amplitude in real stars. For example, it has been suggested that nonlinear effects might couple the unstable fundamental mode to higher radial modes or to nonradial modes. If these other modes are damped, the transfer of energy to them represents a net dissipation, and the amount of this dissipation depends on the strength of the coupling. The generation of shock waves can also help to limit the amplitude, since radiation from a shock front implies a net loss of mechanical energy. Christy (1964) has constructed nonlinear numerical models of pulsating stars, and he notes that, in his models, the energy dissipated in shock waves appears to become significant at an amplitude near to the limiting one.

DISCUSSION OF THE EIGENVALUES

Let us consider in somewhat more detail the roots of equation (29) in the case of small but finite K. The eigenvalue s is then expanded:

$$s = s_0 + s_1 K + s_2 K^2 + \dots \tag{36}$$

We know that for two of the roots (call them $s^{p\pm}$), the zero-order term is

$$s_0 = \pm i\sqrt{B}\,\sigma_0 \tag{37}$$

For the roots $s^{p\pm}$, it is found that the first-order term is real and the second-order term imaginary. Thus, there is a second-order correction to the frequency, and a (positive or negative) damping which is of first-order in K. The result for s_1 is

$$s_1^{p\pm} = \frac{\sigma_0}{2B}(AB - D) = -\frac{3\sigma_0\delta}{2B(\alpha C - \delta)^2}\Lambda \tag{38}$$

From this relation, it is again clear that, so long as $B > 0$, the sign of the damping constant depends on the sign of Λ; $\Lambda > 0$ implies an exponential decrease in the amplitude, $\Lambda < 0$ gives an increasing amplitude ("negative damping"). This agrees with the criterion (35). (Notice that equation (38) is only first order in K, while equation (35) holds to all orders. $s^{p\pm}$ have terms of higher order; all the real terms are proportional to Λ.) The damping time is, in order of magnitude,

$$\tau_d \approx \frac{1}{K\sigma_0} \approx \frac{mP_0}{L_0\rho_0} \approx \frac{m\mathscr{R}T_0}{\mu L_0} \approx \frac{E_{\text{th}}}{L_0} \tag{39}$$

where E_{th} is the thermal energy of the layer. Thus, the damping time is in fact a thermal diffusion time, roughly the time required by the layer to exchange its entire thermal-energy content with the surrounding material. If we were to extend these considerations to the entire star, then from the virial theorem we know that $E_{th} \approx + E_{gr}$, where E_{gr} is the total gravitational energy of the star. Thus, the damping time for the star as a whole is

$$\tau_d^* \approx \tau_{gr} \qquad (40)$$

the gravitational (Helmholtz–Kelvin) contraction time scale. (This point is discussed in a more general case by Ledoux and Walraven, 1958, §65.)

The third root, which is always real, is of the first order in K. Neglecting terms of higher order, it is

$$s^S = K s_1^S = -(K\sigma_0)\frac{D}{B} \qquad (41)$$

If $D/B > 0$, this root is negative and the system is said to be secularly stable, that is, stable against gravitational contraction. If $s^S > 0$, the gas will contract; for an entire star, this time scale is again

$$\tau_s \approx \frac{1}{K\sigma_0} \approx \tau_{gr} \qquad (42)$$

the gravitational time scale. As long as dynamic stability is present $(B > 0)$, the criterion of secular stability is thus $D > 0$, which requires

$$-(4\alpha - 3)(4 - \varkappa_T) + 4\delta(\varkappa_P + 1) > 0 \qquad (43)$$

If $\alpha = \delta = 1$, this becomes

$$\varkappa_T + 4\varkappa_P > 0 \qquad (43a)$$

In slightly different form, this is a well-known criterion for secular stability of a region in which there is no nuclear energy production. For more complete discussions of this point, see Ledoux (1958, §14) and Ledoux (1963, §2,2).

DEPTH DEPENDENCE OF THE NONADIABATIC EFFECTS

We now wish to investigate the relative effect of different layers upon the stability. We consider a layer of mass m somewhere within a star of mass M, which pulsates as a whole. The frequency of pulsation of the whole star is given approximately by

$$\sigma^* = \sqrt{\frac{GM}{R^3}} \qquad (44)$$

We can now look upon the one-zone model as a layer within the star which is subject to an external force (from the rest of the star), driving the layer periodically with frequency σ^*. Then the proper measure of adiabacy for the single layer is no longer K, but rather

$$K' = \frac{K\sigma_0}{\sigma^*} = \frac{2L_0\rho_0}{mP_0\delta\sigma^*} \qquad (45)$$

According to (39), then

$$K' \approx \frac{L_0/E_{th}}{\sigma^*} \approx \frac{\Pi}{\tau_d} \qquad (46)$$

The measure of the departure from adiabacy is thus the ratio of the pulsation period to the damping or thermal diffusion time of the layer. This is simply an expression of the physically obvious point that compression and expansion of a layer of gas will be an essentially adiabatic process if the change is carried out over a time short compared to the time needed for radiative exchange of the thermal energy of the layer with its surroundings. The fact that $\tau_d \approx E_{th}/L_0$ implies that the greater the thermal energy of a layer, the smaller will be its contribution to the nonadiabatic damping or excitation. (We are excluding from discussion the interior region of nuclear-energy production; thus, L_0 = constant.) If we consider two layers of equal mass and having mean temperatures T_1 and T_2, then

$$\frac{K'_1}{K'_2} = \frac{T_2}{T_1} \tag{47}$$

In general, a layer lying deep in the star will be much less effective in producing radiative damping or excitation than a layer of the same mass lying in the surface regions. This means that in practice it is possible to neglect the effect of the interior regions in estimating the damping coefficient. The steeper the fall in temperature from center to surface, the smaller will be the proportion of the total mass which contributes significantly to the damping. In cool supergiants, such as the δ-Cephei stars, it is found that only the outer 5% of the mass, at most, makes any contribution to the damping. Some of the pertinent quantities for an actual stellar model are given in Table I.

The table refers to one of the cepheid models investigated by Baker and Kippenhahn (1965). It has the following properties: $M = 7.0M_\odot$, $L = 4600L_\odot$, $T_e = 5320°K$. This model has a fundamental period of 11.5 days, and the detailed linear calculations show it to be vibrationally unstable. The various quantities are given at different levels in the stellar envelope, starting at the photosphere. In general, C is large in the ionization zones; Λ is seen to be negative in the center of all the ionization zones and tends to be positive in other regions (Λ is negative at the photosphere because of the large $\varkappa_T$ due to H^- ionization). The quantity K'' is the measure of nonadiabacy—it is exactly the same as K', except that we have used the total mass of the star instead of the mass of a single layer,

$$K'' = K'\frac{m}{M} = \frac{2L_0\rho_0}{MP_0\delta\sigma^*} \tag{48}$$

Table I

Region	$\log P_0$	$\log T_0$	C	$\varkappa_T$	$\varkappa_P$	Γ_1	Λ	K''	m'
Photosphere	3.422	3.726	2.51	6.02	0.667	1.67	−2.27	1.3×10^{-3}	—
H-I ionization	3.717	4.004	12.36	6.89	0.367	1.14	−1.77	1.9×10^{-4}	2.4×10^{-6}
	3.907	4.161	4.44	−3.58	0.820	1.34	−0.11	2.1×10^{-4}	5.1×10^{-6}
He-I ionization	3.957	4.236	6.53	−3.95	0.861	1.21	−0.74	1.4×10^{-4}	6.0×10^{-6}
	4.077	4.377	2.99	−3.06	0.792	1.57	+0.74	1.0×10^{-4}	8.5×10^{-6}
He-II ionization	4.677	4.605	5.29	−2.94	0.869	1.23	−0.64	3.6×10^{-5}	3.9×10^{-5}
	5.567	4.838	3.28	−3.83	0.865	1.53	+0.65	2.4×10^{-5}	2.3×10^{-4}
	7.997	5.403	3.08	−3.44	0.740	1.56	+0.83	7.8×10^{-6}	1.4×10^{-2}

The quantity m' is the fraction of the star's mass lying above the layer in question:

$$m' = \frac{M - M_r}{M} \tag{49}$$

The detailed calculations show that the pulsations are strongly adiabatic at the last point given in the table, i.e., at the level given by $P_0 \approx 10^8$ dynes/cm^2. Thus, nonadiabatic effects are confined to the outer 1.4% of the star's mass. The He-II ionization zone lies slightly above the adiabatic interior, but deep enough so that it contains more than 10^{-5} of the total mass of the star.

GENERAL DISCUSSION OF THE CEPHEID PHENOMENON

In discussing the radiative damping or excitation in cepheids, it has proved convenient to speak of three separate regions of a star (Cox, 1961). The first is the inner adiabatic region, which can be neglected in calculating the damping. The second is an outer zone, which may contain a fairly large fraction of the radius and in which the variation in Λ may be quite large due to ionization of abundant elements, but whose mass is so small that it cannot make a significant contribution to the damping. The third region is the intermediate nonadiabatic zone; the behavior of Λ in this region determines whether the damping is positive or negative and thus whether the star as a whole is vibrationally stable or unstable.

In the course of a star's evolution, it will pass through the region of the classical cepheids in the H–R diagram several times (Hofmeister, Kippenhahn, and Weigert, 1964). In each crossing, the luminosity remains nearly constant, while the surface temperature undergoes a large increase or decrease. Let us consider a star lying to the left of the cepheid region, which is beginning one of its crossings from left to right. Initially, the surface temperature is very high ($\approx 8000°$). The H, He, and He$^+$ ionization zones lie in the outermost layers of the star, and very little mass is in a partially ionized state. In the nonadiabatic zone, the abundant elements are fully ionized, C is very small ($\approx \frac{5}{2}$), and the star is strongly stable against pulsation. As the surface temperature decreases, first the He$^+$ ionization zone, then the He and H ionization zones, come into the nonadiabatic region. Thus, there is a range of effective temperatures where stars are vibrationally unstable. At still lower surface temperatures, all three ionization zones are in the adiabatic interior and can effect only a small decrease in the positive damping from the nonadiabatic zone. The existence of a region of linear instability in the H–R diagram, due to the destabilizing effect of the second helium ionization, has been quite well substantiated by detailed numerical computations of Baker and Kippenhahn (1962 and 1965) and Cox (1963).

The actual situation is somewhat more complicated than that sketched above. By the time the surface temperature drops to the point that a significant amount of mass is in the H and He ionization zones, most of the energy transport in these zones occurs by convection. In this case, the above considerations, which assume radiative transport, are clearly invalid. The interaction of convection with the pulsations is not understood; it may well be that the onset of convection determines the right-hand limit of the cepheid strip in the H–R diagram (see Baker and Kippenhahn, 1962, and Baker, 1963).

THE DISTRIBUTED MODEL

Although it is encouraging that the one-zone model leads to physically reasonable results, it is appropriate to attempt to understand the interaction of one layer with its neighboring layers. In the one-zone model, it is assumed that the physical quantities in neighboring layers do not vary; in fact, of course, they do vary and this will affect the layer being considered.

In particular, it would seem to be important to consider the spatial variation of t' and l'; the temperature fluctuation and the net radiative energy gained or lost in a layer will surely depend upon the temperature fluctuation in neighboring layers and upon the fact that the impinging radiative flux is not constant in time ($l'_L \neq 0$). We expect the spatial variation of fluctuations in the mechanical quantities to be less important for the damping.

An approach to the problem may be made using the *quasi-adiabatic* approximation (Rosseland, 1949). Essentially, this amounts to assuming that $K' \ll 1$ throughout the star and evaluating the damping coefficient to first order. Although it turns out that there are important regions in cepheids where this approximation is not valid, it has been very useful in qualitative discussions. In practice, the quasi-adiabatic approximation is carried out as follows: Assume that in zero-th order, $\partial l'/\partial M_r = 0$ (complete adiabacy). This gives a relation between p' and t', and by eliminating t' from the mechanical equations we can solve them for r' and p' throughout the star. Knowing r', $\partial r'/\partial M_r$, and p', we can get t' and $\partial t'/\partial M_r$ everywhere; these are then put into the linearized energy-transport equation to get l'. Since $\partial l'/\partial M_r$ is now nonzero, the damping may be finite. This procedure could be iterated to obtain higher approximations, although it is not clear that it would converge if there are regions of large departure from adiabacy.

Applying this procedure to the set (9 to 12), we find in zero-th order

$$t' = \frac{1}{C}p' \tag{50}$$

Substituting this into equation (10), one finds

$$3r' - 4\pi r_0^3 \rho_0 \frac{\partial r'}{\partial M_r} = -\left(\alpha - \frac{\delta}{C}\right)p' \tag{51}$$

One would then proceed to solve this equation together with (9). Then a first-order value of t' is found from (10) and substituted into (11) to get l'. Instead of doing this, we make the additional simplification of assuming $r' = $ constant for all M_r. Then the second term in equation (51) vanishes, and upon substituting equations (50) and (51) into equation (11), we find

$$l' = -b\frac{\partial t'}{\partial M_r} + \Lambda t' \tag{52}$$

where

$$b = -\frac{1}{T_0}\frac{\partial T_0}{\partial M_r} > 0 \tag{53}$$

The discussion of the stability is most simply made by using Eddington's (1928) expression for the total change in the pulsation energy during one cycle:

$$\Delta W = - \int_0^M \oint t' \frac{\partial L_r}{\partial M_r} dt \, dM_r$$

$$= L_0 \oint dt \int_0^M t' \frac{\partial l'}{\partial M_r} dM_r \qquad (54)$$

where the time integral is taken over one complete period. If $\Delta W > 0$, the pulsation energy is increasing in time and the star is vibrationally unstable, whereas $\Delta W < 0$ implies positive damping. Thus, we wish to evaluate

$$-\int_0^M t' \frac{\partial l'}{\partial M_r} dM_r = -t'l' \Big|_0^M + \int_0^M l' \frac{\partial t'}{\partial M_r} dM_r$$

$$= t'(M)l'(M) - \int_0^M b \left(\frac{\partial t'}{\partial M_r} \right)^2 dM_r + \int_0^M \Lambda t' \frac{\partial t'}{\partial M_r} dM_r \qquad (55)$$

where we have applied the boundary condition $l'(0) = 0$. The first term on the right-hand side of equation (55) depends upon the surface boundary condition; as a good approximation we may take

$$\frac{\partial t'}{\partial M_r} = 0 \qquad \text{at} \qquad M_r = M \qquad (56)$$

[This is a fairly realistic condition, valid if applied at an optical depth $\tau \gg 1$, i.e., if the oscillations of the atmosphere are neglected. The condition (56) was used for one of the models in Baker and Kippenhahn (1962). The correct boundary conditions for cepheids have been derived in detail by Unno (1964). He also has discussed the effect of the boundary conditions upon the stability calculations. A simple condition such as (56) appears to be entirely adequate for our present purpose.]

The last term in equation (55) may be integrated by parts. Then with the definition

$$\langle \Lambda t'^2 \rangle = \tfrac{1}{2}[(\Lambda t'^2)_M + (\Lambda t'^2)_0] \qquad (57)$$

we finally have

$$\Delta W \propto - \langle \Lambda t'^2 \rangle - \int_0^M b \left(\frac{\partial t'}{\partial M_r} \right) dM_r + \tfrac{1}{2} \int_0^M t'^2 \left(\frac{\partial \Lambda}{\partial M_r} \right) dM_r \qquad (58)$$

In the one-zone model, we assumed t' and Λ to be constant in space. In that case, the last two terms in equation (58) vanish and $\Delta W \propto -\Lambda$; this stability condition agrees with that derived from the one-zone model [see equation (35)]. The second term in equation (58) is a correction for the spatial variation in the temperature fluctuation. It might be thought that this differential heating and cooling could be such as to provide either positive or negative damping; in fact, this effect always enhances the stability, since b is always positive. This term also illustrates another effect which is missing in the one-zone model: the "wavelength dependence" of the damping constant.

The function t' has an increasing number of nodes as the frequency increases, and thus $|dt'/dM_r|$ will be, on the average, greater for the higher harmonies (this is especially true since the nodes appear first and are most numerous in the outer nonadiabatic regions, where the sound velocity is much smaller than in the interior). Hence, we expect the positive damping represented by the second term in equation (58) to be the greater, the higher the harmonic considered. This is one reason that the numerical linear calculations usually show that the higher harmonics (although often not the first) are damped, even though the fundamental is unstable. Physically, this comes about because the distance between hotter-than-average and cooler-than-average regions is smaller, leading to an increase in the amount of radiative diffusion between these regions. The last term depends on the variation of Λ. If we ignore the change in t'^2, the term is proportional to $[\Lambda(M) - \Lambda(0)]$. A general decrease in Λ from center to surface contributes to instability, whereas an increase stabilizes. This may possibly be related to the fact that the luminosity fluctuation is zero at the center and tends generally to grow with increasing M_r. Thus, the stabilizing or destabilizing effect of the outer layers would be relatively larger because $|l'|$ is larger there. In any case, it appears that in a simple two-zone model, where only the specific heat C differs from one zone to the other, the configuration in which the layer with high C is on top of the other would be less stable than the reverse situation.

ACKNOWLEDGMENTS

The author wishes to thank L. B. Lucy for many useful discussions. The section on the distributed model arose from discussions with W. Unno, who also made other helpful suggestions which are gratefully acknowledged. The manuscript was completed during a stay at the Max-Planck-Institut für Physik und Astrophysik, Munich, for whose hospitality the author wishes to thank Prof. L. Biermann. This work has received financial support from the United States Air Force Office of Scientific Research under Contract No. AF-AFOSR-62-386.

REFERENCES

N. Baker (1963), in: Star Evolution, Proc. Internat. School Phys. E. Fermi, Varenna, 1962, Academic Press, Inc., New York and London.
N. Baker and R. Kippenhahn (1962), Ztschr. Astrophys. 54: 114.
N. Baker and R. Kippenhahn (1965), Astrophys. J., In press.
S. Chandrasekhar (1939), An Introduction to the Study of Stellar Structure, Chicago University Press, Chicago, Illinois.
J. P. Cox (1961), Report to Commission 35, Eleventh General Assembly I.A.U., Berkeley, Calif.
J. P. Cox (1963), Astrophys. J. 138: 487.
R. F. Christy (1964), Rev. Mod. Phys. 36: 555; this volume, p. 359; and private communications.
A. S. Eddington (1962), The Internal Constitution of the Stars, Cambridge University Press, England.
E. Hofmeister, R. Kippenhahn, and A. Weigert (1964), Ztschr. Astrophys. 59: 215 and 242; 60: 57; and this volume, p. 263.
P. Ledoux (1958), Handbuch der Physik, Vol. 51, Springer Verlag, Berlin-Göttingen-Heidelberg, p. 605.
P. Ledoux (1963), in: Star Evolution, Proc. Internat. School Phys. E. Fermi, Varenna, 1962, Academic Press, Inc., New York and London.
P. Ledoux and Th. Walraven (1958), Handbuch der Physik, Vol. 51, Springer Verlag, Berlin-Göttingen-Heidelberg, p. 353.
L. L. Rauch (1950), in: S. Lefschetz (ed.), Contribution to the Theory of Nonlinear Oscillations, Annals of Mathematics Studies No. 20, Princeton University Press, Princeton, N.J.
S. Rosseland (1949), The Pulsation Theory of Variable Stars, Oxford University Press, England.
W. Unno (1964), In preparation.

EXCITATION AND GROWTH OF RADIAL PULSATIONS*

J. P. Cox

In the preceding paper, Baker has explained how the driving mechanism of the ionization zones works. In this paper, I shall describe the results of some calculations recently carried out in collaboration with A. N. Cox, K. H. Olsen, D. S. King, D. D. Eilers, and others in the J-15 group at the Los Alamos Scientific Laboratory. The physics of the operation of the destabilization mechanism of second helium-ionization has been studied using a hydrodynamic code for solving the complete set of nonlinear, nonadiabatic pulsation equations. A stellar model is followed in the course of time, using realistic equations of state and opacities. The basic purpose of such calculations has been twofold: (1) to check the linearized calculations that have been carried out on stellar pulsations (Baker and Kippenhahn, 1962, and Cox, 1963) to see if the mechanism of second helium-ionization will really work; and (2) to study certain nonlinear effects such as limiting amplitude, light and velocity curves, etc., of pulsating stars. Further details of these calculations will be described in a forthcoming paper.

In our calculations, we have so far considered only stellar envelopes rather than entire stellar models. The mass of the envelope is of the order of 10^{-4} times the mass of the entire star, and the temperature at the bottom of the envelope is of the order of 10^5 °K; the radius of the bottom of the envelope is typically 80 to 90% of the total stellar radius. The crucial region is the region which contains 50% second helium-ionization. The envelopes are chosen deep enough so that they contain the complete ionizing region plus enough material below to provide considerable radiative damping. Therefore, all the basic physics is presumably contained in the model. One of the main disadvantages of using only envelope models rather than complete stellar models is that the natural pulsation periods will be about 50% too short; the fundamental pulsations of the envelopes therefore correspond very roughly to first overtone oscillations of real stars. However, all *e*-folding times are relatively short because of the small mass of the envelopes, and computing time is therefore reduced.

The envelope is divided into a number of discrete mass zones, and the mass of each zone remains fixed with time. The initial conditions, i.e., the positions, velocities, and temperatures of all mass shells at time $t = 0$ must be specified. The bottom boundary of the envelope is kept fixed in time (the core is taken to be a rigid sphere) and a constant luminosity L_0 is fed into the envelope from below. The future time-behavior of the model can now be determined; for example it is possible to adjust the equilibrium radius R_0 of the model, keeping L_0 and M (total stellar mass) fixed.

The first step in a calculation is to set the envelope into a good approximation to hydrostatic equilibrium by using some sort of implicit hydrostatic scheme. The

* Supported in part by the Advanced Research Projects Agency Contract DA-31-124-ARO(D-139) with the University of Colorado.

347

hydrostatic equilibrium requirement is then removed and the time behavior of the envelope is calculated, somewhat as Christy (1962) has done. Two kinds of behavior have been observed. One type of behavior, corresponding to a stable envelope, results from using too little helium or the wrong equilibrium radius for given values of L_0 and M. Following one of the mass zones in time, a typical behavior is as follows: Initially, because the envelope has not been placed at the outset in perfect hydrostatic equilibrium, there are transients, characterized by an irregular, small-amplitude "jostling" of the mass zone; after 10 to 30 periods, however, all the transients have died out and a very-small-amplitude sine wave is left, whose amplitude in the case of a stable envelope decays in time.

The second type of behavior corresponds to an unstable envelope. To date, approximately 10 have been found. They behave as follows: For about the first 10 periods or so there are transients, as before, and the transients decay, leaving a small-amplitude sine wave; in the case of an unstable envelope, however, the amplitude of the sine wave grows in time. Eventually, a limiting amplitude is reached at which the pulsations become essentially strictly periodic. The typical growth time is of the order of a few hundred periods.

Three sets of "thin"-envelope models have been constructed and studied so far. In the first set of models (Cox, Cox, and Olsen, 1963), the composition was a pure helium–hydrogen mix, with the numerical helium–hydrogen ratio 0.15, and the envelope was divided into 20 mass zones. The equilibrium luminosity, i.e., the energy fed in from below, was chosen for computational convenience to be approximately twice that given by the Sandage mass–luminosity relation (Sandage, 1958). The total stellar mass for all models was taken to be $M = 1.073 \times 10^{34}\,\text{g} = 5.395 M_\odot$. Figure 1 shows the location of these envelope models on the H–R diagram.

For radii that were too small or too large, the envelopes were stable. The unstable envelopes lie in the vicinity of the observed instability strip for classical cepheids. This agreement is encouraging, but the closeness of the agreement is not

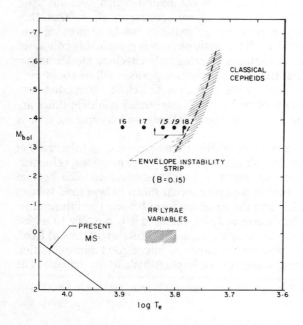

Figure 1. H-R diagram for envelope models with helium/hydrogen ratio = 0.15 (by numbers). Points labeled by slanted numbers represent unstable envelopes.

entirely significant, because our envelopes have the wrong periods. The total kinetic energy of the envelope as a function of time, as the limiting amplitude is attained, is shown in Figure 2. The limiting amplitude was reached, in this case, after about 1000 periods of growth.

The kinetic energy leveled off at about 10^{41} ergs for all envelope models calculated. The limiting pulsation properties of the two unstable envelopes in this series are summarized in Table I.

In order to insure that the limiting amplitude was independent of starting conditions, a large perturbation was applied to one of the envelopes after it had attained limiting amplitude. The perturbation quickly decayed, and, after some 10 periods or so, the original limiting amplitude had been recovered to within a

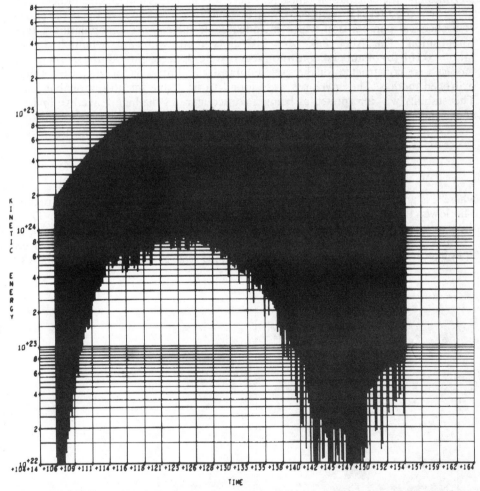

Figure 2. Total kinetic energy of the envelope as a function of time. Ordinates are in units of 10^{16} ergs. The leveling off of the upper envelope of the curve at about 10^{41} ergs represents the attainment of the limiting amplitude. The curious appearance of the lower envelope of the curve is a spurious effect introduced by the method of plotting the data. Approximately 750 periods are shown in the figure.

Table I. Limiting Pulsation Properties of Unstable Envelopes
(Helium–Hydrogen Mix, $B = 0.15$)

PM	Radius semiamplitude $(\delta R/R_0)$	Total bolometric light range	Total velocity amplitude (km/sec)	Maximum kinetic energy (ergs)
15	≈ 0.04	$\approx 0^{m}\!.32$	64	1.0×10^{41}
19	≈ 0.045		61	8×10^{40}

few percent; the detailed properties of the pulsations were also practically the same as before the perturbation.

To test the effect of zoning, one of the unstable envelopes was divided into 50 zones and the calculation repeated. There was no important change in behavior.

The second set of models was constructed in an attempt to investigate the effects of a chemical composition which included some metals, i.e., a population I type composition. The Aller mix, $Z = 0.02$, was used, with the helium–hydrogen ratio = 0.161 (by numbers), and the models were divided into 50 mass zones. The models were constructed for the same luminosity and mass as for the pure helium–hydrogen models. The envelopes were found to be stable for all reasonable radii. This stability was interpreted as being a consequence of the higher average opacity above the ionizing region, caused by the higher metal abundance, which leads to a lower density in the ionizing region. Very roughly, the density in the ionizing region varies as

$$\rho \propto \frac{M}{\varkappa L}$$

where $\varkappa$ is an average value of the opacity in the regions above the ionizing region. The lower the density, the less sensitive is the opacity to the temperature. Thus, the kappa-mechanism, as described by Baker, is less effective in driving pulsations.

In order to raise the densities, a lower luminosity was chosen and a third set of models (with the same composition and zoning) was calculated with a luminosity

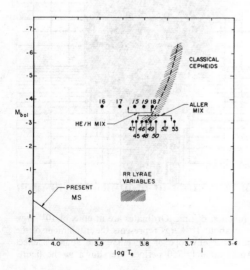

Figure 3. Location of the two sets of envelope models on the H-R diagram. Points labeled by slanted numbers represent unstable envelopes. The arrows indicate the estimated edges of the computed instability strips.

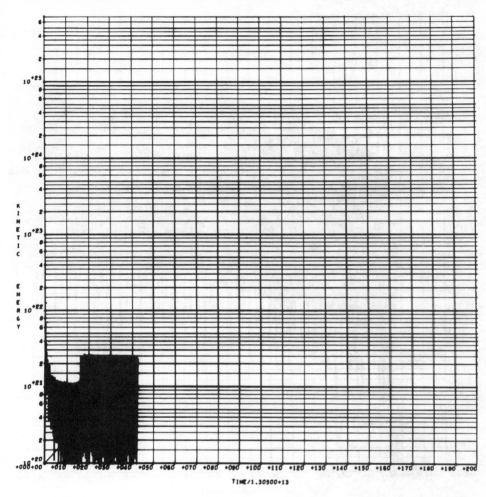

Figure 4. Total kinetic energy of a stable Aller-mix envelope lying just to the left of the instability strip as a function of time. Ordinates are in units of 10^{16} ergs and abscissae are in units of the pulsation period of the envelope. This envelope is barely stable, as is indicated by the very slow rate of decrease of the upper envelope of the curve with time.

about half the previous value, that is, the luminosity was now taken as equal to that given by the Sandage mass–luminosity law (Sandage, 1958). L_0 and M were kept constant, and the equilibrium radius was varied. Again, a group of unstable models was found; those having too small or too large a radius were stable, as in the first set of models. The results are shown in Table II.

The location of the two sets of unstable envelopes on the H–R diagram is shown in Figure 3. The unstable envelopes with the higher value of Z are shifted to the right because of the greater mean opacity in the regions above the ionizing region. This mean opacity is increased by 20 to 30% over that for the pure helium–hydrogen models. This calculated instability strip (for the Aller-mix composition) is close to that observed for the classical cepheids. However, if the fact were taken

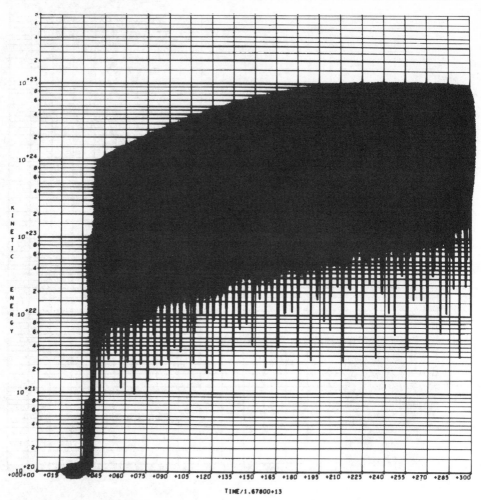

Figure 5. Total kinetic energy of an unstable Aller-mix envelope located near the middle of the instability strip as a function of time. Units are the same as in Figure 4. The steep initial rise in the upper envelope of the curve to about 10^{40} ergs was artificially induced in the calculations to save computer time.

Table II. Results for Envelopes with Aller-Mix Composition (50 Zones)

PM	$M_{bol,0}$	Π_{Nat} (days)	R_0 ($R_\odot = 1$)	$\log T_{e,0}$ (°K)	$\tau_{\dot{x}}/\Pi_{Nat}$	Remarks
44	−3.720	2.359	36.8	3.827	−200	Unstable, He–H mix
47	−3.055	1.444	27.1	3.828	+260	Stable
45	−3.055	1.510	28.5	3.817	+2500	Barely stable
46	−3.055	1.678	30.0	3.806	−150	Unstable
48	−3.055	1.806	31.4	3.796	−100	Unstable
49	−3.055	1.942	33.0	3.787	−125	Unstable
50	−3.055	2.089	34.7	3.774	—	Unstable
52	−3.055	2.411	38.1	3.754	−135	Unstable
53	−3.055	2.782	41.9	3.733	+340	Stable

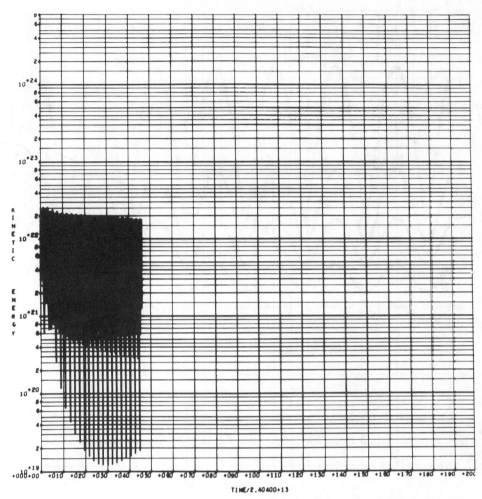

Figure 6. Total kinetic energy of a stable Aller-mix envelope lying to the right of the instability strip. Units are the same as in Figure 5. The kinetic energy decays with time, as is shown by the decrease with time of the upper envelope of the curve.

into account that these envelopes have periods about 50% shorter than the fundamental periods of the corresponding whole stars, the instability strip would be shifted further to the right (by about $\Delta \log T_e \approx -0.036$) *away* from the observed strip. This shift would then bring these results into approximate agreement with those obtained by Baker and Kippenhahn (1962) on the basis of their linearized calculations. Thus, there is still not very good agreement between theory and observations.

The kinetic energies of Aller-mix envelopes lying immediately to the left of, in, and just to the right of the instability strip are shown in Figures 4 to 6. If the radius is too small or too large, the energy decays, while for the unstable envelopes it grows and again levels off at about 10^{41} ergs.

Figure 7 shows radius versus time for some of the mass zones for one of the unstable Aller-mix envelopes after limiting amplitude has been attained. Notice

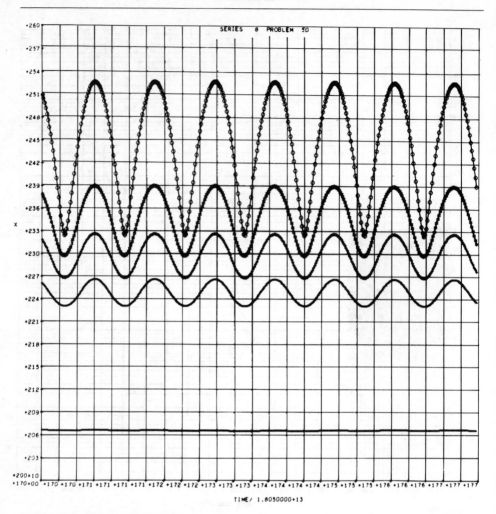

Figure 7. Radius versus time for several mass zones for one of the unstable Aller-mix envelopes after limiting amplitude has been attained. Ordinates are in centimeters. The top curve is for the outermost zone, the bottom for the innermost zone.

that the nonlinearity becomes appreciable only for the outermost few zones. In other words, the pulsations are still quite small, even at the limiting amplitude.

The luminosities and radial velocities of some of the zones as a function of time for one of the unstable Aller-mix models after attainment of the limiting amplitude are shown in Figures 8 and 9. The details of the luminosity curves, especially for the outermost zones, are probably not significant because of some rather crude approximations used in treating the radiation flow in these regions (for example, the entire region of hydrogen ionization was contained in only one or two mass zones).

The limiting pulsation properties of the unstable envelopes for the Aller mixture are summarized in Table III. Note that the limiting values of $\delta R/R_0$, velocity amplitude, and total kinetic energy are at least reasonable compared to

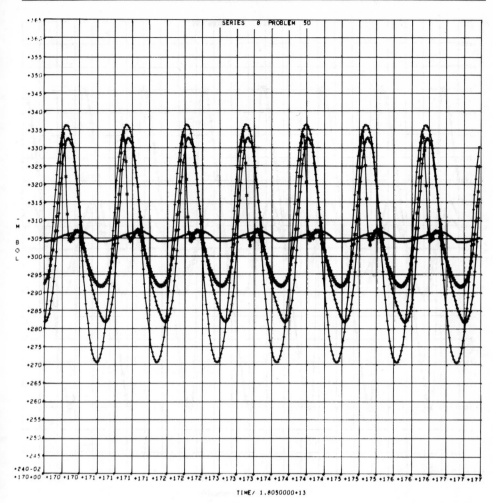

TIME/ 1.8050000+13

Figure 8. Luminosity (in bolometric magnitudes) versus time for several mass zones for the envelope of Figure 7. Luminosity increases toward the top of the figure. The outermost zone is represented by the curve marked with the asterisk (*) symbols; the curve marked with the plus (+) symbols lies the next farthest in; the curve marked with the minus (−) symbols lies still farther in; and the curve not marked with any symbols represents the innermost zone.

Table III. Limiting Pulsation Properties of Unstable Envelopes (Aller Mix)

PM	Radius semiamplitude ($\delta R/R_0$)	Total bolometric light range	Total velocity amplitude (km/sec)	Maximum kinetic energy ($\times 10^{40}$ ergs)
44*	0.047	0.^{m}40	69.5	9
46	0.036	0.^{m}36	63	7
48	—	—	—	8
49	—	—	—	9
50	0.042	0.^{m}41	71	9
52	0.041	0.^{m}40	67	9

* Helium–hydrogen mix.

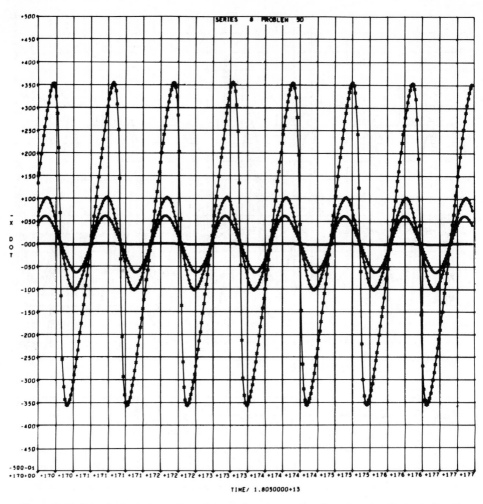

Figure 9. Radial velocity (values *above* the zero line denote *inward* motion, $\dot{r} < 0$) versus time for several mass zones for the envelope of Figures 7 and 8. Ordinates are in kilometers per second. The significance of the symbols on the various curves is the same as in Figure 8.

observed properties of classical cepheids. In addition, the relatively small values of $\delta R/R_0$ at the limiting amplitude favor the suggestion by Ledoux and Whitney (1961) that the limiting amplitude is determined by a "saturation effect" of the second-helium ionization-driving mechanism.

Perhaps the most important conclusion to be drawn from this set of calculations is that "soft" self-excited radial oscillations (see Ledoux and Walraven, 1958) can arise in a stellar envelope (and therefore presumably also in a whole star) from the action of second-helium ionization in the envelope, provided that the parameters of the star (luminosity, mass, radius, and composition) form a favorable combination. Moreover, the attainment of a stable limiting amplitude shows that a "limit cycle" kind of behavior is possible. Because of the good overall agreement with the results of the linearized theory (Baker and Kippenhahn, 1962, and Cox, 1963) (reasonable growth rates, well-defined instability strip in approximately the right location, stability when insufficient helium is present, etc.), the instabilities found here are very likely of an actual physical origin and not merely a reflection of spurious mathematical instabilities.

The stability of the second set of models (Aller-mix composition, L_0 about twice the Sandage value) leads to the following conclusions (these might also have been deduced from a linearized theory): (1) If the density in the critical region becomes too small, instability is likely to disappear. This means, with population I cepheids, that the ratio L/M should not be too large. For a given luminosity, a fairly large mass is required in order that these stars be unstable. (2) Going up along the cepheid-instability strip, the luminosity increases as about M to the third or fourth power; therefore, the density in the critical region should decrease in going up the strip, and it may well be that this decrease is what cuts off the instability strip at the top. (3) Since the ratio L/M is probably larger for population II stars than for population I stars, it may be that low metal abundance is a necessary condition for instability in population II stars. These last three conclusions are based on studying envelope models alone and are therefore rather tentative when applied to whole stars. Models with much deeper envelopes are needed to test these conclusions; work with such models is now in progress.

REFERENCES

N. Baker and R. Kippenhahn (1962), *Ztschr. Astrophys.* **54**: 114.
R. F. Christy (1962), *Astrophys. J.* **136**: 887.
J. P. Cox (1963), *Astrophys. J.* **138**: 487.
J. P. Cox, A. N. Cox, and K. H. Olsen (1963), *Astron. J.* **68**: 276.
P. Ledoux and Th. Walraven (1958), in: S. Flügge (ed.), *Handbuch der Physik, Vol. 51*, Springer-Verlag, Berlin, p. 550.
P. Ledoux and C. Whitney (1961), in: R. N. Thomas (ed.), "Aerodynamic Phenomena in Stellar Atmospheres," *Suppl. Nuovo Cimento* **22** (1): 131 ff.
A. Sandage (1958), *Astrophys. J.* **127**: 513.

DISCUSSION

G. Wallerstein: From your remarks, it seems that a helium-ionization zone is needed to get a population II cepheid to pulsate. If one needs a low metal content combined with small mass, which is reasonable for population II, perhaps it should be concluded that there are stars, namely, these stars, with low metal content but high helium content. This is of considerable interest to the question of the primordial helium content of the galaxy, which might be considerable.

N. Baker: Convection is important in determining the stratification of the equilibrium model. The H$^-$ absorption can change the outer boundary conditions of the convection zone. There can be a fairly large change in the stratification in the case where there is a convection zone. For the stars you are considering, I don't know how important this is.

J. P. Cox: Some of the Los Alamos calculations were carried out using your code for convection. We found that convection may very well become important in some of these stars in the second helium-ionization region, but we have not treated convection adequately yet.

STELLAR VARIABILITY*

R. F. Christy

The general idea behind these calculations is that the observed pulsation motions of stars arise spontaneously because of the particular physical properties of the envelopes. The relevant physical properties are the equation of state and the opacity. The approach is to integrate numerically the time-dependent equations of hydrodynamics (with spherical symmetry) and heat flow. Large-amplitude oscillations and their nonlinear effects are studied.

The calculation of the motion of the envelope is performed omitting the core of the star where nuclear-energy generation occurs. The bottom boundary of the envelope is taken to be a rigid sphere ($\dot{r} = 0$ at $r = R_1$) with a constant luminosity [$L(R_1) = L_0$]. These boundary conditions will not affect the pulsations, because the amplitude of the motion decreases toward the center of a star. The base of the envelope is taken at a radius small enough that further variation in its location was demonstrated to have no effect. This was at a radius varying from a quarter to a sixth of the outside radius, and at a temperature at the base in excess of 10^6 degrees. The periods could conceivably be wrong by a few percent, with similar errors in other quantities. The boundary condition at the top of the atmosphere is taken to be such that the pressure goes to zero and the flux is the solution of the time-dependent diffusion equation, with the boundary condition on the temperature given by the relation

$$T^4 = \tfrac{3}{4}T_e^4(\tau + \tfrac{2}{3})$$

where τ is the optical depth.

The calculation of the pulsations starts from the static solutions for the temperature and density, with a superimposed velocity distribution. The results of a sample calculation are shown in Figure 1. The mass of this particular model was taken to be artificially small, $M = 0.4M_\odot$, so that the amplitude would grow quickly to a limiting amplitude with a minimum amount of computing time. The static luminosity was $L = 1.5 \times 10^{35}$ ergs/sec, and the effective temperature was $T_e = 6500\,^\circ$K. The period was two thirds of a day. The luminosity, velocity, and radius are shown as a function of time. The luminosity shown is the rate at which energy leaves the star. The development in time of the velocity is shown for an optical depth $\tau = 0.1$ to 0.2. This was the third zone in from the surface. The velocity is positive upward. The radius is measured at an optical depth of approximately 1, which was in this case the fourth or fifth zone in from the surface.

The behavior is a little irregular at first, because the technique of initiating motion introduces unwanted harmonics as well as wanted ones; nevertheless, in a case like this it settles down after not too long to a rather regular motion in which even certain peculiar features become quite regular.

---

* For details of the methods, refer to R. F. Christy (1964), "The Calculation of Stellar Pulsation," *Rev. Mod. Phys.* **36**: 555.

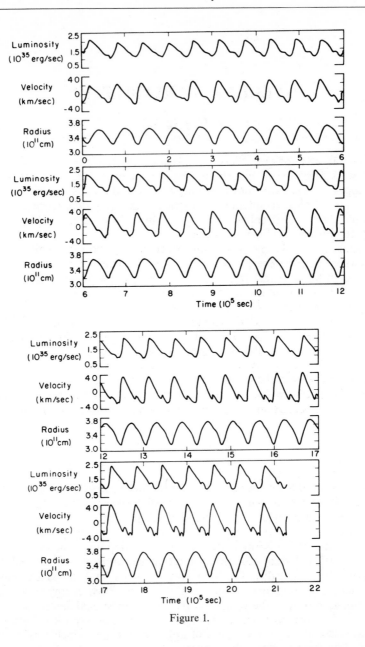

Figure 1.

The structure of the static model which preceded the dynamic calculation is shown in Figures 2 to 5. There were 38 mass points in this particular model. Figure 2 shows the radius of the various mass points in units of 10^{11} cm. The little marks show the amplitude of oscillation that eventually showed up in this model. The amplitude decreases toward the center and is quite inconsequential at the inner boundary, which could have been taken anywhere in the last few mass zones with-out making any essential difference. Figure 3 shows the pressure in the static model

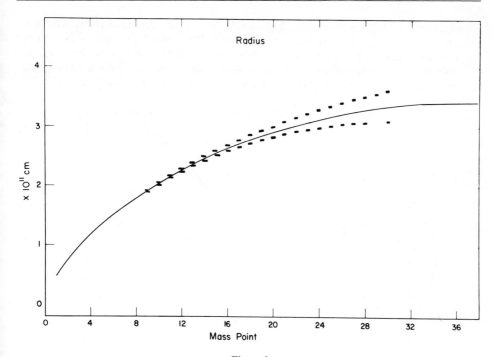

Figure 2.

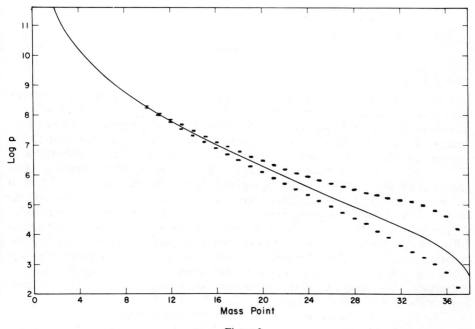

Figure 3.

as a function of the mass points. The amplitude of the pressure is shown; it is large near the surface and again becomes negligible toward the core. The zoning was approximately exponential but deviated toward the inside in an effort to keep the time required for sound to cross each zone roughly constant. Figure 4 shows the temperature as a function of pressure. Convection was ignored because of a belief

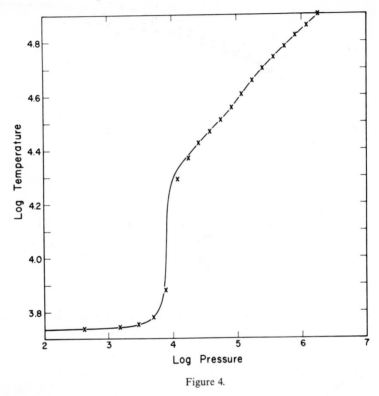

Figure 4.

that for models with sufficiently high temperature and sufficiently low density convection could hardly play an essential role; thus, purely radiative equilibrium was assumed. The solid line indicates the result of a static integration of much finer zoning; the crosses mark the mass points in the particular zoning used in the dynamic calculation. The sudden change in the temperature, in the neighborhood of 10^4 °K, is the most awkward feature of the problem to handle numerically. This jump occurs where hydrogen is ionizing and the opacity increasing rapidly, by a factor of 10^2 to 10^3 across one mass zone, to a maximum at about 2 to 3×10^4 °K. In spite of the sudden change in temperature, the crude zoning was able to reproduce the temperature distribution rather well, because it was possible to make an unusual difference approximation to the radiative diffusion equation, which represented the behavior of radiation flow through this region correctly in the average. Figure 5 shows the temperature as a function of mass point, again showing the jump in temperature. The amplitude of the temperature fluctuations is shown. Helium ionization occurs at mass points around 25 or 26 at 5×10^4 °K.

The details of the variation of the velocity and luminosity as a function of time over a period and with depth are shown in Figure 6. In the deep interior, below

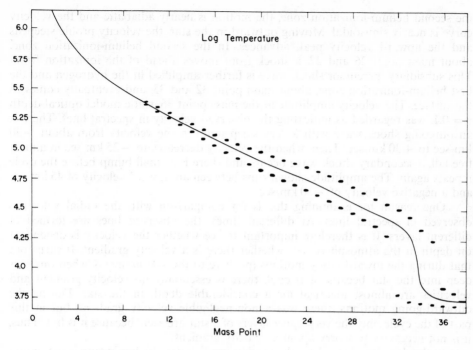

Figure 5.

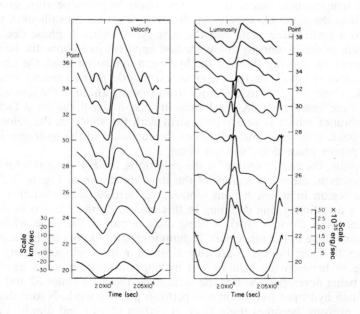

Figure 6.

the second helium-ionization zone, the motion is nearly adiabatic and the velocity curve is nearly sinusoidal. Moving outward in the star, the velocity profile steepens and the upward velocity peak advances. In the second helium-ionization zone, about mass point 26 and 27, a shock front moves ahead of the ionization front. This subsidiary, precursor shock wave is further amplified in the hydrogen and the first helium-ionization zone, about mass point 32 and 33, and eventually comes to the surface. The velocity amplitude at the mass point 36, static model optical depth $\tau = 0.1$, was regarded as indicating the observed velocity in spectral lines. There is an outgoing shock wave with a very steep jump in the velocity from about -30 km/sec to $+30$ km/sec. Then, when the velocity decreases to -25 km/sec in almost free fall, a secondary shock wave occurs and there is a small bump before the cycle repeats again. The amplitude shown runs between an upward velocity of 45 km/sec and a negative velocity of -30 km/sec.

One reason for examining this is for comparison with the radial velocities observed in spectral lines. At different times, the observed lines are formed at different layers. It is therefore important to see whether the velocity is dependent on depth in the atmosphere, i.e., whether there is a velocity gradient. It turns out that during the inward-going motions (positive to the astronomers), when one sees deep into the star because it is cool, there is essentially no velocity gradient and velocities are almost identical for a considerable depth in the star. During the outward-going motions, there was a non-negligible velocity gradient, but in this part of the cycle only the very upper layers of a star are seen, because it is hot. Thus, it is not necessary to worry about a velocity gradient.

In the deep interior, where the motion is adiabatic, the luminosity is maximum when the radius is minimum, i.e., at maximum compression. In the second helium-ionization zone, mass points 22 to 27, the luminosity develops a very great peak of short duration. On contraction, the heat absorption due to the ionization delays the rise in temperature; this causes an abnormally large temperature gradient and heat flux into the zone. On expansion, the opposite effect prevails until the flux is reduced to a small value again and the next cycle begins. A phase delay and the general form of the luminosity curve are first apparent just above the helium zone, at mass point 28. Passing through the hydrogen ionization zone, the phase of the peak is further delayed. There is also an extra little pulse, running ahead of the main peak, which is caused by the shock front running ahead of the ionization front. At the surface, mass zone 38, the luminosity has a sharp rise by a factor of 2.5 (one magnitude), which is nearly coincident with the shock in the velocity curve. After the peak, it decays slowly. Without the inclusion of the hydrogen-ionization zone, the proper phase delay will not occur.

To explore the energy source for the pulsations, the P-V diagrams for a number of mass points in the outer regions of the stars are shown in Figure 7. Zone 22 is below the region in which helium ionizes; the arrows show that this P-V cycle occurs in a direction giving damping in this region. Zone 24 is just at the base of helium ionization. Positive work is being done in part of the cycle, while negative work is being done in the other part. Helium ionization occurs in the range of mass zones 26 to 28. Their P-V cycles are all performed in the work-performing direction. In the region between the helium- and hydrogen-ionization zones, essentially no energy is being developed. Nearer the surface are the mass zones 32 and 33, which cycle through hydrogen ionization and perform positive work. Notice that in these zones the pressure becomes quite large at certain phases and almost vanishes at other phases.

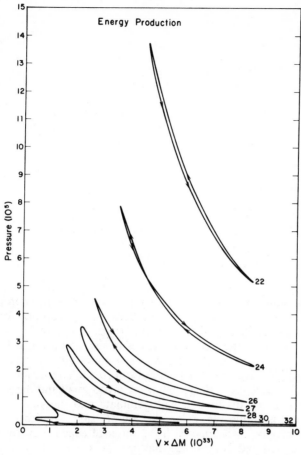

Figure 7.

The sum of the positive contributions to the energy production after 18 periods is 6.3×10^{38} ergs/period, that is, about 11% of the luminous flux is converted into mechanical energy per period. Of this, the hydrogen-ionization zone produces 33% and the second helium-ionization produces 67%. The total dissipation is 4.9×10^{38} ergs/period. Thus, the increase in kinetic energy per period is 1.5×10^{38} ergs/period, which is very sensitive to slight errors in the treatment of the hydrogen zone. Although the hydrogen ionization accounts for only one-third of the total energy production, if it had not been included, this model would not have oscillated. The increase in kinetic energy per period is shown in Figure 8. The fractional increase in kinetic energy per period is initially about 4% and decreases as the limiting amplitude is approached. The amplitude is limited by the damping, which is mainly produced by radiation damping throughout the envelope and partly by the creation of shock waves in the very outer layers.

The basic reason for the energy generation in the ionization zones is their small γ (ratio of specific heat) and high heat capacity, so that the ionization zones are cooler, relative to their surroundings, on adiabatic compression and hotter on

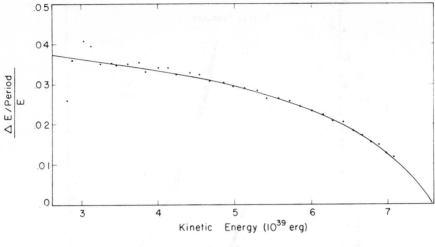

Figure 8.

adiabatic expansion. This means that they will absorb heat when compressed and give it off when expanded. This is just the behavior appropriate for the generation of work and is exemplified in the ionization zones in Figure 7.

The importance of the ionization zones is also conditioned by their depth in the envelope. If T_e is too great, the ionization zones are too near the surface and consequently will involve too little material to provide any significant heat capacity. As a result, the phase delay of the temperature is too small and the energy production is reduced, so that the dissipation will dominate and the star will be stable. The boundary to the unstable region on the low T_e side is more obscure. Estimates suggest that it may be associated with the onset of effective convection. Since convection has been ignored in these calculations, we are unable to settle this question at this time.

In regions of constant ratio of specific heats, it is also known that the dependence of opacity on temperature and density is very important. The usual dependence $\varkappa \sim \rho/T^{3.5}$ leads to dissipation, since the heat transport is greatest when the material is compressed adiabatically. This accounts for the dissipation in the deeper regions of the envelope.

The instability strip is shown as a function of mean effective temperature in Figure 9 for a particular mass $M = 0.9 M_\odot$, luminosity $L = 2.8 \times 10^{35}$, and chemical composition helium = 45%. At the higher temperatures, the first harmonic is unstable, with periods ranging from 0.3 to nearly 0.45 days. Periods in this range are first-harmonic periods. At lower effective temperatures, the fundamental mode becomes unstable, with periods ranging from a little over 0.5 days up to 1 day. The period decreases with increasing effective temperature. These periods, with the gap, correspond closely to what is seen in ω Centauri. Table I shows the properties of a particular model which behaves like SU Draconis.

The effect of varying the helium composition is shown in Figure 10. At 45% helium, the fundamental is unstable up to an effective temperature of 7000° and the first harmonic is unstable to 7500°. At 30% helium, the instability of the fundamental ends at 6750° and at 60% it ends at 6500°. For a much higher luminosity,

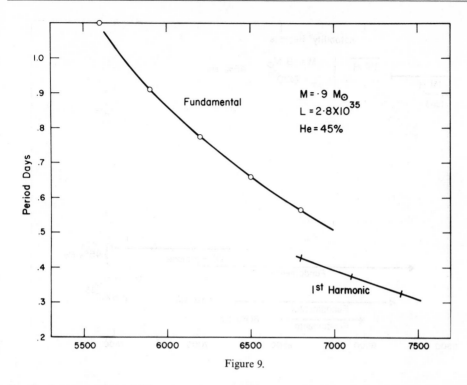

Figure 9.

the fundamental instability extends out to about 6700° in temperature, the first harmonic instability to about 5900°, and even a third harmonic instability showed up in one calculation.

It is possible to use the observations to determine many properties of the stellar envelopes. The radius can be deduced from the velocity curve and the color and luminosity curves; the mass can be deduced from the period, the $P\sqrt{\rho} = $ constant relation, which, since these periods are reasonably accurate, can be rather accurately achieved. It is also possible to use a measure of the ratio of radius amplitude to luminosity amplitude as an additional quantity that characterizes different properties of the envelopes. Figure 11 shows the ratio of this measure of radius to luminosity amplitude as a function of the mean temperature of a model with a certain luminosity, 2.8×10^{35} ergs/sec as used before, and as a function of the helium concentration. Thus, from the mean temperature, it is possible, by measuring the ratio of radius to luminosity amplitudes, to determine something about the helium composition. It was in this way that 45% helium was shown to give the best fit to observations for this model considered, which corresponds to SU Draconis.

Table I. SU Draconis Model

Mass	Radius	Luminosity	M_{bol}	T_e	Composition	Period	$P(\rho/\rho_\odot)^{\frac{1}{2}}$
1.75×10^{33} g $= 0.9(\pm 0.3)M_\odot$	4.7×10^{11} cm $= 6.7(\pm 0.7)R_\odot$	2.8×10^{35} ergs/sec	0.1	6500 °K	H = 55% He = 45% Z = 0.2%	0.66d	0.035

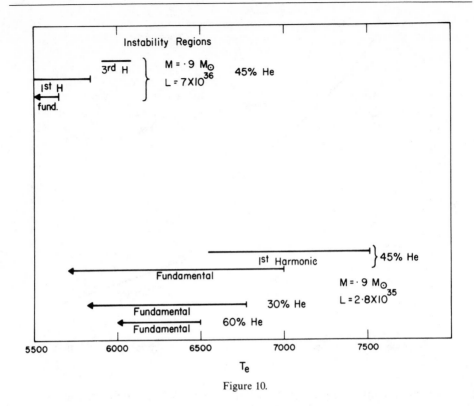

Figure 10.

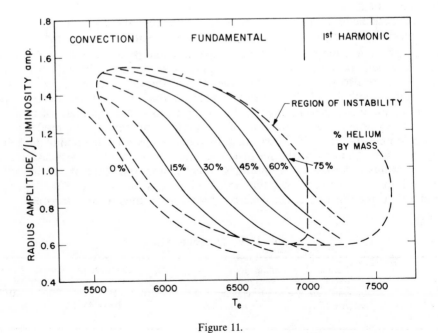

Figure 11.

There is a sizeable uncertainty in such a helium-abundance determination because of uncertainties in the observational values of the radius and effective temperature. The radius is determined from the radial velocities, which are fairly well known, and the colors, which is where the uncertainty comes in. The radius may be uncertain by 10%. This makes a 30% uncertainty in the mass. The temperatures are known to within 100°. Thus, the helium content could vary from 30% to 60%.

Part V. Stellar Mass Loss

T TAURI MASS EJECTION

L. V. Kuhi*

T Tauri stars are believed to be stars in the general region of one solar mass that are still undergoing gravitational contraction toward the main sequence. The observational evidence to support this hypothesis is as follows: (1) T Tauri stars are found only in association with obscured regions of nebulosity, i.e., near concentrations of dust. (2) Since the space density of T Tauri stars is in some regions one to two orders of magnitude above that of field stars of the same luminosity, the T Tauri stars can hardly be chance interlopers from the general field. (3) The T Tauri stars occur in greatest numbers in regions where bright O- and B-type stars are found, such as in the Orion nebula. These early-type stars are believed to be of recent formation, and it seems reasonable that the intimate association with T Tauri stars means that the latter are examples of the same phenomenon at lower masses. (4) Two otherwise inexplicable features of the T Tauri stars—their abnormally wide absorption lines and their elevation above the main sequence in the H-R diagram—find natural explanation if the stars are indeed in contractive evolution toward the main sequence.

T Tauri stars show an emission-line spectrum. The strongest emission lines (hydrogen and ionized calcium) are observed to have violet-displaced absorption components superimposed upon them (Figure 1) with negative velocities ranging from 80 to 230 km/sec, thus indicating the presence of foreground material moving away from the star. There is no sign of the return of this rising material; it has thus been inferred that it actually leaves the star, being driven in some manner from

* Presented by G. H. Herbig.

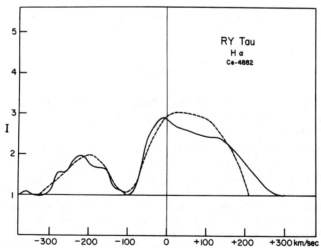

Figure 1. $H\alpha$ in RY Tau: Solid curve is observed. Dashed curve is computed. $I = 1.0 =$ continuum.

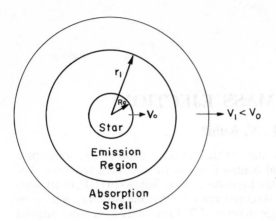

Figure 2. Schematic representation of the model.

below, since the line displacements usually do not exceed the velocity of escape at the surface of the star.

The line profiles were analyzed on the basis of a simple spherically symmetric geometric model. It is assumed that surrounding the star of radius R_0 there is an envelope moving radially outward and subjected only to the deceleration of gravitational forces. This envelope provides the emission lines and is in turn surrounded by a thin shell of material also moving radially outward which effectively produces the absorption component (Figure 2).

The velocity at which the material leaves the surface of the star is v_0. In the shell in which the absorption lines are formed, the velocity has decreased to v_1, which can be directly determined from the velocity shift of the absorption component with respect to the center of gravity of the emission line, or, preferably, with respect to the velocity corresponding to the star itself (which has a separate set of absorption lines). The ejection velocity at the surface v_0 is determined by assuming that the material moves ballistically in the gravitational field and is thus given by the half-width of the emission profile. The effect of pressure pushing it from below is neglected.

The rate of mass loss is obtained from the expansion velocity, the amount of material per unit area required to produce the lines (both emission and absorption), and the surface area of the star.

$$\frac{dM}{dt} = 4\pi R_0^2 \rho(R_0) v_0$$

where $\rho(R_0)$ is the density at the surface of the star and is determined from fitting computed emission profiles to those observed. The results for six T Tauri stars are shown in Table I.

The outward velocities at the surface of the star are probably just a little bit less than the velocity of escape from these stars. Thus, pressure forces must be present in order for the material to escape. Because the absorption components in these stars are never seen on the red side of the emission lines, thus indicating material falling back to the surface, it is believed that the outward-moving material leaves the system. However, this does not prove that it actually does leave the system.

The question of how long the process of mass loss continues is of interest because the calculated instantaneous rates of mass loss are quite large. If these stars are of the order of one solar mass, this process cannot go on for long periods

Table I

Star	Absolute visual magnitude	Emission-line intensity class	r_{shell}/R_{star}	v_0 (km/sec)	v_1 (km/sec)	dM/dt ($10^{-7} M_\odot$/yr)	Mass of envelope (10^{25} g)	Replacement time (days)
Ru Lup	4.0	5	1.4	300	200	1.4	0.0037	0.048
LKH α 120	0.21	4 (peculiar)	1.7	300	180	5.9	0.17	0.55
AS 209	4.33	4u	2.9	325	0	0.65	1.1	31
GW Ori	1.7	(2)	9.0	240	80	0.35	9.5	500
T Tau	3.35	2	2.6	225	140	0.35	4.4	230
RY Tau	3.73	2	10.5	325	100	0.31	4.9	290

of time, because then there would be very serious difficulties with the total mass ejected being comparable to the original mass.

T Tauri stars are observed to range from those with extremely intense emission spectra down to the stars which resemble the sun. It is found that there is a correlation between the rate of mass loss and the general level of intensity of the emission lines as indicated by Herbig's emission-line intensity-class lines (Table I and Figure 3).

The assumption is made that all of the T Tauri stars in the solar neighborhood (out to about 1000 parsecs) represent different stages in exactly the same phenomenon and exactly the same kind of stars. In other words, the stars that have very strong emission lines and a large mass-ejection rate at the present time will later on move down the curve, their emission lines will weaken, and their rate of mass ejection will decrease. It is further assumed that a $1.5M_\odot$ star is representative of the entire class of T Tauri stars and that the relative numbers of stars observed in these various stages of emission line intensity represent the relative length of time a star of this type remains in that stage. The total time of pre-main-sequence contraction for such a star is 1.17×10^7 years, according to recent evolutionary tracks computed by Iben. Thus, the time spent in each stage can be estimated. The corresponding rate of mass loss can be found from the observed correlation (Figure 3), and thus the mass ejected in each stage and the total mass loss can be found (Table II).

During the entire contraction time, the integration gives a total mass loss of $0.20M_\odot$, or approximately 13% of the mass of the star, which indicates that the contraction time-scale will be slightly shorter (a few percent). Although the initial

Table II

Emission-line intensity class	Fraction in each class	1.5$M_\odot$ star	
		Time in each class (10^7 yrs)	Mass ejected ($M_\odot$)
5	0.02	0.02	0.028
4	0.06	0.07	0.056
3	0.10	0.12	0.060
2	0.11	0.13	0.039
1	0.11	0.13	0.013
—	0.60	0.70	0.000
Total	1.00	1.17	0.196

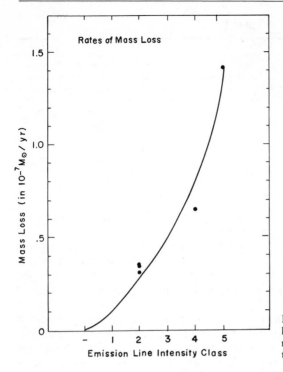

Figure 3. The dots are observed stars; the line is the smoothed curve used in obtaining mean ejection rates for each emission intensity class mass loss in $10^{-7} M_\odot$ per year.

rate of mass ejection may be quite high, the total mass ejected does not amount to a larger fraction, simply because the star passes rather rapidly through this activity. However, the kinetic energy of the ejected material in the initial stages may be comparable to the luminosity of the star; this implies that energy transfer by means of mass ejection may have to be considered in the construction of pre-main-sequence evolutionary models. The problem is even greater for $1.0 M_\odot$ stars, where the mass loss may reach $0.4 M_\odot$ if the same rates apply. However, there is some indication that lower-mass stars eject at a lower rate.

In addition, it must be noted that the distribution among classes is biased against stars showing weaker activity (these being more difficult to observe), thus overestimating the time spent in the violent ejection phase and hence the total mass loss. Therefore, we have obtained an upper limit on the amount of mass lost, and the conclusions must be treated accordingly.

Consider now the significance of the mass loss in T Tauri stars on a galactic scale. Suppose that every star in this mass interval (there are very many) passed through a T Tauri stage early in its career, during which it ejected at a weighted mean rate of $3.7 \times 10^{-8} M_\odot$ per year and returned this mass to the interstellar medium. At any one time, there are probably about a million T Tauri stars in the galaxy, so that they are returning to the interstellar medium about $0.04 M_\odot$ per year. This is insignificant compared to other means by which stellar material may be returned to the interstellar medium. Biermann has estimated that these other sources may contribute between 1 and $10 M_\odot$ per year. In the lifetime of the galaxy, if everything has proceeded at the present rate, then 8% of the present mass of the interstellar medium can then be said to have been inside T Tauri stars at one time or another.

MASS LOSS FROM RED GIANTS

Armin J. Deutsch

There is spectroscopic evidence for the efflux of cool gas from the surfaces of all giant stars with spectral types later than M0 (Figure 1). In the M giants, all the strong absorption lines arising from the ground states of atoms and ions are composed of two components. One component is a broad, relatively shallow absorption line formed in the photosphere. The second component is a deep, narrow line with near-zero central intensity. This component is displaced toward the violet from the center of the photospheric line on which it is superposed. It must be formed in an expanding envelope or shell, where the pressure and excitation temperature are very low. The expansion velocity indicated by the displacement is always quite small, much less than the escape velocity at the photosphere.

These observations could be taken to indicate that a giant M star is surrounded by gas which is seen to rise slowly, and which must then fall back to the star in an unobservable state of ionization. Where an M giant is a member of a visual binary system, however, some deep circumstellar components may also be seen in the spectrum of the companion. In spectroscopic binaries, the circumstellar lines remain fixed, while the photospheric lines shift with the changing position and velocity in the orbit. Thus, in some binaries, the circumstellar envelope of the M giant must surround its companion. The minimum size of the envelope can be

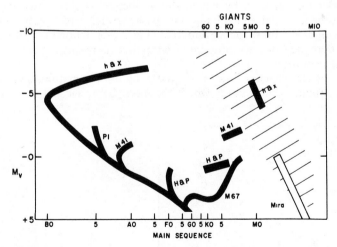

Figure 1. Hertzsprung–Russell diagram adopted from Sandage (1957, *Astrophys. J. 125*: 436). Stars within the ruled area show spectroscopic evidence of mass loss. This diagram was first published in *I.A.U. Symposium No. 10*, 1959, p. 109, and subsequently in the *Liege Symposium*, 16, and in *Stellar Atmospheres* (J. L. Greenstein, ed).

found from the geometry of such a pair. The radius of the envelope is found to exceed greatly the maximum height of a particle in a ballistic trajectory; moreover, at such large distances, the observed expansion velocity exceeds the local escape velocity. Thus, it can be concluded that the M giants are losing mass to the interstellar medium in a quasi-steady outflow that produces the circumstellar spectrum. Circumstellar lines due to cool expanding envelopes may be recognized in the spectra of all normal giants later than type M0, and of supergiants that are somewhat earlier, as well (see Figure 1).

The strongest of the circumstellar lines are always the H and K lines of Ca II. A strong correlation is found between the intensity of these lines and the spectral type (Figure 2). The surface density of circumstellar Ca II is obtained from the curve of growth. The rate of mass loss can then be estimated on the basis of a simple model for the flow. Since the profiles of the circumstellar lines give no evidence of a velocity gradient in the envelope, the equation of continuity for steady flow requires that the gas density will vary as r^{-2}. Assuming the normal (solar) abundance of hydrogen relative to calcium, and that most of the calcium in the envelope is in the singly ionized state, the rate of mass loss is found. For stars in the range of spectral types from M5 to about M8, the most probable value for the rate of mass loss is about $4 \times 10^{-9} M_{\odot}$ per year.

There are many more early M-type stars than there are late M-type stars. However, their circumstellar lines are much weaker, so their rate of mass loss appears to be correspondingly less. The rate of mass loss observed for an M1 star is only about one-thousandth as fast as it is for a star in the range M5 to M8. In the aggregate, the M-type giant stars appear to lose about $1 \times 10^{-12} M_{\odot}/(\text{pc}^2\text{-yr})$ in projection on the plane of the galaxy.

It has usually been supposed that stars of intermediate mass ultimately become white dwarfs when all their nuclear fuels are exhausted. They must therefore reduce their mass to below the Chandrasekhar limit. Mass loss during the red-giant stage may be the essential process by which this reduction is accomplished. Since we have reason to believe that all evolving stars pass through a red-giant stage at some time after leaving the main sequence and following the depletion of central hydrogen, it is possible that the flows we observe in M giants represent the principal loss of mass in stellar evolution.

Does the observed mass loss satisfy the evolutionary requirements? Maarten Schmidt has estimated the rate at which matter is being ejected from stars upon

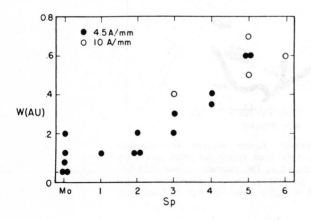

Figure 2. The correlation for giants (M_v fainter than -2.5) of the estimated strength W of the circumstellar K line, and the spectral type. This diagram was first published in *Stellar Atmospheres* (1960, J. L. Greenstein, ed.).

their transformation from main-sequence stars into white dwarfs. The stars of the solar neighborhood, in the aggregate, lose $1 \times 10^{-9} M_\odot/(\text{pc}^2\text{-yr})$ in projection on the plane of the galaxy (this projection takes account of stratification effects). Of this amount, about two thirds is lost from stars that had absolute magnitudes brighter than $+2$ on the main sequence (these are still not very massive stars), and about one third is lost by stars which were on the main sequence at absolute magnitudes fainter than $+2$.

There seems to be a wide discrepancy between the observed rate of mass loss and the evolutionary requirement as calculated by Schmidt. Is it possible to understand this gap? Actually, the estimates of rates of mass loss in red giants are crude. To determine the rate for an M star, I measure the equivalent widths of the Ca II H and K lines, enter the curve of growth for radiative damping, and find the surface density (Ca II ions per square centimeter) of these ions that are coming off the star. The principal uncertainty arises because the H and K lines lie on the flat part of the curve of growth; therefore, a small error in the measurement of the equivalent width—or (which is even harder to avoid) a small error in the Doppler width—can change the derived surface density by a factor of 10.

Also, I make an assumption, which is probably incorrect, that in the flows around these stars the cool matter is seen right down to the photosphere. It is more probable that the temperature is rather high near the star. There is evidence that these flows are the same kind of phenomenon as the solar wind. The material starts near the stellar surface with a relatively small radial velocity, but with rather high temperature. The flow then accelerates outward, becoming supersonic, as in the case of the solar wind. Eventually, the gas expands adiabatically at a speed similar to the one observed. However, in contrast to the solar wind, the temperature in this region is not of the order of a million degrees; since Fe I and Ca I are observed, T is probably very low. But if the gas near the star is too hot to absorb H and K, then we are not seeing the material all the way to the surface of the star. If, as Weymann has shown to be the probable case in α Orionis, for example, we can look down in the circumstellar lines only to a level which is already 10 stellar radii out, then our previous estimate of the rate of mass flow must be increased by another factor of 10.

Finally, while toward earlier spectral types along the giant branch the circumstellar lines get progressively weaker in a systematic way and disappear altogether (with the equipment available to us) at M0, the rates of mass flow need not necessarily diminish at a corresponding rate; this may be only an excitation effect. It is possible that the circumstellar K line gets weaker for the hotter stars mainly because more and more of the calcium resides in the second and higher states of ionization, where it is unobservable. Thus, it is possible that mass loss occurs in all giant stars to the right of the Hertzsprung gap, and that it does not, in fact, set in sharply at type M0, where the circumstellar lines first appear, and then increase steeply as the surface temperature of the star falls. However, there is no observational evidence for mass flows from normal giants of spectral types earlier than M0.

To summarize: My spectroscopic measurement is liable to (1) errors in extracting the surface density of Ca II ions from H and K by reference to the curve of growth; (2) errors in the geometry of the model, which involves the assumption that we see the envelope right down to the surface of the star; and (3) the possibility that substantial mass flows occur at earlier spectral types, albeit with ionization sufficiently high so that we have difficulty observing the lines or cannot see them at all.

Together, these effects could make up the difference between the Schmidt estimate and the most probable rate at which matter appears to come out of the late M giants. It is therefore still possible that a process of this kind will account for most of the return of matter from aging stars into the interstellar gas in the solar neighborhood.

There is an additional point to note. In the sun itself, the solar wind carries away matter at a rate which has been estimated to be about $10^{-14} M_\odot$/per year, a rate which is without significance as far as the evolution of the sun is concerned. However, there are very many solar-type stars per square parsec in projection on the plane of the galaxy. Indeed, in Schmidt's current model, the total mass density is about $75 M_\odot$/pc^2 on the plane of the galaxy, of which about $55 M_\odot$ resides in main-sequence stars. Since most of the main-sequence stars are less massive than the sun, this adds up to something like 300 objects. If they all have winds comparable to that which the sun supports, then together they will return to the interstellar medium about $10^{-12} M_\odot$/pc^2-yr. Therefore, despite the feebleness of their light, the dwarf stars, because of their number, return to the interstellar medium perhaps 10^{-3} times as much as the evolving giants do.

If, in other parts of the galaxy, there is a greater preponderance of dwarf stars than giant stars (Spinrad has found that in M31 and some other galaxies the ratio of late-type dwarfs to giants is about a hundred times greater than in the solar neighborhood), feebly mass-ejecting dwarfs in the aggregate may make a significant addition to the net flow from the vigorously mass-ejecting giants.

DISCUSSION

R. F. Christy: I would like to comment on another process of mass ejection. In the course of carrying out calculations of pulsation, one is necessarily struck by the vigor of shock waves that progress up through the atmosphere on each pulsation, in some of our models at least. These shock waves are strongest where the luminosity mass ratio is high. The momentum transported through the photosphere is clearly more than ample to account for a significant mass loss. Unfortunately, the calculation of the passage of a shock wave through the completely transparent region up to very small optical depths, which is necessary in order to decide whether or not mass is ejected and how much, that is, the physics of optically thin shocks, has not been pursued. However, the possibility clearly exists that mass loss by this mechanism is significant.

M. P. Savedoff: Have there ever been any calculations of expected line profiles and intensity based on the Parker model of the solar wind?

A. J. Deutsch: No. I have made attempts from time to time to make these calculations, but so far have always been thwarted by the difficulty of computing the ionization. Obviously, in these envelopes, conditions depart very far from thermodynamic equilibrium. It may well be found possible to reconcile a Parker-type of flow with the relatively sharp symmetrical circumstellar absorption line profiles which we commonly observe in these stars.

MASS LOSS IN THE PLANETARY NEBULA STAGE

D. E. Osterbrock

Planetary nebulae are very small gaseous emission nebulae that invariably have a hot blue star within them which emits ultraviolet radiation that ionizes and heats the gas. The nebulae are always expanding, and the velocities, and also the galactic distribution of the planetaries, clearly show that the material seen around the star actually comes from that star and represents a shell that was thrown off by it. The distribution of the planetary nebulae is highly concentrated at the galactic center and weakly concentrated at the galactic equator. They have high radial velocities, but not as high as the extreme high-velocity stars, and thus they belong to the disk population. The planetary nebulae are interesting from the point of view of stellar evolution in two different ways. They represent a visible loss of mass from stars and replenishment of the interstellar medium, and also they represent the end-stage of the stellar evolution of at least one kind of star.

The summary I will give of mass loss of planetary nebulae is based largely on an observational study by O'Dell, begun at Wisconsin and completed at Mount Wilson Observatory. The physical picture, due originally to Shklovsky, is that all planetary nebulae are assumed to be objects of the same kind, and in particular it is assumed that the shells all have essentially the same mass. In reality, this cannot be true in detail, for some planetary shells have a high degree of spherical symmetry, while others have fairly good cylindrical symmetry, and still others have quite chaotic structures. Thus, we are certainly making an extreme simplifying assumption. It is further assumed that the shell is expanding or coasting into the vacuum with a constant velocity. The shell is so far away from the star that there are no gravitational effects decelerating it, but it is not so clear whether or not there are other forces, for example, electromagnetic forces, that might affect the shell.

With these assumptions, it is possible to find the distance of a planetary by measuring its surface brightness and angular size. The procedure is as follows: The surface brightness is proportional to the electron density, to the ion density, and to the path through the nebula.

$$SB \sim N_e N_i r \qquad (1)$$

where N_e and N_i are the number densities of electrons and ions and r is a representative length through the nebula. If the surface brightness is measured in a hydrogen line, its strength depends only weakly on the temperature, and this dependence can be neglected. The ion density is proportional to the electron density, so

$$SB \sim N_e^2 r \qquad (2)$$

On the other hand, the assumption of constant mass in the shell means $N_e r^3 = $ constant during the evolution, so r and N_e can each be expressed in terms of the other. Thus

$$SB \sim N_e^{\frac{5}{3}} \sim r^{-5} \qquad (3)$$

There is some check of the assumptions, since for some planetaries the electron density can be measured independently from the [O II] lines and also from the surface brightness, and the two values agree with each other. But all planetary densities cannot be found in this way, because they do not all have strong [O II] lines. Thus, the following procedure must be adopted: The surface brightness is measured, from it the radius of the shell is calculated (the representative length in the planetary), the angular size of shell in the sky is measured, and these two quantities give the distance to the planetary. If the apparent magnitude of the star in the planetary is then measured, its absolute magnitude becomes known. Thus, the relevant properties of the planetary can be determined. This program was carried out by O'Dell with photoelectric measurements of all the planetaries that could be reached, and filled in with photographic measurements, mostly by Abell, of the faintest objects.

It is also possible to measure a sort of color index, between the far ultraviolet and the visible, of the stars in the planetary nebulae. The ionizing ultraviolet radiation of the star ionizes the nebula, and therefore the flux of visible radiation from the nebula is really a measure of the ultraviolet flux of the star. This, together with the visual intensity of the star itself, gives the color index, which is usually expressed as the Zanstra temperature, the temperature of the corresponding black body.

To calibrate the luminosities, the distances of at least one planetary or the mass in at least one shell must be measured independently. In practice, this is done in several ways, all rather poor. There are one trigonometric parallax, two or three expansion parallaxes, one double star with a spectroscopic parallax, and several statistical proper-motion parallaxes. All these calibration methods agree fairly well with one another, although none of them is extremely strong by itself.

O'Dell's results for the absolute magnitudes and effective temperatures (converted to B-V color index) of many nuclei of planetary nebulae are shown in Figure 1. The cross-hatched area is where the planetary nebulae nuclei fall—they are all

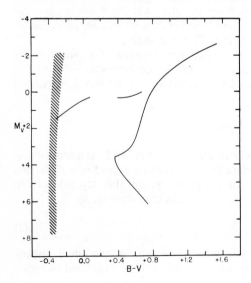

Figure 1. Color-magnitude diagram showing sequences in a typical globular cluster (M5, observed by Arp, corrected for interstellar reddening), with region of planetary-nebula central stars superimposed (cross-hatched).

extremely hot stars with temperatures between 30,000° and 100,000°. It can be seen that there is a systematic trend and that the stars with the highest luminosity have lower effective temperatures, while the stars with the smallest luminosity have the highest effective temperatures.

What deduction can be made about the shells from the results of this study? First, the typical mass of a planetary nebula shell is about $0.2M_\odot$. Second, the sizes of the shells vary between 0.05 parsecs and 0.7 parsecs radius. The ages of the individual nebulae can be determined from their radii, since the velocities of expansion are known. Dating from the time the stars threw off their shells, the oldest observed planetary nebulae have ages of about 2.5×10^4 years. When they become even larger than this, they are too faint to discover, although presumably larger objects do exist, but are not found. In fact, the largest known planetary nebulae, those with radii between 0.3 and 0.7 parsecs, have very low surface brightnesses and almost all of them were found in the course of the National Geographic Society–Palomar Observatory Sky Survey with the 48-inch Schmidt telescope.

The nearest planetary nebula, NGC 7293, is only about 140 parsecs away, while the most distant ones known in our galaxy are at about 5000 parsecs. The statistics are fairly complete out to a distance of about 2000 parsecs, while beyond this distance the discovery probability becomes progressively smaller, and the statistics correspondingly less complete for the larger, fainter planetaries. There are about 700 known planetaries in all, of which about 200 to 300 are within the cylinder of 2000 pc radius centered on the sun. From their individual positions, the average distance of the planetaries from the galactic plane can be found to be approximately 280 parsecs, about the same as the average distance from the galactic plane of stars of mass $1.2M_\odot$. This is the best estimate that it is possible to make of mass of the stars that ultimately become planetaries, and it is very uncertain, of course, because any mass between $1.0M_\odot$ and $1.5M_\odot$ would match almost equally well. Also, the result depends on a statistical argument and cannot exclude the possibility that perhaps 10 or 20% of the planetaries might be completely different kinds of objects.

The space distribution of planetaries seems to be the same as the distribution of mass in the galaxy, so the ratio of the number of planetaries within the known volume centered on the sun to the number in the whole galaxy is the same as the ratio of masses of these two volumes. Therefore (using the galactic model of Schmidt), the total number of planetaries in the galaxy can be estimated to be 5×10^4, and since we have seen that planetaries live 2.5×10^4 years, there must be about two being formed and two dying per year. With $0.2M_\odot$ per planetary shell, it means that about $0.4M_\odot$ per year is returned, by planetary shells, to interstellar space. This is pretty large, but nevertheless shows that planetaries are not the main source of new interstellar matter, since the overall rate of mass return is significantly larger, about $1.5M_\odot$ per year in the whole galaxy, according to the estimate of Schwarzschild.

The abundances of the material in the planetary shells which is ultimately returned to interstellar space are fairly well determined, especially the hydrogen-to-helium ratio. The hydrogen and helium lines are recombination lines, and their relative strengths are nearly independent of temperature and depend only on the relative abundances. In the planetary shells, the ratio of the helium content to the hydrogen content (by mass) is $Y/X = 0.60$, with little if any variation from planetary to planetary, as first found by Mathis. For comparison, according to Gaustad the corresponding ratio in the sun is $Y/X = 0.36$. In other words, the helium abundance in the planetaries is slightly higher than in the sun, which has about the same age. The planetaries start off as stars of $1.2M_\odot$, and in their evolution to the planetary stage they evidently convert only a small fraction of the material in the ejected shell from hydrogen to helium by nuclear reactions. The heavy-element abundance in the shell is just about the same as in typical stars.

Now let us consider the properties of the planetary-nebulae stars. The individual stars can be dated by their nebulae, and the observations show that their luminosities decrease rapidly from 10^4 to $10^2 L_\odot$ in a time of about 10^4 years. The radii decrease very rapidly, and in the final stages are similar to those of white dwarfs. This must mean that the mass is of the order of $1 M_\odot$ or slightly smaller (they cannot be much smaller, or the stars would have larger radii). None of the stellar-interior models that have been calculated in the past for planetary-nebulae central stars can reproduce this rapid evolution, because the models are all based on the idea of nuclear burning in a central core. The last stages of the evolution can be understood in terms of a degenerate star with a rapidly contracting envelope, and there is enough energy in the gravitational contraction to supply all the luminosity, even at the first stage. The whole energy output, the time integral of the luminosity, gives an upper limit to the amount of hydrogen that can possibly be in the collapsing star, because it would undergo nuclear reactions sometime during the evolution if it were there. This upper limit is that the hydrogen content is much less than $0.01 M_\odot$ at the first stage listed in Table I. On the other hand, the observations show that the shell has a relatively high hydrogen content. So apparently in the process by which the planetary nebula forms, there is a clean separation between the hydrogen-rich shell and the nearly completely hydrogen-exhausted stellar remnant.

A nondegenerate gravitationally contracting equilibrium model is, however, not possible. The planetary-nebula stars are so hot that the main source of opacity, particularly in the latter stages of evolution, must be electron scattering. This means the mass-luminosity law takes the form

$$L \sim \mu^4 M^3 \tag{4}$$

independent of radius. Thus, such models will not predict the observed decrease of luminosity with time. Although a model has not been constructed for a homogeneous star with electron-scattering opacity and gravitational contraction as the energy source, the magnitude of the luminosity may be estimated from available integrations of electron scattering and Kramers-law Cowling models by Rogerson and Härm together with the Kramers-law gravitational-contraction model of Levee. The luminosity is $L = 3 \times 10^3$ for a $1.2 M_\odot$ star composed of pure carbon. This is smaller than the observed luminosity in the early stages. Therefore, the

Table I. Typical Properties of Planetary-Nebula Nuclear Stars

M_v	$L/L_\odot$	$R/R_\odot$	Age (years)
−2	1.1×10^4	1.5	0
−1	5.5×10^3	0.7	5×10^3
0	2.5×10^3	0.25	1×10^4
+4	6.0×10^2	0.04	1.5×10^4
+8	1.4×10^2	0.007	2×10^4

nondegenerate gravitationally contracting equilibrium model can be eliminated from consideration.

It is possible that the time scale of the evolution is so short that the planetary-nebula stars are not contracting homologously. Deviations from hydrostatic equilibrium disappear in a time of the order of the pulsation period of the star (a few hours), but deviations from energy equilibrium disappear in a time of the order of the heating or cooling time of the star, which is of the same order as the gravitational contraction time itself. Therefore, a star, soon after the catastrophic loss of a shell containing a significant fraction of its mass, might have an internal structure not in radiative equilibrium. Nevertheless, at its surface there would always be radiative equilibrium, and eventually radiative equilibrium and hence homologous contraction (since the opacity is due to electron scattering) would extend throughout the star. Before the final stage is reached, the inner part of the star might have an arbitrary temperature gradient, depending on its previous history. As the simplest examples of such structures, I have constructed models consisting of electron-scattering (radiative-equilibrium) envelopes and polytropic (nonradiative-equilibrium) cores (with $n + 1 = d \ln P/d \ln T = $ constant) as shown in Table II. Here, $\log C$ is the mass–luminosity parameter. The model with $(n + 1) = 4.0$ in the core and radiative equilibrium only at the surface ($q_1 = 0$) has a luminosity nearly the same as the homologous gravitationally contracting model, which it in fact closely resembles, while the other models with smaller $n + 1$ have smaller luminosities and deviate still further from the observed luminosities of the earliest stages of the planetary-nebula stars. Thus, Table II indicates that only models with $n + 1 > 4$ in the core could have sufficiently large eigenvalues C to match the observations. However, in all normal models, $n + 1$ increases outward and, since the condition of radiative equilibrium at the surface requires that $n + 1 = 4$ there, such models cannot be constructed with $n + 1 > 4$ in the core. The only possibilities for models with large central polytropic index are stars with isothermal cores ($n + 1 = \infty$) surrounded by a region in which nuclear-energy production occurs and $n + 1$ decreases outward, surrounded still further out by a region in which $n + 1$ again increases outward. Physically, this structure seems quite plausible for a star near the end of its lifetime, presumably mostly burned out, with no energy sources at its center.

At the present time, therefore, it seems most likely that a planetary-nebula central star has a central degenerate core containing a large fraction of the total mass, probably composed chiefly of carbon and oxygen, the products of helium-burning. This core is probably surrounded by a helium-burning shell at a temperature of the order of 1.5×10^8 °K or higher at the inner edge of the helium envelope

Table II. Properties of Models with Polytropic Cores and Electron-Scattering Envelopes

Core $(n + 1)$	Log C	Fractional mass in core (q_1)
4.0	−3.02	1.00
3.5	−3.19	0.71
3.0	−3.27	0.48

$L/L_\odot = (4.23 \times 10^4/\varkappa_e)\, C\mu^4(M/M_\odot)^3$

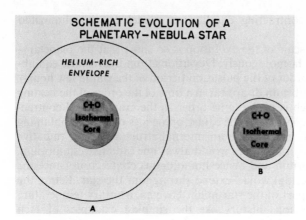

Figure 2. Suggested structure of an evolving planetary-nebula nuclear star, with central isothermal degenerate core, composed chiefly of carbon and oxygen, surrounded by helium envelope burning helium at its inner edge. Stage A is earlier than stage B and has much larger radius, reduced in drawing so that core can be shown.

containing a small fraction of the mass and, as the shell burns outward in mass, the helium-rich envelope contracts rapidly in radius (Figure 2).

Neutrino emission from the hot isothermal core may possibly be important, particularly the plasma-neutrino process. However, these speculations can only be proved or disproved by detailed model calculations; such calculations are now being carried out by J. S. Mathis and L. P. Bautz at Washburn Observatory, and by J. L'Ecuyer at Yerkes Observatory.

The observational data summarized above suggest that the initial masses of the stars that are now observed as planetary nebulae have about the same masses as the evolving stars in globular clusters. In fact, there is one planetary nebula in a globular cluster, K 648 in M15. This observation suggests that stars just at the end of the horizontal branch probably go through the planetary stage immediately before becoming white dwarfs. Furthermore, it shows that the planetary process is probably not connected with the helium flash, which explains the tip of the red-giant sequence in the globular-cluster color-magnitude diagrams, but rather is connected with a still later stage of nuclear evolution.

REFERENCES

G. O. Abell (1955), *Publ. Astron. Soc. Pacific* **67**: 258–261.
G. O. Abell (1961), *Publ. Astron. Soc. Pacific* **73**: 323.
L. H. Aller (1959), *Colloque international d'astrophysique tenu a Liege* (Institut d'Astrophysique, Cointe-Sclessin), pp. 41–50.
J. P. Cox and E. E. Salpeter (1961), *Astrophys. J.* **133**: 764–780.
J. E. Gaustad (1964), *Astrophys. J.* **139**: 406–408.
R. Härm and J. Rogerson (1955), *Astrophys. J.* **121**: 439–444.
R. D. Levee (1953), *Astrophys. J.* **117**: 200–210.
J. S. Mathis (1957), *Astrophys. J.* **126**: 493–502.
C. R. O'Dell (1962), *Astrophys. J.* **135**: 371–384.
C. R. O'Dell (1963), *Astrophys. J.* **138**: 67–78.
M. Schwarzschild (1962), in: L. Woltjer (ed.), *The Distribution and Motion of Interstellar Matter in Galaxies*, W. A. Benjamin, Inc., New York, pp. 266–273.
I. Shklovsky (1956), *Astron. J. Soviet Union* **33**: 222–235.

DISCUSSION

L. H. Aller: I am delighted to see how these numbers check what was found by the Harvard workers some 25 years ago. Whipple, I believe, deduced an age of 30,000 years for the planetaries, and we got average masses of around two tenths of the sun, which a much more extended series of observations confirms.

As for the depletion of hydrogen in the core, it might be pointed out that some of these stars, I am sure, have an excess amount of helium. Others appear perfectly normal. I think the nucleus of NGC 246 has helium, yet other stars like that of NGC 2392 look just like ordinary Of stars.

J. P. Cox: I would like to make one comment. It is known that a thin layer of hydrogen-rich material will increase the radius considerably for a given mass. If this simply blows away, the radius would shrink rapidly.

D. E. Osterbrock: What you are saying is that there might be some very slight mass loss at the end, that the degenerate core stays the same all during the evolution, and that the envelope doesn't collapse, but blows off. That is a good point, because the spectra of the planetary nuclei, insofar as they are available, show that the stars in the very early stages have broad emission lines with the widths corresponding to about the velocity of escape. The spectra suggest that there is some mass loss going on in the early stages, but not in the late stages.

Part VI. Observations Concerning Stellar Evolution

AGE DETERMINATION FOR MAIN-SEQUENCE B, A, AND F STARS

B. Strömgren

The age of a star can be determined from its precise location in the H-R diagram. Through observations, we determine the star's location in a diagram which is, for example, a plot of the visual absolute magnitude against some suitably chosen color index. This diagram must then be converted into one showing bolometric absolute magnitude versus the effective temperature. Finally, on the basis of calculations of evolutionary tracks, this diagram can be calibrated in terms of mass and age. The method can be applied to field stars as well as to clusters of stars.

Since the observed main sequence for spectral classes B to F has a width of about 1.5 magnitudes, an accuracy of $0^m.2$ to $0^m.3$ or better is needed in the absolute-magnitude determinations in order to determine meaningful ages. Let us assume, for simplicity, that the stars are population I stars of the same chemical composition and neglect the effects of rotation and magnetic fields. There are then two basic parameters: mass and age. A one-to-one correspondence therefore exists between mass and age, the location in the main-sequence band of the H-R diagram, and the values of two suitably chosen photometric indices.

The MK classification method separates main-sequence stars into fairly homogeneous age groups, but for age determinations the accuracy must be increased. Two methods of photoelectric narrow-band and medium-band photometry are used. In the $(U - B)_0$, β method of Crawford (1958), the indices are an indicator of the Balmer discontinuity and of the Hβ strength. In photoelectric $uvby$ photometry (Strömgren, 1963), the two indices are c_1 (which again measures the Balmer discontinuity) and a color index $b - y$. Crawford's $(U - B)_0$, β method is insensitive to interstellar absorption, while the $uvby$ method works only for stars that are not affected by interstellar reddening. Because one of the main applications is the determination of space velocities and their correlation with ages, it has been used up to now in programs that are limited to the neighborhood within 80 parsecs of the sun. Interstellar reddening can be neglected in the application of the method in these problems. Crawford's method yields absolute magnitudes with probable errors around $0^m.2$, and for the $uvby$ method the corresponding errors are between $0^m.1$ and $0^m.2$.

The $(U - B)_0$, β method is described in detail by Crawford in the next paper; here, we shall consider the $uvby$ method only. The characteristics of the $uvby$ system of filters are shown in Table I. The transmission curve of the u-filter lies entirely below the Balmer discontinuity while the v is entirely above it and the crowding of the hydrogen lines.

Three indices are determined from these measurements: (1) $b - y$ is a color index that is relatively insensitive to chemical-composition effects; (2) $c_1 = (u - v) - (v - b)$ is a color difference that is a measure of the Balmer discontinuity; and (3) $m_1 = (v - b) - (b - y)$ is a color difference that is a measure of the total intensity of the metal lines in the v-band. The distribution of stars in the c_1, $(b - y)$ diagram

Table I. Characteristics of the Four-Color Filters (Photoelectric *uvby* Photometry)

	Central wavelength (Å)	Half-width (Å)
u	3500	300
v	4110	190
b	4670	180
y	5470	230

is shown in Figure 1. The lower envelope of the distribution is the zero-age line; above it are the stars evolving through the main-sequence band, and above them are a scattering of supergiants. An absolute-magnitude calibration of the c_1, $(b - y)$ diagram for A and F stars in the main-sequence band, based on cluster and trigonometric parallaxes, is shown in Table II (see Strömgren, 1963).

Evolutionary-model sequences for hydrogen-burning stars in the range $1.5M_\odot$ to $10M_\odot$ have been calculated by Kelsall for six different chemical compositions covering the range encountered for population I stars. On this basis, Kelsall and

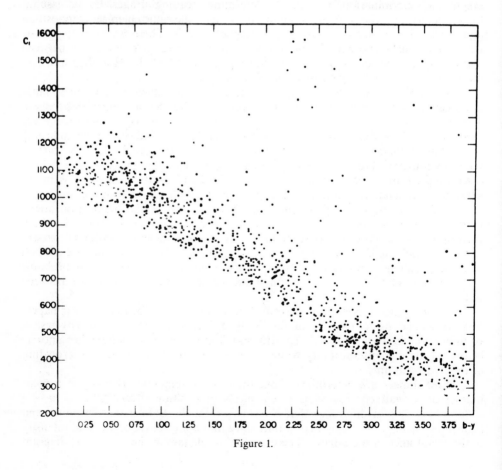

Figure 1.

Table II. Calibration of the c_1, $(b - y)$ Diagram

$b - y$	c_1	M_v	Factor $\Delta M_v / \Delta c_1$	Spectral type
0^m05	0^m96	2^m0	8	A3
0.10	0.88	2.4	8	A6
0.15	0.78	2.7	8	A8
0.20	0.65	3.1	8	F0
0.25	0.52	3.5	9	F2
0.30	0.42	4.0	10	F5
0.35	0.35	4.5	13	F8

Strömgren (1965) have carried out a calibration of the T_e-M_{bol} diagram in terms of mass and age. It is found that for B, A, and early F stars in the upper half of the main-sequence band, the age determined from T_e and M_{bol} is insensitive to the assumed relative hydrogen content X of the star. The sensitivity to the heavy-element content Z is greater, and, from the approximately known range of variation of Z for population I stars, it is estimated that this gives rise to a contribution to the probable age-determination error of at most about 7% of the age.

The conversion of location in the observational $(U - B)_0$, M_v diagram or $(b - y)$, M_v diagram to location in the T_e-M_{bol} diagram requires knowledge of bolometric corrections and of the temperature scale derived from model-atmosphere calculations (see, for example, Strömgren, 1964).

Now let us apply this method of age determination to the Hyades stars in the upper half of the main-sequence band (regarding the $uvby$ photometry used, see Strömgren, 1963). The ages derived are shown in Table III. The spread of the ages in this small sample is $\pm 6\%$ of the age (p.e.) when we restrict ourselves to stars more than 0^m7 above the zero-age line, which is an encouraging result. If all Hyades members more than 0^m5 above the zero-age line are included, the scatter becomes $\pm 15\%$ of the age (p.e.).

Table III. Ages Determined for Members of the Hyades Cluster from Photoelectric $uvby$ Photometry Using the Age Calibration by Kelsall and Strömgren (for $X = 0.70$, $Y = 0.27$, $Z = 0.03$)

Van Bueren No.	Star	Δc_1	log age
24	51 Tau	0^m071	8.79
38	60 Tau	0.084	8.85
47	64 Tau	0.078	8.57
54	65 Tau	0.125	8.61
72	78 Tau	0.137	8.67
74	79 Tau	0.063	8.63
80	80 Tau	0.084	8.86
82	HR 1427	0.067	8.58
104	90 Tau	0.123	8.61
108	92 Tau	0.116	8.65
123	97 Tau	0.085	8.70
129	102 Tau	0.120	8.63

The quantity Δc_1 is equal to the observed value of c_1 minus the zero-age-line value of c_1 that corresponds to $b - y$. According to Table II, the factor by which Δc_1 must be multiplied to yield ΔM_v, the vertical distance from the zero-age line in the Hertzsprung–Russell diagram, is approximately 8.

It should be emphasized that strongly peculiar stars must be excluded in applying the method of age determination just described. The metal index m_1 as determined from *uvby* photometry (see Strömgren, 1963) is of value in this connection.

It is desirable to extend analysis of the type described for the Hyades clusters to many more galactic clusters in order to gain a better estimate of the age-determination accuracy. An extension of the work in this direction is planned.

The results regarding the accuracy of the age determinations obtained from the cluster material can be generalized with some confidence to the case of population I field stars. For the field stars, the effect of scatter in the chemical compositions is larger, but it is possible to allow for this on the basis of the results already mentioned regarding the influence of chemical composition upon age determination.

REFERENCES

D. L. Crawford (1958), *Astrophys. J.* **128**: 185.
T. Kelsall and B. Strömgren (1965), in: A. Beer (ed.), *Vistas in Astronomy, Vol. 6*, Pergamon Press (to be published).
B. Strömgren (1963), *Quart. J. Roy. Astron. Soc.* **4**: 8.
B. Strömgren (1964), *Rev. Mod. Phys.* **36**: 532.

PHOTOMETRY OF B STARS

D. L. Crawford

One photometric method of locating B-type stars is to measure the Balmer discontinuity and an Hβ index. Figure 1 shows the transmission curves of the two filters used for the Hβ determination, a wide filter and a narrow one, both centered on Hβ. The Hβ index is the magnitude difference of these two filters. No corrections are necessary for atmospheric extinction or interstellar reddening, since the two filters have the same effective wavelength and both effects completely drop out when the ratio of the two filters is taken.

There is a close correlation between the Hβ index, called "β," and the measure of the photographic equivalent width of Hγ (Figure 2). The scatter around the mean line in this diagram is about 10% of the Hγ index (which is about the observational accuracy of the Hγ index).

Figure 3 gives a rough idea of the absolute magnitude calibration of the Hβ system. It shows the β-values for the Pleiades-cluster members plotted against their apparent magnitude, corrected for interstellar reddening. The apparent magnitude (corrected for reddening) V_0 can be converted to absolute magnitude by using the distance modulus of the Pleiades (5.55 magnitude). The B-type stars lie on the upper branch. The scatter around the line is ± 0.1 magnitude. The observed mean error of one Hβ measurement varies between ± 0.007 and ± 0.013 magnitudes, depending upon the magnitude of the star.

Figure 4 shows a plot of the Hβ index against a measure of the Balmer discontinuity, in this case $(U - B)_0$, the intrinsic $(U - B)$ color index. The lines divide the MK spectral types. The solid circles are those that agree; the open circles are exceptions. The triangles are giant stars (luminosity class III), and the crosses are luminosity II or I. The spectral classification from such photometry is rather good.

Figure 5 shows the c_1 versus β plot for B-type stars. A main-sequence band is apparent, and there is a falling off into the supergiant region. The mean position for luminosity class V is located near the lower part of the main-sequence band. The mean position for luminosity class III is located near the upper part of the band. Here, therefore, is a way of showing the zero-age line and the evolution from it. The age of the stars can be dated as they move across this main-sequence band before moving rapidly to the right. The observed accuracy of the individual points is ± 0.014 in c_1 and ± 0.013 in β-magnitude. Each point is made up of three observations or more.

Figure 6 shows the same stars in a $(b - y)$ versus β plot. It should be noted that there is a selection effect here. The density of points is not per unit volume, but per unit apparent magnitude.

If one looks at clusters, then there is no large scattering as seen in Figures 4 and 5, but rather a smooth line. In particular, young clusters are observed to lie right on the zero-age line.

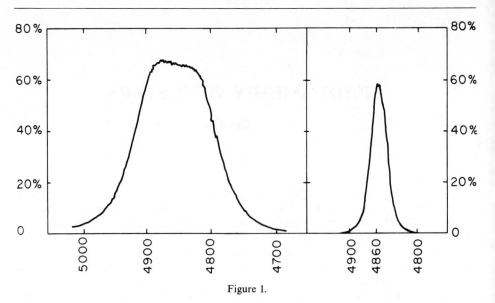

Figure 1.

Figure 2.

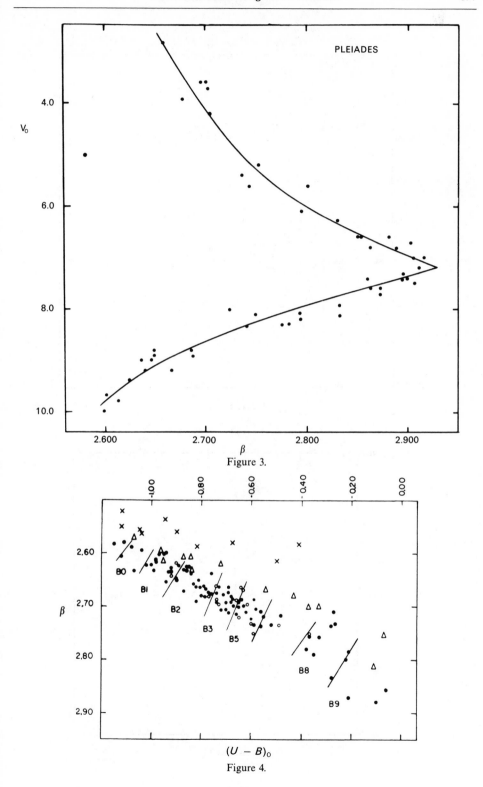

Figure 3.

Figure 4.

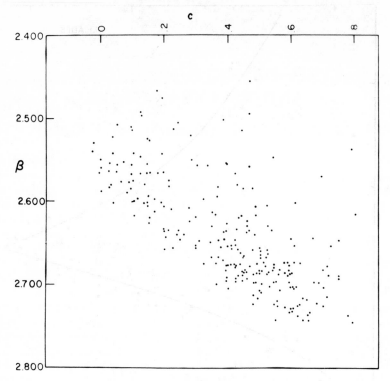

Figure 5.

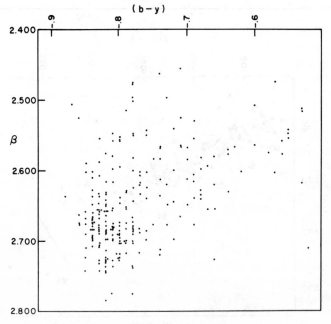

Figure 6.

THE ANALYSIS OF FIELD HORIZONTAL-BRANCH AND RR LYRAE STARS

J. B. Oke, J. L. Greenstein, and J. Gunn

This paper is concerned with stars which are similar to those found on the horizontal branch of globular clusters. The discussion will be confined to stars which are very near or within the variable-star gap. Since these objects must be very highly evolved, it is important to determine their chemical abundances. Also, there is a possibility that mass loss has occurred at some time during the stars' evolution and it is of interest to see if the masses can be determined (Woolf, 1964). It is also possible that mass loss occurs while the horizontal-branch star is pulsating as an RR Lyrae star. If the mass can be obtained for stars on both sides of the variable-star gap, the existence of mass loss can be checked and the direction of stellar evolution along the horizontal branch determined.

The main difficulty in studying horizontal-branch stars in globular clusters is the fact that they are between fifteenth and sixteenth magnitude. Thus, any high-dispersion spectroscopy is impossible, and even moderate-dispersion spectroscopy is severely limited. It is therefore extremely important to find field stars which are much brighter and susceptible to detailed spectroscopic investigation. Possible candidates for horizontal-branch field stars can be found among those stars with large space motions and spectral types earlier than F0. Cooler, horizontal-branch stars are much more difficult to discover. Very large numbers can be found among the faint blue stars near the galactic poles. At the present time, there are approximately 30 suspected relatively bright horizontal-branch stars. One of the best known and brightest of these is HD 161817.

In the range of temperature considered here, one of the best ways to study these suspected horizontal-branch stars is by means of photoelectric spectrum scans, from which absolute energy distributions in the spectrum can be obtained. If the stars have very weak lines, the measurements actually refer to the absolute energy distribution of the continuum. Hence, comparisons with model atmospheres can readily be made. In Figure 1 is shown the emergent flux computed from a model atmosphere with $\theta_e = 5040/T_e = 0.70$ and $\log g = 3.0$. The flux is plotted as $-2.5 \log F_\nu$. The plot covers the range which is also covered by observations. The energy distribution between the Balmer and Paschen discontinuities is a smooth function of the effective temperature, expressed as θ_e, and depends only slightly on the gravity. The size of the Balmer discontinuity is a function of both θ_e and $\log g$. It is therefore very easy to compare the observed absolute energy distribution with a grid of models and determine both θ_e and $\log g$ with high accuracy. Apart from the problem of the absolute calibration of α Lyrae and the effects of absorption lines, neither of which will be discussed here, the main difficulty is the determination of the interstellar reddening; unless it is determined, errors in both temperature and gravity will result. One of the most successful ways to obtain the reddening is to measure the profile of $H\gamma$. In the temperature range in question, it has been shown (Searle and Oke, 1962) that the $H\gamma$ profile is sensitive only to temperature. It has

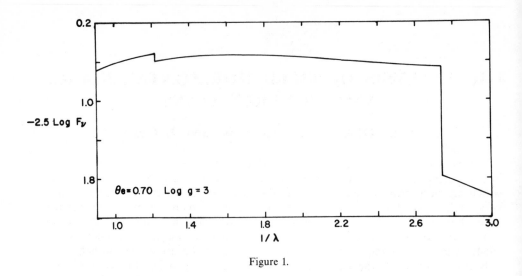

Figure 1.

further been shown by Oke, Giver, and Searle (1962) that the comparison of actual profiles of Hγ with those computed from model atmospheres yields values of θ_e which are in excellent agreement with those obtained from the spectrum-scanner measurements of the continuum. Consequently, the Hγ profile can be used to determine θ_e; the scan observations can then be used to determine the reddening and log g accurately. The results for 10 stars are given in Table I. The values in columns 2 and 3 do not take account of reddening. At least five, and possibly six, stars appear to be very much like horizontal-branch stars; one of these is to the right of the variable-star gap, while the others are just to the left. If it is assumed that these stars are in fact horizontal-branch stars and that their absolute magnitudes are near +0.5, then the values of θ_e and log g give the masses. Accepting the numbers in Table I literally, the masses for the stars to the left of the gap are all substantially less than $1M_\odot$. Using the same assumptions, the one star to the right of the gap has a mass of approximately $1M_\odot$. These masses must be accepted with caution, since the Balmer jump depends to some extent on the detailed structure of the model atmosphere, and systematic errors would result from inadequate models. It should be pointed out, however, that main-sequence stars at these temperatures have values of log g, determined from the Balmer jump, which are very near the expected value of 4.0.

Chemical-abundance analyses have been made of two of the six stars discussed above. The metal-to-hydrogen ratio in HD 161817 is down by a factor of more than 10 from the sun (Kodaira, 1964). In the case of HD 109995, the deficiency is, less certainly, a factor of 10 (Wallerstein and Hunziker, 1964). A comparison of spectra of the other four stars with these two suggests that their metal deficiencies are of the same order. These deficiencies are typical of moderately metal-deficient population II stars.

The field RR Lyrae stars are very similar in spectral appearance to the horizontal-branch stars discussed above. Their effective temperatures range from 5900 °K to 7600 °K around the pulsation cycle, with very little variation in the minimum temperature from star to star. The masses of these stars are difficult to

Table I

Star	Scans θ_e	$\log g$	Hγ θ_e	Remarks
BD $-$ 6° 86	0.68	2.6	0.67	Horizontal branch
BD $-$ 28° 1205	0.73	3.8	0.70	
BD $+$ 25° 1981	0.72	4.1	0.71	
BD $-$ 12° 2669	0.72	4.2	0.71	Variable velocity
HD 86986	0.63	2.9	0.62	Horizontal branch
HD 106223	0.76	3.2	0.70	Horizontal branch. If reddened, $\theta_e = 0.70$, $\log g = 4.0$
HD 109995	0.59	(3)	0.62	Horizontal branch
HD 161817	0.70	2.6	0.68	Horizontal branch
BD $+$ 17° 4708	0.85	3.0	0.83	Horizontal branch
BD $+$ 39° 4926	0.77	~ 1	0.68	Variable temperature?

determine, since the Balmer jump, which measures the surface gravity, measures only the effective gravity, and in a pulsating star the effective gravity will be less than the true gravity by an unknown amount. If one assumes that in the very high layers of the atmosphere there is free fall, which can be measured by obtaining the radial velocity of the hydrogen lines, the gravity is found to be about $\log g = 2.3$. This implies masses in the vicinity of $0.2M_\odot$. From the measured radius and period–density law, one finds masses which range from 0.2 to $1.0M_\odot$.

The remaining four stars which have been scanned but which do not appear to be horizontal-branch stars are themselves very possibly interesting objects. One of them, BD $+$ 39°4926, has an effective temperature of 6500 °K and $\log g < 1.0$. It has the largest Balmer jump of any star known. If it were a horizontal-branch star, its mass would be $0.01M_\odot$. If its mass is taken to be $1M_\odot$, then its absolute magnitude is -5. The radial velocity is small, as is the proper motion; there is therefore no way of determining, or putting limits on, the absolute magnitude. The scans which have been made suggest that the star is slightly variable, over intervals of some weeks. Broad-band photometry covering several hours of observation virtually rules out the possibility that the period is as short as that of a typical RR Lyrae star. The best guess is that the star is very much like a W Virginis star, but sufficiently hot to be just at the boundary of the region of pulsational instability. The three remaining stars have effective temperatures of 7000 °K and $\log g = 4$. Since they are hotter than normal subdwarfs, they may be metal-poor dwarf stars similar to those beyond the normal turn-off point of the globular-cluster main sequence (Sandage, 1953). One of these stars may also be variable in radial velocity.

REFERENCES

K. Kodaira (1964), *Ztschr. Astrophys.* **59**: 139.
J. B. Oke, L. P. Giver, and L. Searle (1962), *Astrophys. J.* **136**: 393.
A. R. Sandage (1953), *Astron. J.* **58**: 61.
L. Searle and J. B. Oke (1962), *Astrophys. J.* **135**: 790.
G. Wallerstein and W. Hunziker (1964), *Astrophys. J.* **140**: 214.
N. J. Woolf (1964), *Astrophys. J.* **139**: 1081.

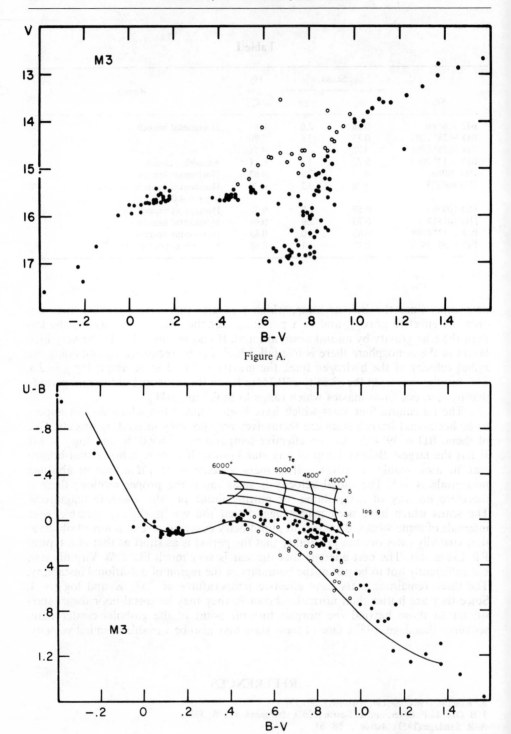

Figure A.

Figure B.

DISCUSSION

M. Walker: In the color-magnitude diagrams of a number of globular clusters, a bifurcation of the red-giant branch is observed. According to some ideas of evolution, the stars of the "bluer" red-giant sequence, together with the horizontal-branch stars, have passed through the tip of the red-giant branch and undergone a helium flash with complete or very deep mixing. Such stars might be expected to show an enrichment of their metal content.

Figure A shows the $C - M$ diagram for the globular cluster M3: Figure B shows the two-color diagram for the same cluster. The stars of the "bluer" red-giant sequence are shown by circles in Figure A. In Figure B, it will be seen that these stars lie below the stars from the "redder" red-giant sequence and from the lower part of the $C - M$ diagram, that is, they have smaller ultraviolet excesses. This difference in ultraviolet excess could arise either from a difference in the hydrogen-to-metal ratio or from a difference in surface gravity.

Theoretical calculations of the U, B, V colors of stars having no absorption lines are shown for different values of the effective temperature and surface gravity in Figure B; these values were calculated from the model fluxes of deJager and Niven. Assuming that the ratio of change of $U - B$ with $\log g$ is the same for the weak-line, metal-poor cluster stars as for the line-free models, the observed difference in $U - B$ between the stars from the two portions of the red-giant branch appears to be too large to be explained by differences in temperature and surface gravity between the two groups. However, refined theoretical calculations will be needed in order to prove definitely whether or not a difference in the hydrogen-to-metal ratio exists for the two red-giant sequences.

ULTRAVIOLET EXCESS IN T TAURI STARS

Merle F. Walker

Among the group of T Tauri stars contracting toward the main sequence, there are a number which have ultraviolet excesses. The spectra are basically similar to normal T Tauri spectra; they have the emission lines of hydrogen, calcium, and sometimes helium, Fe-I, etc., and they have an underlying late-type absorption spectrum. There is, in addition, a very strong ultraviolet excess (also found in some regular T Tauri stars) which covers up some or all of the absorption features. Twenty-three such stars in the Orion nebula cluster and in NGC 2264 have been observed to date and are listed in Tables I and II. The brightest of these is of photographic magnitude 14, and they are arranged in the tables in order of decreasing absolute brightness at the time of observation.

Ten of these stars (listed in Table I) show redward-displaced absorption components at the hydrogen lines and occasionally at the Ca-II K-line (Figure 1). These lines exist primarily among the brighter stars having absolute photographic magnitudes brighter than about $+6.8$ and are not frequently seen in fainter stars. In one case (that of YY Ori), rapid changes occurred in the redward-displaced lines; the absorption component moved from the redward side of the emission line to its center within 24 hours. Evidently, the phenomenon is complicated.

This absorption-line phenomenon might be thought to be some kind of binary effect. But, of a sample of nine stars (single randomly timed observations of nine different objects), only on one plate of one star (YY Ori) is the absorption component not to the red side of the emission line. Thus, the effect is not due to binary motion, because then one would statistically expect to find about half of the absorption components to lie to the violet.

The velocity of the emitting and absorbing material can be measured. The underlying, late-type absorption spectrum, where it can be seen in a few of these

Table I. Ultraviolet-Excess Stars
Showing Redward-Displaced Hydrogen and Ca II Absorptions

Star	Estimated M_{pg}
M0 Mon	$+5.1$
MM Mon	$+5.3$
YY Ori	$+5.4$
SY Ori	$+5.7$
LX Mon	$+6.0$
SU Mon	$+6.2$
Brun 637	$+6.2$
NS Ori	$+6.2$
XX Ori	$+6.7$
CE Ori	$+7.1$

Table II. Ultraviolet-Excess Stars Not Showing Redward-Displaced Hydrogen and Ca II Absorptions

Star	Estimated M_{pg}
LHα 61	+4.0
BC Ori	+6.2
LT Mon	+6.3
Brun 134	+6.4
AL Ori	+6.6
NX Mon	+6.8
LU Mon	+6.8
AU Ori	+7.0
HS Ori	+7.0
VX Ori	+7.2
VY Ori	+7.2, 8.2
π 1946	+7.2
KP Ori	+7.2

stars, gives essentially the expected cluster velocity. The redward-displaced absorption components are shifted to the red by 150 to 300 km/sec.

The emission lines, too, sometimes have an interesting behavior. In some of these stars, the lower members of the Balmer series are shifted to the violet; the shift decreases as one goes up the series and eventually approaches the cluster

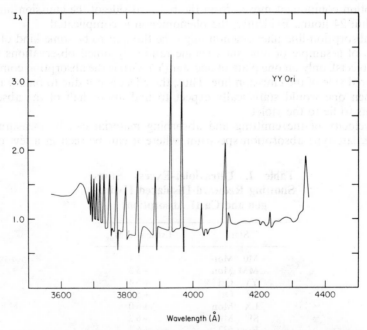

Figure 1. Energy curve of YY Ori obtained by spectrophotometric comparison of YY Ori with BD + 28° 4211 on electronographic plates taken with the Lallemand electronic camera and coudé spectrograph of the 120-inch reflector.

velocity. The shifts at Hγ or Hδ amount to about −90 to − 160 km/sec. The presence
of these emission-line shifts is not correlated with the presence of the redward-
displaced absorption components. Thus, the simplest explanation, that the emission
lines are shifted because an overlying absorption eats away the red side, is not
sufficient.

Let us now consider a model for the atmospheres of these peculiar T Tauri
stars. In the model of a normal T Tauri star, there are several layers. The lowest
layer, the photosphere of the star, is where the late-type absorption spectrum
originates. Above it, is the layer where the emission lines occur, and above that a
layer in which the displaced absorption features are produced. This is the layer of
material which in the normal T Tauri stars is rising and produces the mass loss.
Tentatively, then, the model of these peculiar stars would be the same as above,
except that instead of having material rising in the atmosphere, it is actually falling
in. Also, the infalling material will interact with the star and may increase its brightness, which would be consistent with the fact that the brighter ultraviolet-excess
stars are the ones that show sufficient infalling material to produce a directly
observable redward-displaced line.

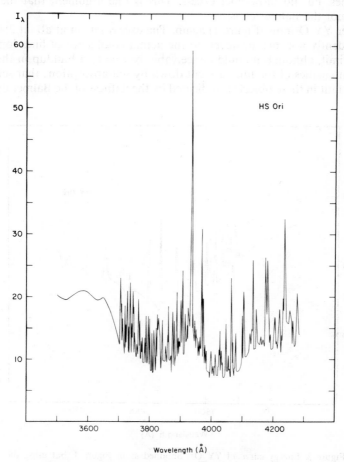

Figure 2. Energy curve of HS Ori, derived as in Figure 1.

Energy curves for a number of the ultraviolet-excess stars have been derived using the Lallemand electronic camera attached to the coudé spectrograph of the 120-inch Lick reflector. The response of the electronic camera is linear, in terms of density on the nuclear-track plates used to record the electronic image, to the intensity of incident light—up to densities of at least 3.4. Thus, simple densitometry of a plate of an unknown star and of a star of known energy distribution directly yields the energy curve of the unknown object. The energy curves shown in Figures 1 to 3 were obtained in this way, using the extremely hot star BD + 28°4211 as the standard.

Having available the energy curves, let us now consider the nature of the ultraviolet excess. In many of the ultraviolet-excess stars, the emission lines grade over into the continuum, which might then simply be the Balmer emission continuum. This is not, however, an adequate explanation for the entire effect.

Figure 2 shows the energy curve of HS Orionis, which has a spectrum very close to that of RW Aurigae. The hydrogen lines in HS Orionis are rather weak and fade out after about Hδ or perhaps Hε. But the star has an ultraviolet excess. Comparing this with RW Aurigae, we see essentially the same fading out of the hydrogen lines, but no ultraviolet excess. This is one argument that the cause of the effect is not the Balmer emission.

Consider YY Orionis (Figure 1) again. The excess sets in at about 3800 Å. The excess is evidently not due, however, to the actual confluence of lines longward of the Balmer limit, although it could conceivably be due to a buildup in the wings if the central intensities of the lines are cut down by self-absorption; that self-absorption is important in these objects is indicated by the flatness of the Balmer decrement.

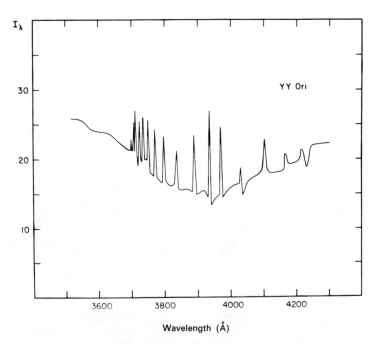

Figure 3. Energy curve of YY Ori obtained as in Figure 1, but using the spectrograph slitless so that there is no light loss as $f(\lambda)$ at the split.

Figures 2 and 3 show more clearly what happens in the ultraviolet beyond the Balmer limit. These observations were made with a grating setting going down to 3500 Å, and Figure 3 was obtained using the spectrograph essentially slitless, so that there is no light loss. In both cases, there is an ultraviolet excess setting in around 3800 Å and increasing up to at least 3500 Å. If the ultraviolet excess were Balmer emission, it would decrease shortward of the Balmer limit. This is a second argument against Balmer emission being completely responsible for this effect. The explanation of the ultraviolet excess is still not known, although it appears possible that it might be explained by two-photon emission, provided that the excitation temperature can be 30,000 °K. The necessary external source of this energy might be the energy of the infalling material.

LITHIUM IN MAIN-SEQUENCE STARS

G. H. Herbig

T Tauri stars have been found to contain a great deal of atomic lithium in their atmospheres. In individual stars, the lithium abundance is between 50 and 400 times that of the sun. The question is: What happens to this lithium after these T Tauri stars eventually arrive on the main sequence at intermediate masses? If the sun is representative, there is very little lithium left in G stars when they are as old as the sun (approximately 5×10^9 years). Is this lithium destroyed early, by the time the stars reach the main sequence, or is it gradually burned later? Do the observations help decide between the quick destruction of lithium in the first 10^7 years or more (during which the sun contracted to the main sequence) and the gradual decay over long periods after stars arrive on the main sequence with a lot of lithium?

Table I shows the statistics of a survey of nearby stars between spectral types F5 and late G on the main sequence. (A deliberate attempt has been made to exclude stars above the main sequence or those that have excessive rotation.) These stars were selected from those within 20 parsecs of the sun and chosen without any bias with respect to space motion or any other kinematic characteristics. Hopefully, this is a sample of stars of all ages, from zero up to the maximum age of G stars on the main sequence, of the order of 1.5×10^{10} years. Curves of growth analyses for lithium abundance are now available for about 100 of these stars. Figures 1 and 2 show typical spectra and the variation in the intensity of the lithium λ 6707 line in G dwarfs in a very small range of spectral type.

The percentage of stars with fairly high lithium abundance (more than 5 times solar) is at a peak of about 70 to 80% at late F and early G types and then decreases toward late G types and becomes zero around K0. There are no known stars on the

Table I

Main-sequence spectral type	$l \equiv (Li/Ca)^*/(Li/Ca)^{\circ}$				Percent of stars in that spectral class with $l > 5$
	$l > 20$	$20 > l > 10$	$10 > l > 5$	$l < 5$	
F5, F6	1	2	—	1	75
F7	—	3	2	1	83
F8, F9	2	1	1	2	67
G0	1	6	2	6	60
G1	1	1	4	2	75
G2	3	1	1	8	38
G3, G4	—	—	2	6	25
G5, G6	—	—	5	6	45
G7, G8	1	—	—	12	8
K0					0

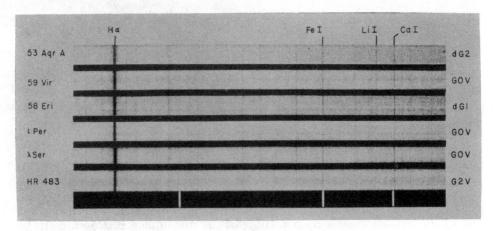

Figure 1.

main sequence later than K0 with lithium that is detectable at this dispersion. The break near K0 can most plausibly be interpreted as being due to the destruction of lithium by convection, because the convective zone in K0 and later dwarfs is quite deep.

The discussions by Hayashi and by Ezer and Cameron do not establish whether the convection during the contractive stage of a $1M_\odot$ star is adequate to destroy

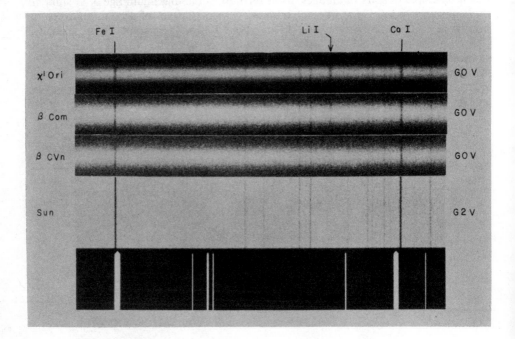

Figure 2.

most of the lithium initially present. The general variation in lithium abundance indicated in Table I appears to be consistent with the destruction of lithium on a long time scale. But can one prove that this really is an age effect?

If the destruction process goes at a slow rate, the lithium abundance as a function of age can be predicted for solar-type stars by assuming that the sun destroyed its lithium exponentially with a time scale of $1/e \approx 1.3 \times 10^9$ years and that this rate is typical for all stars of about one solar mass. A depletion factor of Li(0)/Li(t) = 300, where $t = 4.7 \times 10^9$ years for the sun, comes from the abundance of lithium contained in the Suess–Urey table and the Goldberg–Muller–Aller solar determination. But the value resulting from the abundance of chondritic lithium recommended by Shima and Honda and the Mutschlecner solar value is 35 instead. Comparison of prediction with the observational data for lithium abundance in F-G dwarfs in the Hyades, Pleiades, and Coma does not, at least, contradict the idea of a slow lithium-destruction process at our exponential rate (Figure 3).

What other evidence bears on this question? Is there any dependence of the lithium abundance on mass? Figure 4 shows Wallerstein, Herbig, and Conti's results for the Hyades cluster. The arrows indicate upper limits. These Hyades stars are presumably all of nearly the same age. There does not seem to be any systematic trend in this diagram, as might be expected if lithium destruction depended on the mass of the star. But the intrinsic uncertainty of the measures is not well known at the moment.

Now consider the correlation of lithium abundance with space motion. The space motions of the stars with high lithium abundance tend to be lower than with those with no lithium at all (Figure 3). There is also a tendency for the stars with the higher lithium abundances to possess as well one other characteristic of the T Tauri stars: They have unusually strong H and K chromospheric-type emission which could be considered evidence of surface activity left over from the T Tauri stage, and still present to a lesser degree in stars that have recently emerged from it.

There is no correlation of lithium abundance with the conventional metal-dependent ultraviolet excess and metal abundance in the F stars, but there is a tendency for such a correlation in the G stars. This fact may or may not be significant,

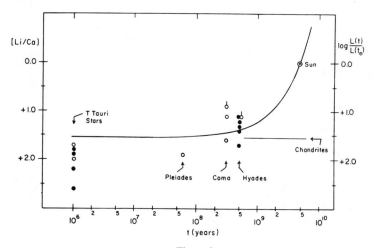

Figure 3.

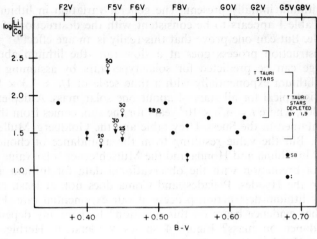

Figure 4.

because Pagel has discussed evidence that small differences in metal abundance are not correlated with age.

In summary: The data support the idea that there is a progressive destruction of lithium with time, at least in the early G-type stars.

In principle, spectroscopic observations also provide some evidence on the mechanism responsible for the slow depletion of lithium. Lithium may be lost by simple dilution with completely lithium-free material from below, or lithium from the surface may be circulated down to an intermediate depth where Li^6, which has a greater cross-section for (p, α) reactions, will be destroyed much more rapidly than Li^7, and so the surface material gets relatively richer in Li^7 as time goes on. Figure 5

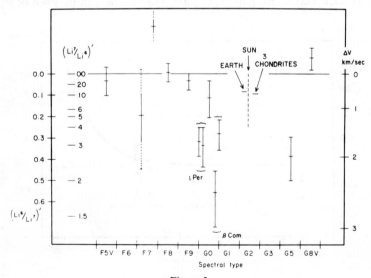

Figure 5.

shows the data now available on the "apparent" lithium-isotope ratio in lithium stars. The Li^7/Li^6 ratio in the sun is probably between 4 and infinity. If lithium is made by surface spallation reactions in T Tauri stars then, it has been claimed, a star will start with comparable amounts of Li^6 and Li^7. (I might mention, in passing, that there is evidence from one star at the top of its Hayashi track that the lithium is present within no more than a few decades of its luminous "beginning," so there is clearly very little time to do a great deal of nucleosynthesis after the star becomes luminous.) In β Comae, where the total lithium abundance is down from the T Tauri value by a factor of about 5, the ratio is 2 or 3. If in the beginning, $Li^7/Li^6 \approx 1$, it seems unlikely that in this star the method of destruction of lithium is by the circulation of atmospheric material to an intermediate depth, with Li^6 vanishing at a rate 100 times faster than Li^7, and the subsequent return of the same material to the surface, relatively enriched in Li^7. A possible alternative is through the dilution of originally comparable parts of Li^6 and Li^7 with completely lithium-free material from underneath. But there are many such possibilities, and it is difficult to choose from among them without either more observational information or some theoretical guidance.

RECENT ABUNDANCE DETERMINATIONS

H. L. Helfer

Helfer, Wallerstein, and Greenstein (1963) have recently performed detailed abundance analyses on a few stars from the set for which abundances of selected elements had been determined by Wallerstein (1962). There was very good agreement between the determinations. Let us introduce the notation

$$[X] = \log X \text{ (star)} - \log X \text{ (sun)}$$

For the thirty abundance determinations in which the two papers overlap, fifteen agree within something like $\Delta[X] \approx 0.07$. This means that the determinations are consistent in the errors performed. The accidental errors have been kept low and, furthermore, the few elements that are listed in Wallerstein are truly representative of the various nuclear-genesis groups. Thus, Wallerstein's results (Table I) for G dwarf stars may be used to study abundance variations. The stars listed in the table are collected in five groups ordered by values of [Fe/H], which is our index of metal to hydrogen content. In a very crude sense, this index may be used as an age indicator.

Some systematic variations can be noted. First, as the iron (metal)-to-hydrogen ratio decreases, the manganese-to-iron ratio decreases. This is known as the "manganese anomaly." The other e-process elements do not show any variation relative to iron. The barium–iron variation may or may not be significant. Second, in the last two groups of stars of roughly the same iron-to-hydrogen abundance ratio, that is, both deficient by a factor of 3 in the metals, one group has the solar α-element abundance while the other group is definitely overabundant in the α-group of

Table I

Group	Number of stars in group	[Fe/H]	[α/Fe]	[Mg/Ca]	[Sc/Ca]	[Mn/Fe]	[e/Fe]	[Ba/Fe]
1	6	+0.25	−0.05	−0.02	+0.06	−0.10	−0.02	+0.13
2	7	+0.01	0	−0.03	−0.05	−0.08	0	+0.13
3	6	−0.23	+0.08	+0.07	+0.01	−0.37	+0.01	−0.06
4	5	−0.44	+0.04	+0.02	−0.09	−0.34	0	+0.01
5	5	−0.54	+0.21	+0.07	−0.23	−0.37	+0.06	−0.11

α refers to the elements magnesium, silicon, sulfur, argon, potassium, calcium, and titanium.
e refers to equilibrium-process elements near iron.
Group 1 consists of HD 86728, 102870, 34411, 186408, 10307, and 114710.
Group 2 consists of HD 19373, 186427, 28099, 187923, 28068, 28344, and 115043.
Group 3 consists of HD 30649, 143761, 55575, 90508, 30455, and 142267.
Group 4 consists of HD 110897, 13974, 152792, 16508, and 114762.
Group 5 consists of HD 157214, 148816, 224930, 22879, and 160693.
The α-rich star HD 106516 has been omitted; it probably belongs in group 5.

elements. The α-rich group of stars has galactic orbits with higher eccentricities than the others. Also the α-rich group of stars has a lower scandium abundance. Third, in the metal-deficient stars, the abundances (relative to iron) of the various groups of metals formed by different processes (equilibrium-process, slow-neutron capture, and fast-neutron capture) are independent of the iron-to-hydrogen abundance. Considering the diverse conditions of pressure, temperature, and density necessary for each of the mechanisms, this uniformity from one metal group to another implies that there must have been significant large-scale mixing of the interstellar matter in the galaxy in order to maintain this sort of uniform composition.

The mechanism of the enhancement of the α-elements but not of scandium is a problem. The abundances in G dwarfs depend on the products of nucleosynthesis in preceding generations incorporated in the stars on their formation. Noting that observationally titanium behaves as an α-element and that the lighter α-elements may be produced by carbon and oxygen burning, the investigation by Hayashi *et al.* (1959) strongly suggests that the observed relative abundances of the α-elements can only be produced at those high temperatures and low densities presently hypothesized for non-core regions of a supernova flash-heated by the explosion. The variation of the abundance of the α-group of elements relative to iron may then reflect relative differences in the frequency of occurrence of supernovae of different masses in different parts of the galaxy. Presently, one of the possible mechanisms for the production of scandium is the absorption of neutrons on a slow time scale starting from A^{36} and Ca^{40}, present as initial contaminators. The uncertainty in neutron cross-sections permits such an s-chain to be competitive with the s-chain using Ne^{22} as seed nucleus (Fowler and Tuttle, 1957), even if all the initial CNO contamination is converted into Ne^{22}. The normal value of [Sc/Fe] observed in the α-rich group of stars would not be expected if scandium is produced by an s-process operating on seed nuclei which were α-elements present as initial contaminators.

Some extremely old stars have been observed with iron (metal) abundances 0.01 of that of the sun (Wallerstein *et al.*, 1963). These are probably first-generation stars formed within 10^7 to 10^8 years of the formation of the galaxy. Thus, from a naive point of view, 100 generations, or about 10^{10} years, are needed to produce the present metal abundances.

REFERENCES

Fowler and Tuttle, quoted in: E. M. Burbidge, G. Burbidge, W. Fowler, and F. Hoyle (1957), *Rev. Mod. Phys.* **29**: 584.

C. Hayashi, M. Nishida, N. Ohyama, and H. Tsuda (1959), *Progr. Theoret. Phys.* **22**: 101.

H. L. Helfer, G. Wallerstein, and J. L. Greenstein (1963), *Astrophys. J.* **138**: 97.

G. Wallerstein (1962), *Astrophys. J. Suppl.* **6**: 407.

G. Wallerstein, J. L. Greenstein, R. Parker, H. L. Helfer, and L. H. Aller (1963), *Astrophys. J.* **137**: 280.

DISCUSSION

A. G. W. Cameron: The so-called α-elements are probably produced partly by carbon-, neon-, and oxygen-burning and partly by the approach to equilibrium. Scandium is also produced in the approach to equilibrium. There are many different possibilities for producing the enrichment in the so-called α-group and for increasing or decreasing the amount of scandium. This particular part of the nucleid chart will likely be the last to be understood in full quantitative detail because of the variety of processes which can be a function of age as well as of stellar mass and structure.

TRANSPORT OF s-PROCESS ELEMENTS TO THE SURFACES OF STARS

Vern L. Peterson and Marshal H. Wrubel

The simplest explanation for the appearance of technetium in stellar atmospheres is that this element is produced in the deep interior and transported to the atmosphere before it β-decays completely (Burbidge, Burbidge, Fowler, and Hoyle, 1957). Technetium is presumed to be formed by the s-process along with other elements, and if the technetium is transported to the atmosphere, so must the other s-process elements. Observations of the relative abundance of these elements in stellar atmospheres might yield some information about the transport process.

The observations, unfortunately, cannot be made in general for individual isotopes, but have to be made for all the isotopes of an element as a group. In the discussion which follows, we will examine one model by means of which the abundances of adjacent elements in the periodic table might be used to explore the transport of s-process elements. A requirement is that one of the elements has an isotope produced by the s-process with a β-decay lifetime on the order of, or longer than, the time of transport from core to atmosphere.

The s-process is a neutron-addition process. A nucleus of atomic number Z absorbs neutrons on a slow time scale, forming isotopes with larger and larger numbers of neutrons until, through a rapid β-decay, it becomes an isotope of the element $Z + 1$. Some isotopes with long β-decay lifetimes are produced along the way. Except near magic-number nuclei, adjacent stable or long-lived isotopes will usually have abundance ratios inversely proportional to their cross-sections for neutron capture; that is, their abundances will be in relative equilibrium.

One model of the mixing process is shown in Figure 1. The core of the star is assumed to be an unobservable reservoir of s-process material in which the isotopes of adjacent elements occur in their equilibrium ratios. Some kind of transport process carries a fraction of the core material through the interior up to the atmosphere; during this transport, some of the isotopes β-decay. The material is observable only after it reaches the atmosphere.

The material in the atmosphere was not removed from the core at a unique time, but rather it is a mixture of material which reached the atmosphere at different

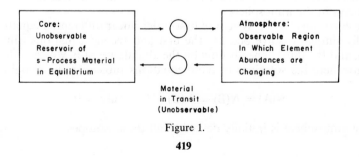

Figure 1.

419

times. Furthermore, some isotopes β-decay within the atmosphere, and, in addition, the atmosphere loses material back into the interior as part of the circulation pattern. Thus, at any time we observe only a snapshot of material in the atmosphere, some of which is old and some newly arrived.

The differential equations of this model are as follows: There are three types of isotopes that must be considered:

Isotopes Unchanged by β-Activity

All that affects the abundance of such isotopes is that they are brought up into the atmosphere and carried back down to the core. Calling isotopes of this kind type A, we have

$$\frac{dN(A)}{dt} = f_1 N_c(A) - f_2 N(A)$$

where f_1 is the fraction of the material leaving the core that reaches the atmosphere per unit time and f_2 is the fraction of atmospheric material exchanged with the interior per unit time. $N_c(A)$ is the equilibrium core abundance and $N(A)$ the (observable) atmospheric abundance of isotope A. Thus, the first term on the right is the addition to the atmosphere from below, and the second term is the loss to downward circulation.

Isotopes Which Decay

We call these isotopes type B, and find

$$\frac{dN(B)}{dt} = f_1 N_c(B) \exp(- \tau_t/\tau_\beta) - f_2 N(B) - \frac{N(B)}{\tau_\beta}$$

Here, the first term on the right is the addition to the atmosphere from below, taking into account the decay during transit, where τ_t is the transit time from core to atmosphere, and τ_β is the β-decay half-life. The second term is the loss due to downward circulation, and the last is the loss due to β-decay in the atmosphere itself.

Products of β-Decay

Isotopes of this type obey the equation

$$\frac{dN(C)}{dt} = f_1 N_c(C) + f_1 N_c(B)[1 - \exp(- \tau_t/\tau_\beta)] - f_2 N(C) + \frac{N(B)}{\tau_\beta}$$

Here, the first and third terms are similar to those for type A, the second term is the production of isotope C by β-decays in transit, and the last term the production of C by β-decays in the atmosphere.

These equations form a system of first-order linear differential equations which is especially simple to solve because the first and second equations can be solved separately, and the third depends only on the abundance of B, which is then known.

We give here the solutions of these equations subject to the initial conditions

$$N(A) = N(B) = N(C) = 0 \quad \text{at } t = 0$$

that is, the atmosphere is initially devoid of all these isotopes.

The abundance of type-A isotopes as a function of time is easily found to be

$$N(A) = \frac{f_1}{f_2} N_c(A)[1 - \exp(- f_2 t)]$$

If we define

$$N_c^*(B) \equiv N_c(B) \exp(- \tau_t/\tau_\beta)$$

which is the abundance of type-B isotopes after decay in transit, and let

$$f_B = f_2 + \frac{1}{\tau_\beta}$$

then the abundance of type-B isotopes is

$$N(B) = \frac{f_1}{f_B} N_c^*(B)[1 - \exp(- f_B t)]$$

If we further define

$$N_c^*(C) = N_c(C) + N_c(B)[1 - \exp(- \tau_t/\tau_\beta)]$$

the abundance of type-C isotopes is found to be

$$N(C) = \frac{f_1}{f_2} N_c^*(C)[1 - \exp(- f_2 t)] + \frac{N(B)}{1 - \exp(- f_B t)}$$
$$\left[\frac{1 - \exp(- f_2 t)}{f_2 \tau_\beta} + \exp(- f_B t) - \exp(- f_2 t) \right]$$

The nature of the solutions is indicated by the limiting cases.

For $t \to 0$,

$$N(A) \to f_1 N_c(A) t$$
$$N(B) \to f_1 N_c^*(B) t$$
$$N(C) \to f_1 N_c^*(C) t$$

For $t \to \infty$,

$$N(A) \to \frac{f_1}{f_2} N_c(A)$$

$$N(B) \to \frac{f_1}{f_B} N_c^*(B) = \frac{f_1}{f_2} \frac{N_c^*(B)}{1 + (1/f_2 \tau_\beta)}$$

$$N(C) \to \frac{f_1}{f_2} \left[N_c^*(C) + \frac{N_c^*(B)}{f_B \tau_\beta} \right] = \frac{f_1}{f_2} \left[N_c^*(C) + \frac{N_c^*(B)}{f_2 \tau_\beta + 1} \right]$$

Now consider the abundance ratios of two adjacent elements. Each pair of adjacent elements, Z and $Z + 1$, normally contains some isotopes of type A (that do not decay and are not decayed into, i.e., are unchanged by β-decay); element Z

(the feeder element) contains an isotope of type B, and element $Z + 1$ (the receiver element) contains an isotope of type C. The abundance ratio of elements Z and $Z + 1$ is

$$\frac{\Sigma N_Z(A) + N_Z(B)}{\Sigma N_{Z+1}(A) + N_{Z+1}(C)}$$

The mixing factor f_1 disappears in taking this ratio.

Three times are unknown: the time since the onset of mixing t, the time of transit τ_t, and the removal time from the atmosphere $(1/f_2)$. Consider several cases:

(A) If mixing has been going on only a short time (t small), the ratio is determined by the transit time τ_t, which would be found by measuring the abundance ratios. However, the time elapsed since the beginning of the mixing process would remain unknown.

(B) If the time the material remains in the atmosphere is short compared to its decay time (or, equivalently, $\tau_\beta f_2$ is large), the ratio is determined by the transit time τ_t for all t.

(C) If the material has a chance to decay in the atmosphere ($\tau_\beta f_2$ is small) the abundance ratio after time $t \sim f_2 + (1/\beta)$ changes to a value which depends upon f_2 as well as τ_t (Figure 2) (Peterson, 1963).

For stars in which the mixing process has been going on for a long time (t large), both the transit time τ_t and the time in the atmosphere $(1/f_2)$ can be determined. For adjacent elements, a constant abundance ratio corresponds to a curve in the $f_2 - \tau_t$ plane. Since f_2 and τ_t are independent of the element involved, abundance ratios for two pairs of elements should intersect at the proper f_2 and τ_t, determining both quantities.

To summarize: The abundance ratios of s-process elements observed in the atmospheres of stars will not always be the same as the equilibrium abundance ratios in the core. To determine the time scales of the mixing process, elements satisfying the following conditions are needed:
1. The element is formed by the s-process.
2. An abundant isotope with long β-decay lifetime exists.
3. The element produces lines in the visible spectrum for which good f values are available.
4. The abundance ratios are independent of τ_n.

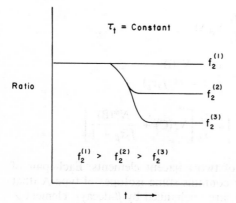

Figure 2.

Likely candidates in addition to technetium worthy of further attention are

$$Zr \rightarrow Nb \rightarrow Mo$$

with decay times 1.37×10^6 and 2.9×10^4 years, respectively, and

$$Pd \rightarrow Ag$$

with a decay time 1.0×10^7 years.

ACKNOWLEDGMENT

The authors wish to express their indebtedness to Dr. John N. Bahcall for many stimulating discussions during the course of which the model described in this paper was evolved and explored.

REFERENCES

E. M. Burbidge, G. R. Burbidge, W. A. Fowler, and F. Hoyle (1957), *Rev. Mod. Phys.* **29**: 547.
V. L. Peterson (1963), Indiana University Ph.D. Thesis, University Microfilms, Ann Arbor, Michigan.

DISCUSSION

A. G. W. Cameron: The problem of technetium is particularly instructive, and may put strong boundary conditions on the nature of the mixing process. The laboratory lifetimes for β-decay for Tc^{99} are about 2×10^5 years, but in the interior of a star, under the conditions of neutron capture in which the technetium is produced, that lifetime is drastically reduced because the β-decay can go through excited states. Using neutrons from the carbon-burning stage, the lifetime of the Tc^{99} is only a few days. Using neutrons from Ne^{22} (α, n) at temperatures around 2×10^8 at the end of helium-burning, the lifetime is increased only to about a year. Thus, to transport technetium from a region in which it is produced to the surface, it must be abruptly removed from the neutron flux. There cannot be storage in the interior. This consideration may severely limit the nature of possible mixing processes.

CH STARS AND NEUTRON-ADDITION PROCESSES

G. Wallerstein and J. L. Greenstein

Among the carbon stars are a few with very distinct group characteristics (Keenan, 1942); CH and the resonance line of barium II at λ 4554 are very strong, while bands of C_2 and CN are not generally as strong as in other carbon stars. All of these CH stars have high space velocities (about 200 km/sec). Many are located at substantial galactic latitudes, while almost all of the other carbon stars are located at low galactic latitudes, except, of course, for a few bright ones, which are close by. The fact that the CH stars are actually population II carbon stars was confirmed about a year ago when a CH star was found in ω Centauri, a metal-poor globular cluster (Harding, 1962).

We have investigated these stars using Palomar spectra at dispersions from 6.7 to 13.5 Å/mm. Figure 1 shows the spectra of four stars in the yellow region. The top one is γ Leo B, a slightly metal-poor G8 giant; shown next are the two CH stars which we have investigated, HD 26 and HD 201626; finally, at the bottom for comparison is HD 122563, the most metal-poor giant star we have been able to find. These spectra are arranged in order of increasing metal deficiency: γ Leo B may be metal deficient by a factor of 2, HD 26 by a factor of 6, HD 201626 by a factor of 30, and HD 122563 by a factor of over 100. Interstellar D-lines are visible in the CH stars, because they are faint and therefore distant. The spectra are arranged so that the stellar lines are lined up. In this way the interstellar lines, which are actually at a radial velocity near zero, appear to be displaced, and the high radial velocity of the star can be seen by the displacement of these sodium D-lines. The extremely metal-poor star HD 122563 is very bright, however, and has a low space velocity. It has no C_2, evidence that not all metal-poor stars have excess carbon.

An abundance analysis was made by the curve of growth method using ε Virginis, of type G8III, as the standard star (Cayrel and Cayrel, 1963). From six-color photometry, the temperatures of the CH stars are almost exactly the same as ε Virginis.

Figure 2 shows the curve of growth for ionized elements compared to neutral iron. The general strength of the ionized lines shows that the electron pressure is lower in HD 201626 than in ε Virginis. The relative overabundance of some elements is shown by the great strength of their lines. Much of the scatter is caused by blending with other features. A reasonable curve of growth is obtained from the bottom envelope of either the open figures or the closed figures.

Table I gives the results of the abundance analysis compared to the abundance in ε Virginis. From the table it can be seen that the stars are metal deficient, but that carbon, barium, and the rare earths (excluding europium) are enhanced relative to iron. The enhancement is sufficiently large to make some of the absolute abundance ratios (element/hydrogen) larger than in the sun. No C^{13} is seen in either star; thus $C^{12}/C^{13} > 20$ in HD 26 and $C^{12}/C^{13} > 50$ in HD 201626 (Climenhaga, 1960).

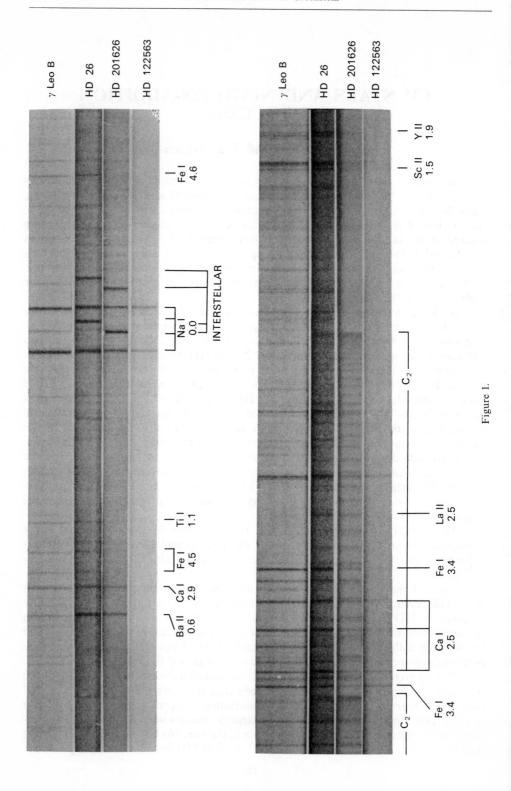

Figure 1.

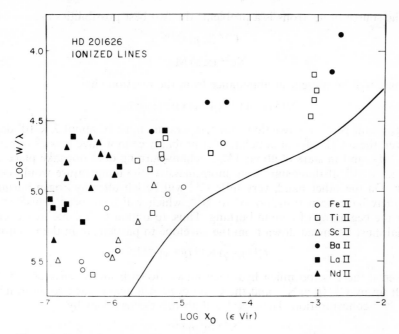

Figure 2. Curve of growth for ionized lines. The open figures are iron or lighter elements. The closed figures are barium, lanthanum, and neodymium. The continuous line is for Fe I in HD 201626.

In normal carbon stars, C^{12}/C^{13} varies from 2 to more than 40.

What evolutionary history could produce these abundances? Since these are population II stars, it is reasonable to assume that the stars started with low metal abundance. The excess of carbon could be formed by the triple alpha-reaction, $3 He^4 \rightarrow C^{12}$, and any original C^{13} would be destroyed during this helium-burning. Barium and the rare earths could be produced by neutron-capture processes (Burbidge and Burbidge, 1957, and Cameron, 1955). Although there is a large abundance of rare earths, no great enhancement of copper or zinc has been found. Thus, the neutron-addition process has gone well beyond iron. On the other hand, the seed elements have not been all processed up to lead, at which point they would not be seen at all. The number of neutrons per seed iron nucleus needed to make a large amount of rare earths without first making an extremely large amount of zinc and copper, or without pushing the nuclei all the way to lead, is roughly 100 neutrons per initial Fe^{56} nucleus (Clayton, Fowler, Hull, and Zimmerman, 1961).

Table I

	HD 26	HD 201626
Fe/H	$\frac{1}{6}$	$\frac{1}{30}$
C/Fe	5	6
(Ba + rare earths, excluding europium)/Fe	20	20
Eu/Fe	5	< 2

The source of neutrons is a problem; the two best possibilities are

$$C^{13} (\alpha, n) O^{16}$$

$$Ne^{22} (\alpha, n) Mg^{25}$$

Ne^{22} may well be present in abundance from the reaction chain

$$N^{14} (\alpha, \gamma) F^{18} (\beta^+ v) O^{18} (\alpha, \gamma) Ne^{22}$$

For significant $Ne^{22} + \alpha$ reactions, the temperature must be about 2×10^8 degrees. However, these CH stars of population II probably cannot have masses larger than 1 or $1.2 M_\odot$, and in stars of about $1 M_\odot$, helium-burning does not take place above about 1.5×10^8. If these stars were more massive, the temperature would be high enough. On the other hand, very little C^{13} will be left after hydrogen-burning; it will largely have been transformed to N^{14}, which will have been transformed to O^{18} at the beginning of helium-burning. Thus, to obtain C^{13}, some hydrogen-rich material must be mixed down from the envelope to participate in the reaction

$$C^{12} (p, \gamma) N^{13} (\beta^+ v) C^{13}$$

However, if there is too much hydrogen mixed down from the envelope, N^{14} will be built up by $C^{13} (p, \gamma) N^{14}$ and the CNO cycle will again reach equilibrium with a high N^{14} concentration. Then, N^{14} will absorb the neutrons by

$$N^{14} (n, p) C^{14}$$

since $C^{13} (\alpha, n)$ will burn C^{13} faster than $N^{14} (\alpha, \gamma)$ will burn N^{14}.

The $C^{13} (\alpha, n)$ mechanism has been considered in some detail by tracing the evolution of the interior of the star up to the point at which these processes will occur. Since the goal is to produce about 100 neutrons to be captured by each iron nucleus, 100 nuclei of C^{13} per iron nucleus are needed. To avoid the N^{14} neutron-poison problem, mixing with the envelope was assumed to be small, so that, while there would be enough protons to convert some of the C^{12} to C^{13}, the carbon cycle would not go to completion and there would be no great buildup of N^{14}. We have assumed the initial abundances to be the solar values (Goldberg, Müller, and Aller, 1960), except that all heavier elements are reduced by a factor of 30. The terrestrial ratio of C^{12}/C^{13} was used; the initial helium abundance is unknown.

Table II. Nucleosynthesis of Carbon and Barium with Incomplete $C^{12} (p, \gamma) N^{13} (\beta^+, v) C^{13}$

Isotope	I	II	III	IV	V	VI
H^1	12.0	—	—	—	—	12.0
He^4	—	11.4	11.4	11.4	11.4	9.9
C^{12}	7.1	6.3	9.3	9.3	9.3	7.8
C^{13}	5.1	5.7	5.7	7.1	—	5.1
N^{14}	6.5	7.7	7.7	7.7	—	6.5
O^{16}	7.5	5.3	5.3	5.3	7.1	7.5
O^{18}	4.5	4.5	4.5	4.5	7.7	6.2
Fe	5.1	5.1	5.1	5.1	—	5.1
Ba	0.6	0.6	0.6	0.6	3.6	2.1

The evolution of abundances in the star is shown in Table II. Abundances are shown as $\log N_x$ normalized to $\log N_H = 12$ initially. First, within the core of the star, all of the hydrogen is converted into helium, $4H^1 \to He^4$. The carbon, nitrogen, and oxygen isotopes come into equilibrium through the CNO cycle, although O^{16} does not quite reach equilibrium. This results in most of the carbon, nitrogen, and oxygen going to N^{14}. A high value of the temperature, 6×10^7 °K, was used because it was felt that this equilibrium would be reached rather late in the hydrogen-burning stage; as the hydrogen-burning dies out, the temperature rises. The resulting abundances were obtained from Caughlan and Fowler (1962). In the second stage, helium-burning starts and consumes 3% of the helium in the core, transforming it to C^{12}.

The next steps occur simultaneously, but for simplicity of presentation they are discussed one at a time. The C^{13}/Fe ratio is much less than 100, not nearly enough to produce 100 neutrons per initial iron nucleus. Third, some hydrogen from the envelope is mixed into the core. The protons react with the C^{12} to produce C^{13}. The number of protons is assumed to be small enough so that there is no buildup of N^{14}. Fourth, the C^{13} reacts with the α-particles by $C^{13}(\alpha, n)O^{16}$, to produce 100 neutrons per initial iron nucleus, which produces a large buildup in a heavy element like barium. It was assumed that 3% of the iron goes to barium. This is a reasonable guess as to the distribution and number of the products of neutron absorption. The N^{14} reacts with both the neutrons and α-particles, producing C^{14} and F^{18}; the latter decays to O^{18}. The O^{18} may go on to Ne^{22}, depending on temperature.

Finally, the core material is mixed to the surface in the proportions 3% core material and 97% original composition surface material (Table II, column VI). The hydrogen abundance remains the same. C^{12} is slightly enriched, while C^{13} returns to its original low value. The N^{14}, O^{16}, and Fe abundances are normal, and the Ba abundance is increased by a factor of 30.

Details of this work can be found in Wallerstein and Greenstein (1964) and Caughlan and Fowler (1964 a and b).

REFERENCES

E. M. Burbidge and G. R. Burbidge (1957), *Astrophys. J.* **126**: 357.
A. G. W. Cameron (1955), *Astrophys. J.* **121**: 144.
G. R. Caughlan and W. A. Fowler (1962), *Astrophys. J.* **136**: 453.
G. R. Caughlan and W. A. Fowler (1964a), *Astrophys. J.* **139**: 1180.
G. R. Caughlan and W. A. Fowler (1964b), *Astrophys. J.* **140**: 380.
G. Cayrel and R. Cayrel (1963), *Astrophys. J.* **137**: 431.
D. D. Clayton, W. A. Fowler, T. E. Hull, and B. A. Zimmerman (1961), *Ann. Phys.* **12**: 331.
J. L. Climenhaga (1960), *Publ. Dominion Astrophys. Obs. Victoria* **11**: 307.
L. Goldberg, E. A. Müller, and L. H. Aller (1960), *Astrophys. J. Suppl.* **5**: 1.
G. A. Harding (1962), *Observatory* **82**: 205.
P. C. Keenan (1942), *Astrophys. J.* **96**: 101.
G. Wallerstein and J. L. Greenstein (1964), *Astrophys. J.* **139**: 1163.

DISCUSSION

A. G. W. Cameron: We are uncertain as to the ratio of carbon and oxygen produced at the end of the helium-burning. Do you see the oxygen in these stars?

G. Wallerstein: There is no observed excess of oxygen, but only an upper limit on its abundance, about the solar ratio of O/Fe, can be obtained. Probably, little oxygen is produced, although some could be produced without our seeing it.

W. P. Bidelman: I have made some observations of the NH bands in the far ultraviolet in carbon stars and two CH stars. The NH bands are very similar in strength to those in ordinary giant stars. I doubt very much that nitrogen is either overabundant or underabundant. Furthermore, there is no relation between the strength of the NH bands and the presence or absence of C^{13}. This would, I believe, imply that the carbon-cycle processes have not gone to equilibrium.

H. Reeves: Red-giant stars show very different compositions. The ratio of iron to hydrogen is related to the activity of the *e*-process and the age of the star, the ratio of Ba-La-Ce-Nd to Fe is related to the activity of the *s*-process, and the ratios of C and O to Fe are possibly related to the moment of occurrence of the *s*-process.

Some red giants having a low ratio of N_{Fe}/N_H (apparently very old) are quite similar in many abundances, except for their *s*-process abundance. This may reflect the fact that neutron processes are not an automatic consequence of helium stage (and of more advanced stages), but depend, among other things, upon the stellar masses. The two CH stars considered here present evidence of intense neutron activity (increase in the *s*-process elements). The enhancement of C without enhancement of O suggests that the process took place in the early days of helium-burning, possibly from proton admixture or from N^{14} burning at high temperature in the peak of the helium flash. Such analyses are very preliminary and serve mostly as illustrations.

ABUNDANCE DIFFERENCES AMONG POPULATION I STARS

B. E. J. Pagel

Some abundance features of the late-type stars of population I are discussed. These features confirm the conclusions that have been drawn previously, especially by Wallerstein (1962) and by Eggen, Lynden-Bell, and Sandage (1962), from observations of stars of earlier type or with ultraviolet excess. Our conclusions are: (1) Increasing metal deficiency is correlated with increasing dispersion in velocity (W) perpendicular to the galactic plane; in the plane, stars with a large orbital eccentricity are mildly deficient in metals, but the converse does not always hold. (2) Individual abundances of different metals show no correlation with anything, except perhaps luminosity.

OVERALL METAL ABUNDANCE (Fe/H)

Consider first the subgiants, which are particularly interesting in view of the fact that evolutionary theory gives some clue as to their ages. Reliable information on overall metal abundances of both K dwarfs and G and K subgiants is obtained from the results of six-color photometry, e.g., comparing $(B - V)$ with $(G - I)$ or $(R - I)$.

Figure 1 indicates that moderately metal-deficient subgiants (shown by open circles) have the normal large dispersion in z-velocity (perpendicular to the galactic plane) compared to subgiants of solar composition (shown by black circles), but that neither the overall composition nor the dispersion in z-velocity shows any marked correlation with age.

Next, consider the overall metal abundances among K dwarfs. This class is less interesting than the subgiants because there are no direct clues as to the age of the stars. Information on mild differences in composition among K dwarfs can be obtained both from spectral types estimated at high dispersion and from photoelectric colors in the infrared.

Wilson (1961) found that there was a significantly large dispersion in the $(P - V)$ colors for each K subtype and that at any one subtype the bluer stars had stronger Balmer lines and weaker Ca II H and K emission than the redder stars. Revised classifications by Wilson (1962) eliminated the two spectroscopic effects, but confirmed the dispersion in color. This effect is due to differences in metal abundance which lead to differential line-blanketing effects. Metal deficiency leads to blue colors and metal overabundance to red colors. This conclusion has been suggested before (Pagel, 1962) and is quite widely accepted, but the abundance differences must now be considered as small, by a factor of 2 or 3. It is further strengthened by new observations of a few crucial stars according to Kron's (R, I) system (Kron and Smith, 1951), along with the data already available from the work of Kron, Gascoigne, and White (1957).

Figure 2 shows a plot of $(P - V)_E$ against spectral type. Known mild subdwarfs (Sandage and Eggen, 1959) are represented by crosses. Their tendency to be

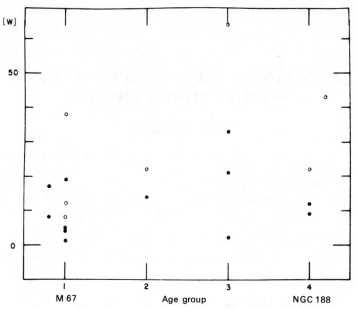

Figure 1A. z-velocity age diagram for subgiants. • normal subgiants;
○ deficient subgiants.

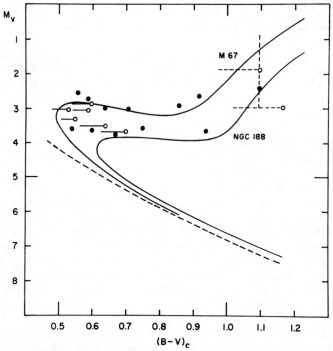

Figure 1B. H-R diagram for normal and metal-deficient subgiants.
• normal subgiants; ○ deficient subgiants. The horizontal lines indicate
the correction for blanketing that has been applied on the basis of
$(G - I)$ (full lines) or $\delta(U - B)$ (broken line).

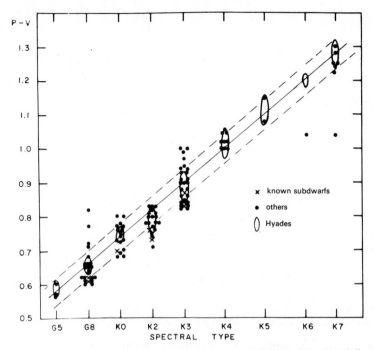

Figure 2. $(P - V)_E$ against spectral type by O. C. Wilson. The dotted lines represent the inevitable effect of bunching due to the discreteness of the spectral types, so that a star can only be considered significantly "blue" or "red" if it lies outside the range of the broken lines.

"blue" at G8, K0, K2, and K3 is quite obvious. It is also seen that certain stars are extremely "red," especially at G8 and K3. One has to ask whether these stars are exceptionally metal rich. Classifying the stars from Wilson's diagram as "blue," "normal," "red," or "subdwarf" it can be seen whether these effects show up as a blanketing effect when we plot $(R - I)$ against $(P - V)_E$ (Figure 3). The expected trend is present to some extent, since the open triangles ("blue") are somewhat on the high side of the diagram toward the subdwarfs, and the black triangles ("red") are somewhat on the low side. We conclude that Wilson's effect is basically a real one, arising from differences in metal abundance by factors of 2 or 3 in either direction from the sun, in accordance with what is already known in the case of the G dwarfs. There is, however, a considerable spread, and observational errors are large.

The distribution of velocity in the galactic plane of G8–K7 stars with varying metal abundances as judged from colors is shown in Figure 4, which closely resembles the diagram for G dwarfs by Wallerstein (1962). There is relatively little correlation between metal abundance and the galactic orbit, except when the orbits have a rather high eccentricity ($v \leq -80$ km/sec relative to the sun).

Figure 5 shows the velocities of these stars at right angles to the galactic plane. Again, we see very much the same result as was found by Wallerstein (1962) and by Eggen, Lynden-Bell, and Sandage (1962)—the increase in dispersion of W with increasing metal deficiency—although no stars with W much greater than 50 km/sec are present in the sample. In view of the age independence of this effect, one should

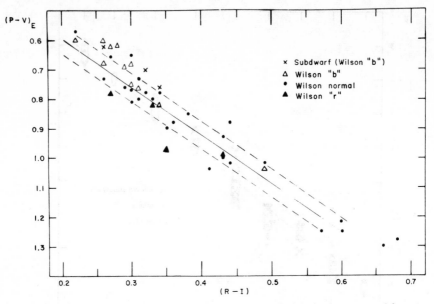

Figure 3. $(P - V)_E$ against $(R - I)$ for G8–K7 dwarfs. The limits of colors expected for stars of normal composition are indicated rather roughly by the dotted straight lines.

consider very seriously whether the hydrogen-to-metal ratio could be a function of the distance above the plane at which the star was born, e.g., through the gas-density distribution rather than as a function of advancing epoch.

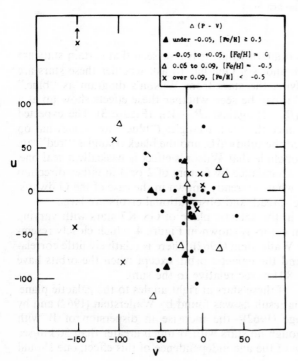

Figure 4. Haas–Bottlinger diagram of velocities of G8–K7 dwarfs in the galactic plane.

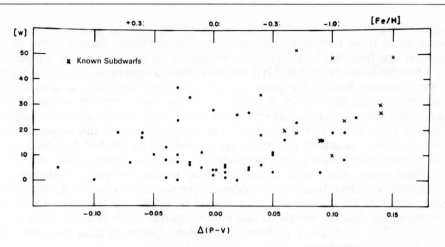

Figure 5. Speeds of G8–K7 dwarfs at right angles to the galactic plane.

We conclude that the overall metal abundances of K dwarfs are similar to those found already for the G dwarfs. Both groups seem to share the same evolutionary history.

INDIVIDUAL METAL ABUNDANCES

Individual metal abundances for stars can also be determined. A great variety of results are obtained which have little or no obvious relationship with the kinematic properties of the stars involved. Table I gives a brief summary of the incidence of the various individual abundance effects as deduced from existing and reasonably reliable analyses of the curve of growth. It is difficult to discern any sort of clear pattern in these results. The red giants, when deficient, appear to have the same

Table I. Individual Abundance Effects in Cool Stars

		(Fe/H)	(α/Fe)	(Mn/Fe)	(s/Fe)	(r/Fe)
Red giants						
ε Cyg	K0 III	−0.31	+0.02	−0.02	−0.11	−0.03
γ Leo A	K0 III	−0.26	+0.03	−0.03	−0.07	−0.03
α Boo	K2 IIIp	−0.35	+0.04	−0.03	−0.09	0.00
Subgiants						
γ Ser	F6 IV–V	−0.4	+0.2:	+0.2	+0.25	—
β Hyi	G1 IV	−0.30	+0.25	−0.03	−0.22	−0.44
HD 185657	dG6	−0.5	+0.1	−0.07	0.0:	—
α-rich dwarfs						
85 Peg	G2 V	−0.60	+0.25	−0.28	−0.06	−0.2:
τ Cet	G8 Vp	−0.40	+0.25	−0.21	−0.10	0.0
o² Eri	K1 V	−0.15	+0.2	−0.05	—	—
Other dwarfs						
99 Her	F7 V	−0.45	0.00	−0.35	−0.02	−0.3:

α refers to elements produced by the α-process; s refers to elements produced by the slow-neutron-capture s-process; r refers to elements produced by the fast-neutron-capture r-process.

degree of deficiency in all elements, except perhaps for an additional marginal deficiency in those formed by the s-process. The subgiants, covering most of the range in age from M67 to NGC 188, all appear to be α-rich; what happens to the s-process is not clear, but β Hyi seems to show a definite additional deficiency in the r-process. Neither the giants nor the subgiants show the manganese effect (manganese being about twice as deficient as iron) that is usually so conspicuous among the dwarfs. For the dwarfs themselves, there is only a rather marginal effect in both the s- and r-processes. The interpretation of this sequence from giants to dwarfs as one of increasing age is inhibited by the fact that each group also contains many stars having precisely the same abundances as the sun. These individual effects show little correlation with motion in space; furthermore, if all the metal-deficient dwarfs are considered to be very old because of the manganese effect, then it is hard to see why they should be subdivided into an α-rich group resembling the deficient subgiants and another group which is metal deficient but not α-rich. The α-rich group of dwarfs has higher eccentricity than the others, and there is a large spread in v.

Recently, Helfer, Wallerstein, and Greenstein (1963) have suggested the formation of titanium by the α-process because this element is more abundant than iron in the α-rich stars which they studied. Our analyses tend to confirm this conclusion and also suggest that sodium might be added to the list, in accordance with the idea that it was produced during the α-process by the $C^{12}(C^{12}p)$ reaction (Reeves and Salpeter, 1959, and Reeves, 1962). One star (RY Sag) is found by Danziger to be very rich in carbon; its atmosphere is mostly carbon and helium, and it is also very rich in sodium.

It is possible that some of these effects are connected with surface reactions that took place in an earlier phase of a star's history, as in the case of lithium, beryllium, and boron. There does seem to be one crucial type of observation that has not yet been made, namely, a study of individual element abundances in sub-dwarfs that lie high up on the main sequence and would therefore seem to be considerably younger than M67, say, at least as young as NGC 752. If these have the α-rich and manganese effects, then one would be able to say that the effects cover a very considerable spread in time of formation of the star.

REFERENCES

O. J. Eggen, D. Lynden-Bell, and A. R. Sandage (1962), *Astrophys. J.* **136**: 748.
H. L. Helfer, G. Wallerstein, and J. L. Greenstein (1963), *Astrophys. J.* **138**: 97.
G. E. Kron and J. L. Smith (1951), *Astrophys. J.* **113**: 324.
G. E. Kron, S. C. B. Gascoigne, and H. S. White (1957), *Astron. J.* **62**: 205.
B. E. J. Pagel (1962), *Roy. Obs. Bull.* No. 55.
H. Reeves and E. E. Salpeter (1959), *Phys. Rev.* **116**: 1505.
H. Reeves (1962), *Astrophys. J.* **135**: 779.
A. R. Sandage and O. J. Eggen (1959), *Monthly Notices Roy. Astron. Soc.* **119**: 278.
G. Wallerstein (1962), *Astrophys. J. Suppl.* **6**: 407.
O. C. Wilson (1961), *Astrophys. J.* **133**: 457.
O. C. Wilson (1962), *Astrophys. J.* **136**: 793.

DISCUSSION

D. Clayton: Evolutionary models have not shown conclusively how much of the products of carbon- and oxygen-burning phases can ever get to the interstellar medium, because these processes occur on a rapid time scale (due to the neutrino luminosity) on the way toward the equilibrium buildup. We do not know how much of anything is produced by these reactions, unless they occur in shells that are blown off, and we do not know how much can get out.

ELEMENT ABUNDANCES IN THE PECULIAR A STARS

William P. Bidelman

There are two varieties of stars in which certain elements are definitely very overabundant. The first class consists of late-type giants: the carbon, CH, S-type, and Ba II stars. The abundance anomalies of these objects seem to be well accounted for by α- and s-process production of heavy elements in the stellar cores and subsequent mixing of these products to the surface layers. The second class consists of dwarf stars with spectral types between B5 and F0, most of which show evidence of large magnetic fields. What is happening in this group is much less clear.

Recently, I have studied a number of stars of the latter type on high-dispersion coudé spectrograms obtained at the Lick Observatory. A number of interesting element identifications, some made for the first time in stellar spectra, have resulted from this work (Bidelman, 1965). Abundance determinations have now been made by various investigators for a considerable number of peculiar A stars. Table I presents a very rough summary of some of the more remarkable results obtained. The values given are taken from a great variety of determinations; in some cases, they refer to only a single star, while in others they are averages of individual values showing a considerable dispersion. Thus, the table is merely indicative of the general situation. It should be emphasized that the over- and underabundances certainly differ greatly among the stars of the group.

There are, it appears, two distinct problems posed by the peculiar A stars: (1) the structure of the stars, and (2) the extraordinary chemical composition indicated by their spectra. Many of these objects are also variable in spectrum and in

Table I. Some Abundances in the Peculiar A Stars

Element	Over- or underabundance (log)	Comments
He^3	—	$He^3/He^4 \sim 5$ in 3 Cen
Be	$+2.0$	
O	-1.2	Only in some stars
Si	$+2.0$	
P	$+1.7$	
S	-1.6	
Ca	-1.5	
Mn	$+2.0$	Mn/Fe = 3 in 53 Tau
Ga	$+3$:	
Kr	$+2.5$	
Sr, Y, Zr	$+1.7$	
Xe	$(+2.5)$	Rough estimate
Rare-earth elements	$+2.2$	
Hg	$(+2.5)$	Rough estimate

magnetic-field intensity (and also slightly variable in light, as well), but these additional phenomena are probably only incidental to the main problems cited above.

The peculiar A stars are, in general, fairly near the main sequence. This naturally has led to the belief that they are young stars. Furthermore, they are quite frequent, constituting about 10% of all the A stars near the main sequence. This is another fact that has made their explanation very difficult. It seems very doubtful that the overabundances of the heavy elements observed in the atmospheres of these stars can be produced by surface reactions operating on material of normal chemical composition. Thus, one is forced to the conclusion that these stars must already have undergone extensive nucleosynthesis in their interiors which has resulted in an abnormal chemical composition—one rich in all of the heavy elements —throughout the *entire* star. I believe that the abnormal abundances characteristic of the peculiar A stars are therefore the result of two processes: (1) an initial heavy-element production, largely by the *s*-process, and (2) subsequent modification of this composition by surface spallation reactions that produce large overabundances of certain elements and deficiencies of others. Reactions such as $(p, 2p)$ and $(p, p\alpha)$ can transform sulfur into phosphorus, iron into manganese and chromium, and lead into mercury, to cite only a few possibilities.

The concept that the peculiar A stars are nuclearly evolved objects makes their interpretation no less difficult than before. At one time, I considered that these stars were the cores or remnants of red giants, lingering near the main sequence on their way toward the white-dwarf stage (Bidelman, 1960). There are certainly some difficulties in this interpretation, and it now seems equally probable that these objects have arrived at their present state as a consequence of binary nature—or something else—without ever having departed substantially from their present position in the H-R diagram.

REFERENCES

W. P. Bidelman (1960), *Transactions I.A.U.*, Vol. 10, Cambridge University Press, p. 677.
W. P. Bidelman (1965), "Abundance Determination in Stellar Spectra," *I.A.U. Symposium No. 26* (in press).

DISCUSSION

G. Wallerstein: Deutsch pointed out in 1947 that the A0 peculiars, and also most of the others, have colors which are too blue for their spectral types, that is, they have colors of about B8 stars. Yet they do not have any helium lines. I think this indicates that He^4 is deficient in these stars, in fact, that all helium is deficient in these stars. It is hard to see how we could be looking at the core of the star and yet not have any helium in it.

A. J. Deutsch: Another possible explanation, certainly no better understood than those you have proposed, was put forward long ago and recently revived by Babcock, namely, the possibility that some kind of a mass-spectrograph or similar process is operating in the envelopes of these stars and separating the elements. There is good evidence that the surfaces that we see of these stars are not uniform. In some of the stars, it is likely that there is a physical separation between the chromium and certain other elements that go with it on the one hand and the rare earths on the other. They probably do not occupy the same volume. If this kind of process operates, it is conceivable that there may be a separation of the elements in a substantial part of the material of the envelopes in these stars; this might provide an explanation for this phenomenon, without any nucleosynthesis at all.

STELLAR GROUPS AND THE MASS–LUMINOSITY RELATION

Olin J. Eggen

There are two lines of evidence that point to a separation of the disk stars, within about 200 parsecs of the sun, into two kinds: (a) those which have a chemical composition (H, He, Fe) like that of the Hyades and Pleiades clusters, and which may have been formed from the same gas cloud within the last few times 10^8 years; and (b) those with only approximately half the helium content of the Hyades stars, and probably formed at distances of greater than 10 kiloparsecs from the galactic center.

KINEMATIC EVIDENCE

The A-type stars (B8 to F0) brighter than visual magnitude 5.5 represent a valuable sample of young stars (younger than about 5×10^8 years) having accurately known apparent motions and being close enough to the sun (within about 100 parsecs) to have colors which are essentially free of interstellar reddening. The data for the stars in this sample have been published elsewhere (Eggen, 1963a), where the distribution of the (U, V) vectors of the space motion, relative to the sun, shown in Figure 1 has been derived. The limiting ellipse that would enclose most of these stars in the (U, V) plane, which is of course related to the velocity ellipsoid, can be

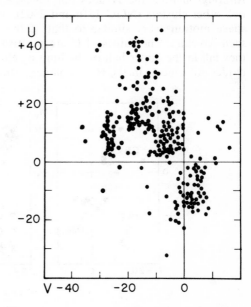

Figure 1.

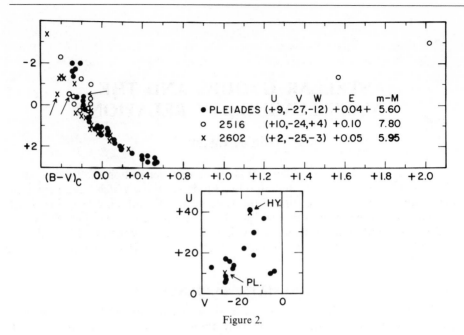

Figure 2.

divided into two nearly equal segments by a line joining $(U, V) = (+20, 0)$ and $(0, -20)$. This dichotomy among the youngest stars is illustrated by the kinematics of two groups of peculiar A-type stars (Ap): (1) those of the "manganese and silicon" types, which are all bluer than $B - V = -0.05$, and (2) the redder Ap stars, which are of the europium, chromium and strontium types. The distribution in the (U, V) plane of the brighter Ap stars of the manganese and silicon types is shown in Figure 2, where the positions of the Hyades and Pleiades clusters are also indicated. Although neither the Hyades nor Pleiades clusters happen to have Ap members, two other clusters (NGC 2516 and 2602), which have color–luminosity arrays and space motions very similar to that of the Pleiades cluster (Figure 2) do contain such Ap stars. The values of U are positive for all of these stars and clusters, and they fall in the upper half of the limiting ellipse in Figure 1. The distribution of the redder Ap objects in the (U, V) plane is shown in Figure 3, where the positions of

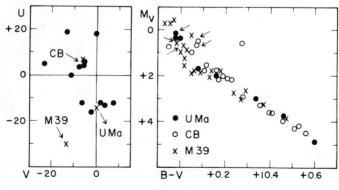

Figure 3

three clusters which contain objects of this kind (M39, Coma Berenices, and U Ma) are also indicated. The color–luminosity arrays of these clusters are also shown in Figure 3, where the Ap stars are indicated by arrows. These stars and clusters are, in general, limited to the lower half of the limiting ellipse in Figure 1. It should also be noted that the stars of type earlier than B8 and within about 300 parsecs of the sun are confined to the upper half of this limiting ellipse (Eggen, 1961).

This dichotomy in the (U, V) distribution of the younger stars (less than about 5×10^8 years) occurs very sharply at an age somewhere between 1 and 5×10^8 years, with all of the youngest objects in the solar neighborhood falling in the upper half of the limiting ellipse in Figure 1.

CHEMICAL COMPOSITION

Distant, physical companions to many of the A-type stars represented in Figure 1 and the Ap stars represented in Figures 2 and 3, as well as later-type main-sequence stars in the relevant clusters, have been observed photoelectrically in three colors (e.g., Eggen, 1963b, and unpublished studies). When the observations of F- and G-type dwarfs physically connected with the A-type star fall on the $(U - B, B - V)$ relation for the Hyades and Pleiades stars, the (U, V) vectors place them in the upper half of the limiting ellipse in Figure 1. Contrariwise, the stars in the lower half of the ellipse are connected with companions showing an ultraviolet excess with respect to the Hyades–Pleiades stars.

The stars in the solar neighborhood satisfy two discrete mass–luminosity relations: (1) the "Hyades–Pleiades" relation, which holds for members of the Hyades and Pleiades clusters and groups (Figure 4) and stars which fall on the Hyades $(U - B, B - V)$ relation (Figure 5); and (2) the "sun–Sirius" relation, which holds for the sun, members of the Sirius group, and all stars showing an ultraviolet excess with respect to the Hyades stars. The shape of these relations in the $(M_B, \log m)$ plane is shown in Figure 6. The few objects in Figures 4 and 5 that follow an extension of the sun–Sirius mass–luminosity relation for luminosities brighter than $M_V = + 2.5$ are evolved stars whose main-sequence luminosities place them on the Hyades–Pleiades relation (Eggen, 1965). All of the stars that fall on the Hyades–Pleiades mass–luminosity relation also lie in the upper half of the

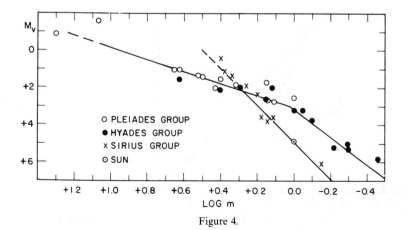

Figure 4.

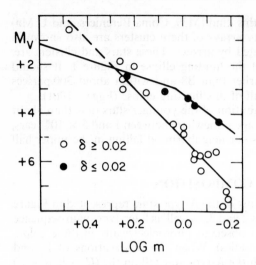

Figure 5.

limiting ellipse in Figure 1. These objects are, therefore, all as young as the A V and Ap stars that also occupy this half of the limiting ellipse.

The objects falling on the sun–Sirius mass–luminosity relation spread over a much larger area of the (U, V) plane than do the A-type stars in Figure 1, and they therefore represent a much wider range of ages than do the stars in that figure. However, if we confine ourselves to stars which follow the sun–Sirius relation and which are also connected, in clusters or in wide binaries, with objects bluer than

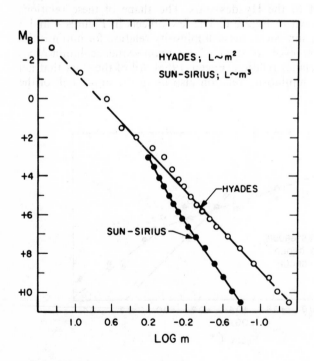

Figure 6.

$B - V = + 0.03$ (i.e., are confined to the same age spread as represented by the stars in Figure 1), we then find that they all lie in the lower half of the limiting ellipse in Figure 1.

The relationships between the values of the ultraviolet excess and Fe/H are extensively discussed elsewhere (e.g., Sandage and Eggen, 1959, and Eggen, 1964). The two mass–luminosity relations can be understood as representing stars with different H/He ratios. Taking the solar ratio as $\frac{3}{2}$, that for the Hyades–Pleiades stars becomes $\frac{2}{3}$ (Eggen, 1963c, 1965).

These results can be summarized as follows:

A. The objects near the sun that are younger than between 1 and 5×10^8 years have values of Fe/H and H/He similar to those of the Hyades and Pleiades cluster members and fall in the very limited region of the (U, V) plane represented by the upper half of the limiting ellipse in Figure 1.

B. Objects only 1 or 2×10^8 years older (about 1 cosmic year) have values of Fe/H from about 0.5 to less than 0.1 and values of H/He twice that of the Hyades–Pleiades stars, and they fall in the lower half of the limiting ellipse in Figure 1.

Although the stars of population B, above, all have smaller values of Fe/H than those in population A, as indicated by the ultraviolet excess, this does not mean that they all have smaller amounts of iron. For example, the ultraviolet excess of the sun indicates that $(Fe/H)_\odot = 0.5(Fe/H)_{Hy}$ (Eggen, 1964), but if $H_\odot = 2H_{Hy}$, as indicated by the mass–luminosity relations, then $Fe_\odot = Fe_{Hy}$.

SOLAR MOTION

The dichotomy in the motions of the young stars is shown in the (U, V) diagrams of Figure 7 for (a) wide binary systems in which the fainter component is a main-sequence star with $B - V$ between $+0\overset{m}{.}3$ and $+0\overset{m}{.}8$ and an ultraviolet excess, with respect to the Hyades stars, of less than $0\overset{m}{.}02$, (b) a repetition of Figure 1; and (c) the wide binary systems in which the brighter component is bluer than $B - V = +0\overset{m}{.}3$ and the fainter has $B - V$ between $+0\overset{m}{.}3$ and $+0\overset{m}{.}8$, with an ultraviolet excess greater than $0\overset{m}{.}02$. The so-called basic solar motion, which is derived from a sample of stars similar to that in Figure 7c, gives values of (U, V)

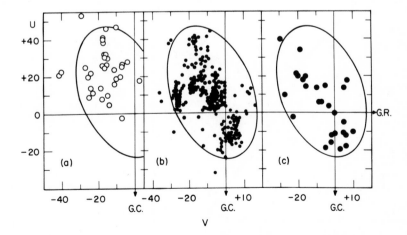

Figure 7.

near $(+10, -10)$ for the center of rest, i.e., the motion of the sun with respect to an object with the circular velocity at the sun's present distance from the galactic center is $(U', V') = (-10, +10)$. However, if we confine ourselves to the youngest stars, Figure 7a, the value of (U', V') is more nearly $(-20, +17)$, whereas the objects in Figure 7c indicate a value closer to $(0, 0)$. Furthermore, solar motions obtained from Figure 7b may have little significance, because, as discussed in detail elsewhere (Eggen, 1963a), the stars in this figure not only represent a fortuitous combination of those in Figures 7a and 7c, but also represent the mean motion of a half-dozen stellar groups, and not of several hundred randomly moving stars. Stars in Figure 7a are predominantly members of the Hyades $(U, V = +40, -17)$ and Pleiades $(U, V = +9, -27)$ groups, and many of those in Figure 7c are members of the Sirius $(U, V = -14, 0)$ group.

If it is assumed that the youngest stars (upper half of the limiting ellipse in Figure 1) represent nearly circular motion, then none of these objects have wandered more than about a kiloparsec from their present distance from the galactic center, and all could have been formed from the same gas cloud at different epochs. Therefore, because the Hyades are old enough to have made at least one, and perhaps two, galactic orbits, it is not surprising to find them now among younger objects, such as the Pleiades and the B-type stars in the Gould-belt complex, of the same composition. On the other hand, the sun and the stars in the lower half of the limiting ellipse in Figure 1 have spent most of their lives outside the region occupied by the Hyades and Pleiades stars.

REFERENCES

O. J. Eggen (1961), *Royal Obs. Bull.* No. 41.
      (1963a), *Astron. J.* **68**: 483.
      (1963b), *Astron. J.* **68**: 697.
      (1963c), *Astrophys. J. Suppl.*, No. 76.
      (1964), *Astron. J.* **69**: 570.
      (1965), *Astron. J.* (in press).
A. R. Sandage and O. J. Eggen (1959), *Monthly Notices Roy. Astron. Soc.* **119**: 278.

COMPOSITION DIFFERENCES BETWEEN THE GALAXY AND THE MAGELLANIC CLOUDS*

L. H. Aller

The great importance of the Magellanic clouds is that they are stellar systems similar to our own in stellar content in which the various types of objects are all conveniently situated at the same distance from us (see discussion in Kerr and Rodgers, 1964). It is unfortunate that they are in the Southern hemisphere, where there are no really adequate modern telescopes. The distance modulus of the Magellanic clouds is about 19.5 magnitudes (which means that a star like the sun would be at about the limit of a 200-inch telescope). The brightest stars in the clouds are observed at about magnitude 10.5, which means that their absolute magnitudes are around -9.

We thus obtain in the Magellanic clouds the top of the Hertzsprung–Russell diagram, i.e., exactly that domain in which the evolution of the stars is proceeding at the most rapid pace. The Magellanic clouds offer great opportunities for studies of stellar evolution, since an apparent-magnitude diagram becomes, in fact, an absolute-magnitude diagram as soon as small corrections for space absorption are made. There exist small but well-established differences in the color-magnitude arrays between the field stars in the Magellanic clouds and our own galaxy. The brighter stars in the cloud are often found to fall not in the regions which are filled with stars in our own galactic system, but into the gaps. Likewise, the clusters are not the exact analogs of those in our own system. There are objects in the clouds which look like globular clusters but have different color-magnitude arrays, with a distribution of stars resembling that of an open cluster in our own system. These effects may be due to a difference in chemical composition between the stars in the Magellanic clouds and those in our own galaxy (Kerr and Rodgers, 1964).

Differences in chemical composition can be checked in several ways. The brightest stars can be observed with moderate dispersions with the existing telescopes in the Southern hemisphere. Thackery and Feast found that these stars resemble their counterparts in our own galaxy. However, there are very few stars in our own galaxy around magnitude -9 or -8.5 for which reliable data on abundances are known. Furthermore, a small error in the assigned temperature can lead to a considerable error in the metal-to-hydrogen ratio, so that a slightly metal-deficient star, for example, could masquerade as a normal star with a slightly different temperature. Hence, the result for the composition is uncertain; it will be necessary to study the stars which we can observe with appropriate spectrographic equipment.

The gaseous nebulae can also be used to determine chemical composition. They have one very great advantage, namely, that the stronger lines of the more abundant elements like oxygen, neon, and helium (and perhaps eventually carbon

* Supported in part by the Air Force Office of Scientific Research, Office of Aerospace Research, United States Air Force, under AFOSR Grant No. 83–63.

Figure 1. The diffuse nebula 30 Doradus in the large Magellanic cloud, photographed in the light of Hα with the 74-inch reflector at Mt. Stromlo Observatory, 1960.

and nitrogen) can be observed. Helium-to-hydrogen ratios can be found by comparing the stronger lines of helium with Hβ. The oxygen-to-hydrogen ratio can also be found (but perhaps not quite so accurately) by examining the emission lines due to singly and doubly ionized oxygen:

$$\frac{N(O)}{N(H)} \approx \frac{N(O^+) + N(O^{++})}{N(H)}$$

The number of hydrogen atoms follows from the surface brightness as measured in Hβ. It is possible that in some objects neutral or triply ionized oxygen, which is not observed, exists. A histogram of the above ratio can be examined to find the actual ratio of oxygen to hydrogen. Neon abundances can be obtained in a similar fashion, except that neon is observed only in one stage of ionization, Ne^{++}, and the corrections that must be made for the unobserved stages of ionization become disagreeably large.

Abundance analyses have been performed for about 40 emission nebulosities in the large cloud and for about 15 or 20 in the small cloud (Dickel, Aller and Faulkner, 1964). The average ratio of oxygen to hydrogen appears to be about 70% of what it is in our own galactic system (as found from the results for the bright B stars, which are believed to have recently been formed from the Orion nebula and from the analysis of the Orion nebula itself). The oxygen-to-hydrogen ratio is (by number)

$$O/H \cong 3 \times 10^{-4} \qquad \text{Magellanic clouds}$$

$$O/H = 4 \times 10^{-4} \qquad \text{Orion nebula}$$

The neon abundances are also about 70% of the galactic abundance, but the result is much more uncertain. The helium-to-hydrogen ratio is found to be of the order of He/H $\approx$ 0.088 by number (Faulkner and Aller, 1965). This is to be compared with He/H = 0.105 for the galaxy (see also the discussion by Mathis, 1965). Thus, there appears to be only a small difference (if any) between the compositions of the Magellanic clouds and our own galaxy insofar as oxygen, neon, and helium are concerned.

A similar change of 30% in the metal abundances would not normally be observed by ordinary spectroscopic techniques, particularly in dealing with abnormal supergiants whose absolute magnitudes are -9 to -8.5. Perhaps, by using the techniques which have been developed by Strömgren, the colors of the stars would provide the needed evidence, but the light-gathering power of present telescopes is not sufficient.

REFERENCES

H. R. Dickel, L. H. Aller, and D. J. Faulkner (1964), *International Astronomical Union Symposium No. 20*.

D. J. Faulkner and L. H. Aller (1965), *Monthly Notices Roy. Astron. Soc.* (in press).

F. J. Kerr and A. W. Rodgers (eds.) (1964), "The Galaxy and the Magellanic Clouds," *International Astronomical Union Symposium No. 20*, Australian Academy of Sciences, Canberra.

J. Mathis (1965), *Publ. Astron. Soc. Pacific* **77**: 189.

DISCUSSION

Bidelman: I do not believe that too much should be made of the difference in the frequency distribution in the diagram of the brighter stars in the Magellanic clouds and our own system, principally because the data on the clouds are not very good and the data on our own system are probably even worse. Only the brightest stars are observed in the clouds. These are the F and G supergiants. I am not at all convinced that the ratio of G supergiants to M supergiants is really any different in the clouds than it is in our own system.

AN EVOLUTIONARY-SIGNIFICANT GROUP OF ECLIPSING VARIABLES

Jorge Sahade

The investigation of close binary systems has disclosed the existence of several groups among them that must be significant from the point of view of their evolution. These groups include those binaries in which one component is a Wolf–Rayet or an Of star, the Algol systems, the W Ursae Majoris systems, and the cataclismic variables.

In this paper, I wish to call attention to the existence of another group; first, however, I shall make a few comments with regard to the first group mentioned above.

THE BINARIES WITH A WR OR AN Of COMPONENT

These systems are characterized by the facts that:

A. The companions to the WR or Of stars are usually O or B main-sequence or supergiant objects.

B. The observational evidence suggests that the WR or the Of components eject matter probably with spherical symmetry.

C. The WR and Of stars are emission-line objects.

D. Irrespective of the periods, the mass of the WR or Of star is always smaller than that of the companion.

E. The WR stars are smaller in diameter than their companions but have a very large envelope the size of which is larger than that of the latter (V444 Cygni). Regarding the Of components, in the case of 29 UW CMa, the Of star appears to be larger than the O companion, but we do not know whether this relative size comes about because of the existence of a large envelope similar to the envelopes that are characteristics of the WR objects. The results by R. Wilson in Edinburgh would seem to support this possibility.

Principally because of reason D, these systems have been interpreted by the writer (Sahade, 1958) as close binaries where one component (the WR or the Of object) is still in gravitational contraction, in close approach to the main sequence. Commonly, these components are considered as evolutionary old objects but the writer's hypothesis seems to find further support in the following considerations:

1. WR and Of stars are found to be associated with the youngest aggregates.

2. The WR and Of stars are located not below the main sequence but rather at the "turn-off point" of the main sequence (Westerlund, 1961, 1964, and Feinstein, 1964).

3. Their characteristics are quite different from those stars that are located in the same region of the HR diagram and are evolving away from the main sequence.

4. The WR and Of objects are not underluminous for their masses, as appears to be true for old stars that are presumably evolving to the left in the HR diagram (see next section).

The mean density that is assigned to V444 Cygni on the basis of the dimensions that result from the combination of Kron and Gordon's photometry and Münch's spectroscopic elements (see Sahade, 1958a) depends heavily on the luminosity ratio of the two components (Russell, 1944) of the system, although, of course, it also depends on the dimensions that the Wolf-Rayet core actually has. With regard to the present accepted luminosity ratio, it appears that it may need revising, in view of the possible effect of electron scattering on the emission profiles and the distortion of the H absorption lines in the blue by the blending effects of neighboring emission features (Münch, 1950).

The available data on γ_2 Velorum and 29 UW Canis Majoris yield mean densities of about 0.3 and $0.004 \, g/cm^3$, respectively, but perhaps these values should not be used in the argument until further observational material becomes available.

THE SYSTEMS WITH UNDERLUMINOUS,
MASSIVE COMPONENTS

One of the groups of binaries for which references are found more often in the literature is that of the Algol systems, which are characterized by a configuration that Kopal has described by the term "semidetached system." These systems are formed by a main-sequence star of spectral type about A and a giant or subgiant object of later spectral type that fills its lobe of the first critical equipotential surface. The important point in the Algol systems is that, at least for those with periods smaller than, say, 10 days, the mass of the larger component is the smallest (in some cases actually quite small). These large, small-mass components are overluminous for their masses and seem to show a correlation between mass and luminosity in the sense that the greater the luminosity the greater the mass.

Investigations of β Lyrae made in the last five years have made it clear that, at least in this peculiar system, we are dealing with a secondary that in a sense has opposite characteristics to those of the secondaries in the Algol systems. It is now accepted that β Lyrae is formed by a supergiant B8 primary and a smaller, more massive secondary that is underluminous for its mass.

Actually, β Lyrae is not the only case with such type of components. Very recently, a new investigation of the eclipsing variable V453 Scorpii (Sahade and Frieboes-Conde, 1965) has disclosed another similar case, while an interpretation that was advanced (Sahade, 1962) a few years ago in regard to the strange behavior of the radial velocities of a number of systems has brought up some other cases that increase the number of members of the group. Table I lists the presently known systems with secondary components that are *underluminous for their masses* and *more massive than their primaries*. Except in the case of HD 47129, these secondaries are definitely *smaller in size than their companions*. The spectral characteristics are, however, not completely similar in all the cases.

It should be emphasized that the values shown in Table I for the masses are exact only for the order of magnitude and that for the last three cases (HD 698, ε Aur, W Cru) reasonable values for the masses as suggested by the mass function and/or by the possible evolutionary track of the primary indicate the secondaries to be more massive than the primaries, which seems reasonable.

Such underluminous, more massive (smaller) secondaries are interpreted as being objects evolving toward the left in the H–R diagram after having reached the end of their evolutionary track to the right. This interpretation takes into specific account the fact that we are dealing with underluminous objects. They are certainly

Table I. Systems with an Underluminous, Massive Component

System	Primary component	Secondary component	$m_P \sin^3 i$ $(M_\odot)$	$m_S \sin^3 i$ $(M_\odot)$	P (days)	Light curve*	Notes
HD 47129	O8 V		45	60	14		1
A0 Cas	O9 III		15	21	3.5	β	2
V453 Sco	B1 Iab		30	32	12	β	3
V448 Cyg	B1 Ib-II	B0	14	20	6.5	β	4
β Lyr	B8 II		8	18	13	β	5
HD 698	B9 III		(5)	(9)	56		6
ε Aur	F2 Ia	(O9-B3 III)	(12)	(13)	9883	A	7
W Cru	G2 Iab	(G8)	(14)	(18)	180	β	8

* $\beta = \beta$ Lyrae, A = Algol type.

Notes:

    1. Not eclipsing. The secondary mimics the absorption spectrum of the primary; in 1957, the lines of the former were absent in the photographic region and present in the red region. This would suggest that the secondary component has the energy distribution of an F star, a conclusion which requires that $R_S > R_P$.

    2. The secondary component displays the same absorption spectrum as the primary. However, $R_S < R_P$.

    3. The secondary component appears to display only emission lines. $R_S < R_P$.

    4. The secondary component displays an absorption spectrum. $R_S < R_P$.

    5. The secondary component appears to display only emission lines. $R_S < R_P$.

    6. Single-lined noneclipsing binary. The masses were derived by considering the mass function and tracing back the evolutionary track of the primary on the assumption that no drastic mass loss has taken place since the star was on the main sequence.

    7. Single-lined eclipsing variable. The masses were derived from the mass function by tracing back the evolutionary track of the primary on the assumption that no drastic mass loss has taken place since the star was on the main sequence. $R_S < R_P$.

    8. Single-lined eclipsing variable. The mass function suggests that the mass of the secondary component must be larger than that of the primary and the values written down in the table were derived by tracing back the evolutionary track of the primary on the assumption that no drastic mass loss has taken place since the star was on the main sequence. $R_S < R_P$.

not main-sequence or contracting or giant stars, because they do not conform to the radius–mass–spectrum relationships that hold for such objects. Furthermore, they appear to be surrounded by extended envelopes.

AN ADDITIONAL REMARK

    As an additional remark, I would like to call attention to the possibility of a new group of early-type short-period binaries suggested by the results that Ringuelet-Kaswalder has obtained in the cases of the V/R variables 27 Canis Majoris (1962) and 48 Librae (1963), where the stellar lines suggest velocity variations with periods of 0.115 and 0.262 days, respectively. If these results are confirmed, we would be dealing with close systems having strange types of components.

REFERENCES

A. Feinstein (1964), *Bol. Asoc. Argentina Astron.* (in press).

G. Münch (1950), *Astrophys. J.* **112**: 266.

A. Ringuelet-Kaswalder (1962), *Astrophys. J.* **135**: 755.

A. Ringuelet-Kaswalder (1963), *Astrophys. J.* **137**: 1310.

H. N. Russell (1944), *Astrophys. J.* **100**: 213.

J. Sahade (1958), *Obsèrvatory* **78**: 79.

J. Sahade (1958a), *Étoiles à raies d'émission*, Institut d'Astrophysique, Cointe-Sclessin, Belgique, p. 46.

J. Sahade (1962), in: J. Sahade (ed.), *Symposium on Stellar Evolution*, Observatorio Astronómico, La Plata, Argentina, p. 185.

J. Sahade and H. Frieboes-Conde (1965), *Astrophys. J.* **141**: 652.

B. E. Westerlund (1961), *Uppsala Obs. Ann.* **5**: No. 1.

B. E. Westerlund (1964), in: F. J. Kerr and A. W. Rodgers (eds.), *The Galaxy and the Magellanic Clouds*, Australian Academy of Science, Canberra, p. 316.

Part VII. Summary

SUMMARY

G. R. Burbidge

This is the first conference on stellar evolution in a number of years in which nearly all aspects of stellar evolution have been discussed.

A fundamental part of stellar evolution is star formation. This problem was not discussed during this conference, but has been discussed by Spitzer (1963); not much progress in understanding the details has been made since. The mechanisms of star formation have considerable bearing on the early stages of stellar evolution. There are problems associated with the angular momentum, with the magnetic fields in the interstellar matter, and with removing the magnetic flux from the gas and dust so that it can condense. Mestel and Spitzer (1956) and Mestel (1960) have shown that it is just marginally possible that stars could form. These apparent theoretical difficulties probably only reflect the limitations of our theories at the present time.

Consider next the problem of gravitational contraction. The initial conditions in the interstellar gas, its rotational and magnetic properties, are generally supposed to have little effect on the evolution of a star once it has condensed. There is a tendency to suppose that stars go through their gravitational contraction phase by the Hayashi mechanism; this may very well be true. Hayashi and Ezer and Cameron discussed such evolutionary tracks. However, Temesváry described experiments with model stars done in order to see the effect of mass accretion on the evolution. Also, Faulkner, Griffiths, and Hoyle (1963) have recently published a paper considering the problems of the magnetic fields in the early history of contraction. They argued that the magnetic fields may inhibit the convection, causing the time scales and the evolutionary tracks to be quite different from those described by Hayashi. Because such factors have to be taken into account, we cannot immediately conclude that the gravitational contraction does take place as described classically by Hayashi.

The observational arguments concerning such tracks are, as usual, indirect. There are two fundamental questions involved. The first concerns the early history of the solar system and the time taken for the solar nebula to condense and planets to form. The other is the question of the color–magnitude diagrams for young clusters. If stars were gravitationally contracting on the radiative Helmholtz–Kelvin time scale, they would not have got as close to the nuclear-burning main sequence as they appear to have done.

Spitzer proposed that stars lose mass or possibly break up in the contraction process. Thus, stars would start contracting with larger masses than they have on the main sequence and the time scales would be correspondingly shorter. A second proposal, made recently by Herbig, is that the simple concept that all the stars in a cluster formed essentially coevally is not correct; he argues that there is a spread in formation times. It would seem to me that there is no reason to believe that the stars in a given cluster should all form and start to contract within a very short space of time. A third possibility, and perhaps the most popular, is that the stars really do follow the Hayashi tracks. However, all these possibilities exist. It is very difficult to determine which is correct from an observational standpoint alone.

Involved in the early history of the forming stars is the problem of the production of the light elements. Whereas some people prefer to believe that the deuterium, lithium, beryllium, and boron are formed at this stage, others think that these elements are formed elsewhere. Observational evidence from the T Tauri stars somewhat favors the idea that light elements are, for the most part, synthesized at this stage. But again, the arguments are not conclusive.

Consider now evolution off the main sequence. Fundamental work was done by Schwarzschild and Hoyle (1955) and later by Hoyle and Hazelgrove (1958). There have been continual attempts to fit the observed color–magnitude diagrams to the theoretical color–magnitude diagrams for the globular clusters and for the galactic clusters. Sandage, in particular, made estimates of the ages of these objects by a process of interpolation and extrapolation.

There has been no attempt made at this conference to date a globular cluster. The reason is that, on the one hand, an effort is being made to include more detailed physics in the models of a star, and to improve on what is now known to be a very crude mixing-length theory. Spiegel showed how difficult this is. On the other hand, Sandage, for example, is making a serious attempt to improve the observations, because it is obvious that the measurements of the fainter stars in the clusters are not very good. So, while more accurate dating is probably not far off, discrepancies within factors of 2 between different estimates should not be overemphasized. Dicke is essentially putting another (as yet unwarranted) parameter into the problem by discussing the possibility of evolution with varying G. Verification or nonverification by astronomical methods is a long way off.

Tremendous interest has been shown in the advanced stages of evolution. Highly evolved stars involve important questions of nuclear physics, as well as the understanding of evolution off the giant branch. Understanding of the later stages of evolution can be advanced only to a limited extent by the interplay of theory and observation. The difficulty arises because as stars get older they evolve faster and faster, and the time scales for evolution get shorter and shorter. This means that in practice we can actually observe less and less of these phases because there is a smaller number of stars that can be seen at any given stage of evolution. On an H-R diagram there is a very small number of stars in the more evolved parts of the evolutionary tracks. Thus, accurate theoretical physics and stellar models are extremely important. Attempts to construct models using estimated rates for nuclear processes in highly evolved stars are interesting, but not too much weight should be given to the numerical results. It has not been clear to what extent there is agreement among the different authors in following the tracks of such stars, and to what extent the differences are due to different techniques, different energy-generation rates, or different opacities.

Regarding pulsating stars, much progress has been made, particularly through the work of Baker, Kippenhahn, Christy, Cox, and Whitney. These authors now seem to have models which agree quite well with the observations, and this is a major advance. Luckily, in considering models of pulsating stars, one does not have to worry about the previous evolution or the structure of the interior; one is dealing only with an envelope, and an envelope which is not highly evolved. The models for the pulsation are insensitive to the conditions throughout the stars.

The question of mass loss from stars involves the problem of the end-point of stellar evolution. It has been known for more than thirty years that there are two critical masses, associated with the support of stars by degenerate-electron pressure and degenerate-neutron pressure, respectively, above which no stable configurations exist. These critical masses lie around $1M_\odot$. The mass function for stars goes all the way from about $0.1M_\odot$ to the order of $100M_\odot$. A tradition seems to have grown up that stars must end up as stable white dwarfs and must therefore lose all of the mass down to the white-dwarf limit in the process of evolution prior to the collapse phase. While we do not know the final state of all stars, there is no reason within the framework of physics to suppose that a star necessarily knows what is going to happen and therefore ejects the appropriate amount of mass to avoid an unstable state. This has, however, been tacitly assumed in supposing that all stars end as white dwarfs.

If mass remains in the object above the critical mass, then, although we do not know what the final state is, it appears possible that the star may vanish as far as the emission of the electromagnetic radiation is concerned; although, providing the mass energy is not radiated, it will still be detectable through its gravitational field. This means, therefore, that mass can go into an unobservable state from the point of view of the astronomer who observes the light from it. Thus, we must take into account the possibility that some fraction of the mass of our own and other galaxies may end up in such a state.

Nucleosynthesis was also discussed. We realize now from the observations that very much of the synthesis of elements in stars must have taken place in the early history of the galaxy. There was a short period of rapid production of elements which was followed by a period of steady but very slow production which has gone on until the present. The oldest stars seen show compositions which are in some cases not very different from the composition of the sun.

Cameron pointed out the difficulty of obtaining all the helium in the galaxy from hydrogen if the galaxy has evolved at a steady and fairly constant rate for about 10 billion years. There are several ways out of this apparent dilemma. One is to suppose that the galaxy did not, in fact, condense out of a cloud of pure hydrogen; rather, there was some admixture of helium. Whether or not this point of view is adopted depends to some extent on one's cosmological prejudices. The second possibility is that the vast bulk of the mass of the galaxy is in the low-mass stars which have not evolved, and that those stars have a much lower helium-to-hydrogen ratio than the material from which we determine the average H/He ratio. A third possibility is that there was in the early phase of the galaxy a far higher degree of activity than there is now. Yet a fourth possibility is that mentioned several years ago by the Cambridge cosmologists, that there are many stars imbedded so deep in gas and dust that we do not see them, so that there is much more energy being dissipated than that which is estimated from the luminosity of the galaxy. The energy they emit is absorbed and goes into the kinetic energy of motion of the surrounding interstellar medium.

Finally, a comment on Bidelman's discussion of the question of the magnetic stars and the origin of their compositions. Not all of the stars that show these anomalous compositions are known to contain strong surface magnetic fields. There are two possibilities in explaining these phenomena. The details of the nuclear physics are not understood in either case. First, we may suppose that the abundance anomalies are a result of surface nuclear reactions produced by the acceleration of particles in changing magnetic fields. The details of such processes have yet to be explained. If this point of view is adopted, then this is a skin effect and there is no mixing of the surface layer (a very small fraction, 10^{-6} or 10^{-7} of the mass of a star) with the interior. If there were any mixing over any reasonable time scale, the dilution would destroy the anomalies and they would not be seen. Thus, there can be no convective zone, not even an incipient one. The alternative is to suppose that the compositions seen are representative of the stars as a whole. In this case, the stars are highly evolved, as stated by Bidelman. How such stars evolved is a difficult question. One possibility is that they have evolved with mixing, and so stayed close to the main sequence, perhaps starting out with a somewhat larger mass and ejecting some of it. It is very hard to believe that the star moved off the main sequence, went through a normal evolutionary path in the limited time scale available to it from its occurrence in young clusters, and returned back to the main sequence in the place where it is now seen. However, it may be that such processes have occurred.

REFERENCES

J. Faulkner, K. Griffiths, and F. Hoyle (1963), *Monthly Notices Roy. Astron. Soc.* **126**: 1.
F. Hoyle and Hazelgrove (1958), *Monthly Notices Roy. Astron. Soc.* **118**: 519.
L. Mestel (1960), in: A. Beer (ed.), *Vistas in Astronomy*, Vol. 3, Pergamon Press, New York, p. 296.
L. Mestel and L. Spitzer (1956), *Monthly Notices Roy. Astron. Soc.* **116**: 503.
L. Spitzer (1963), in: R. Jastrow and A. G. W. Cameron (eds.), *Origin of the Solar System*, Academic Press, Inc., New York.
M. Schwarzschild and F. Hoyle (1955), *Astrophys. J. Suppl.* **1**.

INDEX